PROCEEDINGS OF THE 13th

ASIAN LOGIC CONFERENCE

PROCEEDINGS OF THE 13th

ASIAN LOGIC CONFERENCE

Guangzhou, China 16 – 20 September 2013

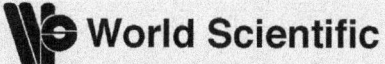

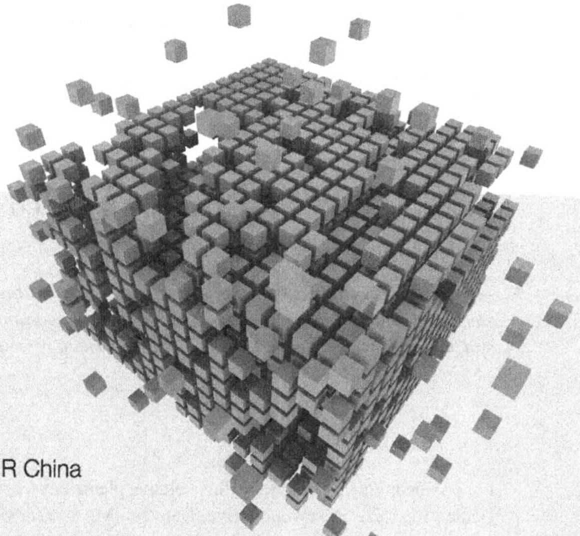

edited by

Xishun Zhao
Sun Yat-sen University, P R China

Qi Feng
Chinese Academy of Sciences, Beijing, P R China

Byunghan Kim
Yonsei University, Korea

Liang Yu
Nanjing University, P R China

World Scientific

NEW JERSEY • LONDON • SINGAPORE • BEIJING • SHANGHAI • HONG KONG • TAIPEI • CHENNAI

Published by

World Scientific Publishing Co. Pte. Ltd.
5 Toh Tuck Link, Singapore 596224
USA office: 27 Warren Street, Suite 401-402, Hackensack, NJ 07601
UK office: 57 Shelton Street, Covent Garden, London WC2H 9HE

British Library Cataloguing-in-Publication Data
A catalogue record for this book is available from the British Library.

PROCEEDINGS OF THE 13TH ASIAN LOGIC CONFERENCE

ISBN 978-981-4675-99-4

Printed in Singapore

Preface

The present volume is the proceedings of the *Thirteenth Asian Logic Conference* (ALC2013) held in Guangzhou, China, from September 16 to September 20, 2013.

The Asian Logic Conference started in 1981, developed into a major international event in logic and related areas. It rotates among countries in the Asia Pacific region and is regularly sponsored by the Association for Symbolic Logic. From 1981 to 2008, the Asian Logic Conference has been held triennially. The previous meetings took place in Singapore (1981), Bangkok (1984), Beijing (1987), Tokyo (1990), Singapore (1993), Beijing (1996), Hsi-Tou (1999), Chongqing (2002), Novosibirsk (2005) and Kobe (2008). In year 2008, the East Asian and Australasia Committees of the Association of Symbolic Logic decided to shorten the three-year cycle to two. The new cycle of ALC started in Singapore (2009), Wellington (2011).

ALC2013 was organized and mainly sponsored by Institute of Logic and Cognition at Sun Yat-sen University. The program consists of 7 plenary lectures, 20 invited lectures in four special sessions, and a number of contributed talks. Plenary speakers were Joerg Brendle (Japan), Yijia Chen (China), Adam Day (New Zealand), Su Gao (USA), Ivo Herzog (USA), Andre Nies (New Zealand), and Kazuyuki Tanaka (Japan). Special sessions on the following topics were organized by: Model Theory (Byunghan Kim), Recursion Theory (Yue Yang), Set theory (Qi Feng), Theoretical Computer Science (Hongde Hu), and Philosophic Logic (Xuefeng Wen).

We are very pleased to find that *World Scientific* were enthusiastic to publish proceedings of ALC2013. Authors were invited to submit articles to the present volume, based on around talks given at the meeting. In order to make sure that the planned volume was of high quality, all the submitted papers were fully refereed, with 13 of them accepted for this volume.

We thank the authors for their contributions and the reviewers for their creful and thorough work. Preparation of the proceedings volume was also partially supported by NSFC grant 61272059 and NSSFC grant 13&ZD186.

Xishun Zhao, Qi Feng, Byunghan Kim, Liang Yu

Organizing committees

Series editorial board

for the series on advances in quantum many-body theory

Xishun Zhao	– Sun Yat-sen University, China
Qi Feng	– Chinese Academy of Sciences, China
Byunghan Kim	– Yonsei University, South Korea
Liang Yu	– Nanjing University, China

International advisory committee

for the Series of International Conferences on Recent Progress in Many-Body Theories

R. F. Bishop (Chairman)	– UMIST, Manchester, UK
S. Fantoni (Secretary)	– SISSA, Trieste, Italy
C. E. Campbell (Treasurer)	– University of Minnesota, USA
H. Kümmel (Hon. President)	– Ruhr-Universität Bochum, Germany
J. A. Carlson	– Los Alamos National Laboratory, USA
S. A. Chin	– Texas A & M University, USA
J. W. Clark	– Washington University, St. Louis, USA
P. Fulde	– Max-Planck-Institut für Komplexer Systeme, Dresden, Germany
A. Kallio	– University of Oulu, Finland
E. Krotscheck	– Johannes Universität Linz, Austria
C. Lhuillier	– Université Pierre et Marie Curie, Paris
A. MacDonald	– Indiana University, Bloomington, USA
E. Manousakis	– Florida State University, USA
D. Neilson	– University of New South Wales, Sydney, Australia
M. L. Ristig	– Universität zu Köln, Germany

Contents

An Analogy between Cardinal Characteristics and Highness Properties of Oracles *

Jörg Brendle

Graduate School of System Informatics
Kobe University
Rokko-dai 1-1
Nada, Kobe, 657-8501, Japan
Email: brendle@kurt.scitec.kobe-u.ac.jp

Andrew Brooke-Taylor

School of Mathematical Sciences
University of Bristol
University Walk
Bristol, BS8 1TW, United Kingdom
Email: a.brooke-taylor@bristol.ac.uk

Keng Meng Ng

Division of Mathematical Sciences
School of Physical & Mathematical Sciences
Nanyang Technological University
21 Nanyang Link, Singapore
Email: selwyn.km.ng@gmail.com

André Nies

Department of Computer Science
Private Bag 92019
University of Auckland
Auckland, New Zealand
Email: andre@cs.auckland.ac.nz

We present an analogy between cardinal characteristics from set theory and highness properties from computability theory, which specify a sense in which

* Jörg Brendle is supported by Grant-in-Aid for Scientific Research (C) 24540126, Japan Society for the Promotion of Science. Andrew Brooke-Taylor is currently supported by a UK Engineering and Physical Sciences Research Council Early Career Fellowship, and was previously supported by a JSPS Postdoctoral Fellowship for Foreign Researchers and JSPS Grant-in-Aid 23 01765. Keng Meng Ng is supported by the MOE grant MOE2011-T2-1-071. André Nies is supported by the Marsden fund of New Zealand.

a Turing oracle is computationally strong. While this analogy was first studied explicitly by Rupprecht (*Effective correspondents to cardinal characteristics in Cichoń's diagram*, PhD thesis, University of Michigan, 2010), many prior results can be viewed from this perspective. After a comprehensive survey of the analogy for characteristics from Cichoń's diagram, we extend it to Kurtz randomness and the analogue of the Specker-Eda number.

1. Introduction

Mathematics studies abstract objects via concepts and corresponding methods. Metamathematics studies these concepts and methods. A common scheme in metamathematics is duality, which can be seen as a bijection between concepts. For instance, Stone duality matches concepts related to Boolean algebras with concepts related to totally disconnected Hausdorff spaces. A weaker scheme is analogy, where different areas develop in similar ways. An example is the analogy between meager sets and null sets. While one can match the concepts in one area with the concepts in the other area, the results about them may differ.

We systematically develop an analogy between

(a) cardinal characteristics from set theory, which broadly speaking measure the deviation from the continuum hypothesis of a particular model of ZFC

(b) highness properties from computability theory, which specify a sense in which a Turing oracle is computationally strong.

One of the simplest examples of a cardinal characteristic is the unbounding number. For functions f, g in Baire space ${}^\omega\omega$, let $f \leq^* g$ denote that $f(n) \leq g(n)$ for almost all $n \in \omega$. Given a countable collection of functions $(f_i)_{i\in\omega}$ there is g such that $f_i \leq^* g$ for each i: let $g(n) = \max_{i\leq n} f_i(n)$. How large a collection of functions do we need so that no upper bound g exists? The unbounding number $\mathfrak{b}$ is the least size of such a collection of functions; clearly $\aleph_0 < \mathfrak{b} \leq 2^{\aleph_0}$.

In computability theory, probably the simplest highness property is the following: an oracle A is called (classically) high if $\emptyset'' \leq_T A'$. Martin[1] proved that one can require equivalently the following: there is a function $f \leq_T A$ such that $g \leq^* f$ for each computable function g.

At the core of the analogy, we will describe a formalism to transform cardinal characteristics into highness properties. A ZFC provable relation $\kappa \leq \lambda$ between cardinal characteristics turns into a containment: the highness property for κ implies the one for λ.

The analogy occurred implicitly in the work of Terwijn and Zambella[2], who showed that being low for Schnorr tests is equivalent to being computably traceable. (These are lowness properties, saying the oracle is close to being computable; we obtain highness properties by taking complements.) This is the computability theoretic analog of a result of Bartoszyński[3] that the cofinality of the null sets (how many null sets does one need to cover all null sets?) equals the domination number for traceability, which we will later on denote $\mathfrak{d}(\in^*)$. Terwijn and Zambella alluded to some connections with set theory in their work. However, it was not Bartoszyński's work, but rather work on rapid filters by Raisonnier[4]. Actually, their proof bears striking similarity to Bartoszyński's; for instance, both proofs use measure-theoretic calculations involving independence. See also the books Ref. 5, Section 2.3.9, and Ref. 6, Section 8.3.3.

The analogy was first observed and studied explicitly by Rupprecht[7,8]. Let $\mathrm{add}(\mathcal{N})$ denote the additivity of the null sets: how many null sets does one need so that their union is not null? Rupprecht found the computability-theoretic analog of $\mathrm{add}(\mathcal{N})$. He called this highness property "Schnorr covering"; we prefer to call it "Schnorr engulfing". A Schnorr null set is a certain effectively defined kind of null set. An oracle A is Schnorr engulfing if it computes a Schnorr null set that contains all plain Schnorr null sets. While $\mathrm{add}(\mathcal{N})$ can be less than $\mathfrak{b}$, Rupprecht showed that the Schnorr engulfing sets are exactly the high sets. Thus, we only have an analogy, not full duality.

2. Cardinal characteristics and Cichoń's diagram

All our cardinal characteristics will be given as the unbounding and domination numbers of suitable relations. Let $R \subseteq X \times Y$ be a relation between spaces X, Y (such as Baire space) satisfying $\forall x\ \exists y\ (xRy)$ and $\forall y\ \exists x\ \neg(xRy)$. Let $S = \{\langle y, x\rangle \in Y \times X \colon \neg xRy\}$. We write

$$\mathfrak{d}(R) = \min\{|G| : G \subseteq Y \wedge \forall x \in X\ \exists y \in G\, xRy\}.$$

$$\mathfrak{b}(R) = \mathfrak{d}(S) = \min\{|F| : F \subseteq X \wedge \forall y \in Y \exists x \in F\, \neg xRy\}.$$

$\mathfrak{d}(R)$ is called the *domination number* of R, and $\mathfrak{b}(R)$ the *unbounding number*.

If R is a preordering without greatest element, then ZFC proves $\mathfrak{b}(R) \leq \mathfrak{d}(R)$. To see this, we show that any dominating set G as in the definition of $\mathfrak{d}(R)$ is an unbounded set as in the definition of $\mathfrak{b}(R)$. Given y take a z such that $\neg zRy$. Pick $x \in G$ with zRx. Then $\neg xRy$.

For example, the relation $\leq^*$ on ${}^\omega\omega \times {}^\omega\omega$ is a preordering without maximum. One often writes $\mathfrak{b}$ for $\mathfrak{b}(\leq^*)$ and $\mathfrak{d}$ for $\mathfrak{d}(\leq^*)$; thus $\mathfrak{b} \leq \mathfrak{d}$. (Another easy exercise is to show that if R is a preordering, then $\mathfrak{b}(R)$ is a regular cardinal.)

2.1. *Null sets and meager sets*

Let $\mathcal{S} \subseteq \mathcal{P}(\mathbb{R})$ be a collection of "small" sets; in particular, assume $\mathcal{S}$ is closed downward under inclusion, each singleton set is in $\mathcal{S}$, and $\mathbb{R}$ is not in $\mathcal{S}$. We will mainly consider the case when $\mathcal{S}$ is the class of null sets or the class of meager sets. For null or meager sets, we can replace $\mathbb{R}$ by Cantor space or Baire space without changing the cardinals.

The unbounding and the domination number for the subset relation $\subseteq_\mathcal{S}$ on $\mathcal{S}$ are called additivity and cofinality, respectively. They have special notations:

$$\begin{aligned} \operatorname{add}(\mathcal{S}) &= \mathfrak{b}(\subseteq_\mathcal{S}) \\ \operatorname{cofin}(\mathcal{S}) &= \mathfrak{d}(\subseteq_\mathcal{S}) \end{aligned}$$

Let $\in_\mathcal{S}$ be the membership relation on $\mathbb{R} \times \mathcal{S}$. The unbounding and domination numbers for membership also have special notations:

$$\begin{aligned} \operatorname{non}(\mathcal{S}) &= \mathfrak{b}(\in_\mathcal{S}) = \min\{|U| : U \subseteq \mathbb{R} \wedge U \notin \mathcal{S}\} \\ \operatorname{cover}(\mathcal{S}) &= \mathfrak{d}(\in_\mathcal{S}) = \min\{|\mathcal{F}| : \mathcal{F} \subseteq \mathcal{S} \wedge \bigcup \mathcal{F} = \mathbb{R}\} \end{aligned}$$

The diagram in Fig. 1 shows the ZFC relationships between these cardinals. An arrow $\kappa \to \lambda$ means that ZFC proves $\kappa \leq \lambda$. The only slightly nontrivial

$$\begin{array}{ccc} \operatorname{non}(\mathcal{S}) & \longrightarrow & \operatorname{cofin}(\mathcal{S}) \\ \uparrow & & \uparrow \\ \operatorname{add}(\mathcal{S}) & \longrightarrow & \operatorname{cover}(\mathcal{S}) \end{array}$$

Fig. 1. Basic ZFC relations of characteristics for a class $\mathcal{S}$.

arrow is $\operatorname{non}(\mathcal{S}) \to \operatorname{cofin}(\mathcal{S})$: Suppose we are given $\mathcal{F} \subseteq \mathcal{S}$ such that for every $C \in \mathcal{S}$ there is $D \in \mathcal{F}$ with $C \subseteq D$. Using the axiom of choice, for each D pick $u_D \notin D$. Now let $V = \{u_D \colon D \in \mathcal{F}\}$. Then $V \notin \mathcal{S}$

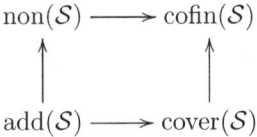

and $|V| \leq |\mathcal{F}|$. (Note that we have used the notations "cover" and "cofin" instead of the standard "cov" and"cof", because the latter two look very much alike.)

2.2. *The combinatorial Cichoń diagram*

In a somewhat nonstandard approach to Cichoń's diagram, we will consider the smaller "combinatorial" diagram in Fig. 2 which describes the ZFC relations between the cardinal characteristics $\mathfrak{d}(R)$ and $\mathfrak{b}(R)$ for three relations R. The first relation is $\leq^*$. The second relation is

$$\{\langle f, g\rangle \in {}^\omega\omega \times {}^\omega\omega\colon \forall^\infty n\, f(n) \neq g(n)\},$$

which we will denote by $\neq^*$. For instance, we have

$$\mathfrak{d}(\neq^*) = \min\{|F| : F \subseteq {}^\omega\omega \wedge \forall e \in {}^\omega\omega \exists f \in F \forall^\infty n \in \omega(e(n) \neq f(n))\}$$

For the third relation, let Y be the space of functions from ω to the set of finite subsets of ω. Recall that $\sigma \in Y$ is a *slalom* if $|\sigma(n)| \leq n$ for each n. We say that a function $f \in {}^\omega\omega$ is *traced* by σ if $f(n) \in \sigma(n)$ for almost every n. We denote this relation on ${}^\omega\omega \times Y$ by $\in^*$. We have for example

$$\mathfrak{b}(\in^*) = \min\{|F| : F \subseteq {}^\omega\omega \wedge \forall \text{ slalom } \sigma \exists f \in F \exists^\infty n \in \omega(f(n) \notin \sigma(n))\}.$$

$$\begin{array}{ccc}
\mathfrak{b}(\neq^*) & \longrightarrow & \mathfrak{d}(\in^*) \\
\uparrow & & \uparrow \\
\mathfrak{b}(\leq^*) & \longrightarrow & \mathfrak{d}(\leq^*) \\
\uparrow & & \uparrow \\
\mathfrak{b}(\in^*) & \longrightarrow & \mathfrak{d}(\neq^*)
\end{array}$$

Fig. 2. Combinatorial ZFC relations.

The nontrivial arrows such as $\mathfrak{b}(\in^*) \to \mathfrak{d}(\neq^*)$ follow from the full diagram discussed next.

2.3. *The full Cichoń diagram*

We are now ready to present (a slight extension of) the usual Cichoń diagram. As a first step, in Fig. 1 we take the equivalent diagram for $\mathcal{S} = \mathcal{N}$

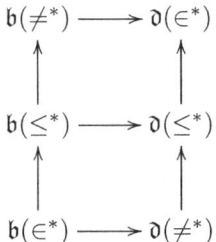

where $\mathrm{cover}(\mathcal{N})$ and $\mathrm{non}(\mathcal{N})$ have been interchanged. We join it with the diagram for $\mathcal{S} = \mathcal{M}$ and obtain the diagram in Fig. 3. The new arrows

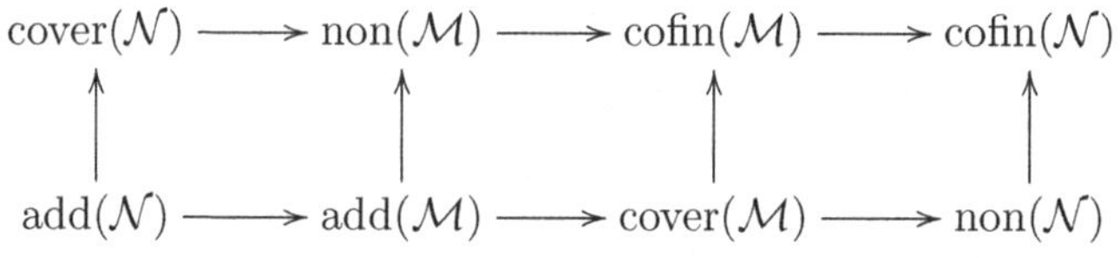

Fig. 3. The diagrams for $\mathcal{N}$ and $\mathcal{M}$ joined.

joining the two 4-element diagrams, such as $\mathrm{cofin}(\mathcal{M}) \to \mathrm{cofin}(\mathcal{N})$, are due to Rothberger and Bartoszyński; see Ref. 5 Sections 2.1.7 and 2.3.1 for details.

Finally in Fig. 4 we superimpose this 8-element diagram with the combinatorial 6-element diagram in Fig. 2. For all its elements except $\mathfrak{b}$ and $\mathfrak{d}$, ZFC proves equality with one of the characteristics in the 8-element diagram. These four ZFC equalities are due to Bartoszyński[3,9] and Miller[10]. Below we will mainly rely on the book Ref. 5 Sections 2.3 and 2.4.

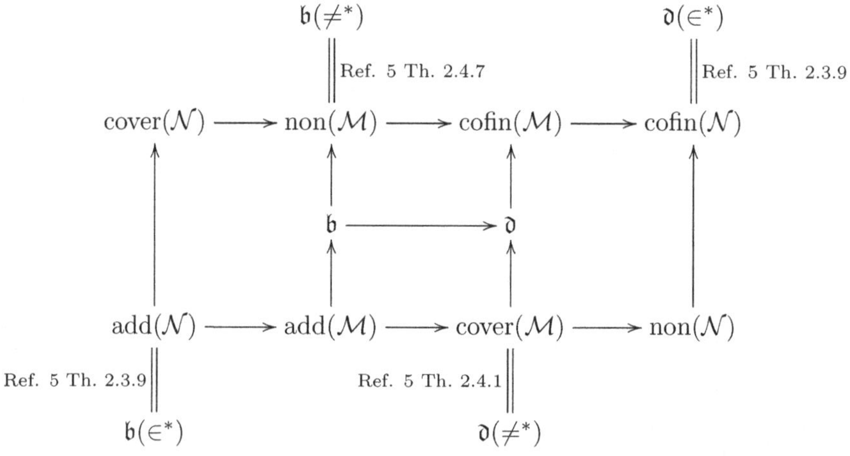

Fig. 4. Cichoń's diagram.

Fig. 4 shows *all* ZFC-provable binary relationships: for any two cardinal characteristics $\mathfrak{v}$ and $\mathfrak{w}$ appearing in the diagram, if there is no arrow $\mathfrak{v} \to \mathfrak{w}$, then there is a known model of ZFC in which $\mathfrak{v} > \mathfrak{w}$. See Subsection 4.2 for more discussion related to this.

Two ZFC-provable ternary relationships will be of interest to us:

$$\mathrm{add}(\mathcal{M}) = \min(\mathfrak{b}, \mathrm{cover}(\mathcal{M})), \quad \text{Miller, Truss; see Ref. 5 Section 2.2.9}$$
$$\mathrm{cofin}(\mathcal{M}) = \max(\mathfrak{d}, \mathrm{non}(\mathcal{M})) \quad \text{Fremlin; see Ref. 5 Section 2.2.11.}$$

3. Highness properties corresponding to the cardinal characteristics

We now present a scheme to describe highness properties of oracles which is analogous to the one in Section 2. As before, let $R \subseteq X \times Y$ be a relation between spaces X, Y, and let $S = \{\langle y, x\rangle \in Y \times X : \neg xRy\}$. Suppose we have specified what it means for objects x in X, y in Y to be computable in a Turing oracle A. We denote this by for example $x \leq_T A$. In particular, for $A = \emptyset$ we have a notion of computable objects. For instance, if X is Baire space and $f \in X$, we have the usual notion $f \leq_T A$.

Let the variable x range over X, and let y range over Y. We define the highness properties

$$\mathcal{B}(R) = \{A : \exists y \leq_T A \forall x \text{ computable } [xRy]\}$$

$$\mathcal{D}(R) = \mathcal{B}(S) = \{A : \exists x \leq_T A \forall y \text{ computable } [\neg xRy]\}$$

If R is a preordering with no greatest computable element, then clearly $\mathcal{B}(R) \subseteq \mathcal{D}(R)$. We will give some examples of such preorderings in Subsection 3.2.

Comparing these definitions with the ones of $\mathfrak{d}(R)$ and $\mathfrak{b}(R)$ at the beginning of Section 2, one notes that, ignoring the domains of quantification, we use here direct analogs of the *negations* of the statements there. For example, the formula defining $\mathfrak{b}(\leq^*)$ is of the form $\forall y \exists x \neg(x \leq^* y)$, and the defining formula for $\mathcal{B}(\leq^*)$ takes the form of its negation, $\exists y \forall x (x \leq^* y)$. The main reason for doing this is the connection to forcing. The cardinal characteristics we consider are each defined as $\min\{|\mathcal{F}| : \varphi(\mathcal{F})\}$ for some property φ, where $\mathcal{F}$ is a set of functions $\omega \to \omega$, or meager sets, or Lebesgue null sets. In each case, there is a forcing that introduces via a generic object a function (or meager set, or null set) such that in the extension model, φ no longer holds of the set of all ground model functions (respectively, meager sets, null sets). In the $\mathfrak{b}$ ($= \mathfrak{b}(\leq^*)$) case, for instance, after adding a function y_0 that dominates all functions from the ground model, the defining formula $\forall y\, \exists x\, \neg(x \leq^* y)$ no longer holds for the ground model functions x, as witnessed by $y = y_0$. Thus, iterating this procedure increases the value of $\mathfrak{b}$. Building a generic object is analogous to building an oracle that is computationally powerful in the sense specified by the analog of φ.

A notational advantage of taking the negations is that, as the analog of $\mathfrak{b}$, we obtain classical highness, rather than the lowness property of being non-high.

3.1. *Schnorr null sets and effectively meager sets*

We find effective versions of null and meager sets. For null sets, instead of in $\mathbb{R}$ we will work in Cantor space ${}^\omega 2$. Let λ denote the usual product measure on ${}^\omega 2$. For meager sets we will work in Cantor space, or sometimes in Baire space ${}^\omega\omega$.

A *Schnorr test* is an effective sequence $(G_m)_{m\in\omega}$ of Σ^0_1 sets such that each G_m has measure λG_m less than or equal to 2^{-m} and this measure is a computable real uniformly in m. A set $\mathcal{F} \subseteq {}^\omega 2$ is called *Schnorr null* if $\mathcal{F} \subseteq \bigcap_m G_m$ for a Schnorr test $(G_m)_{m\in\omega}$.

An *effective F_σ class* has the form $\bigcup_m \mathcal{C}_m$, where the $\mathcal{C}_m$ are uniformly Π^0_1. A set $\mathcal{F} \subseteq {}^\omega 2$ is called *effectively meager* if it is contained in such a class $\bigcup_m \mathcal{C}_m$ where each $\mathcal{C}_m$ is nowhere dense. (In this case, from m and a string σ we can compute a string $\rho \succeq \sigma$ with $[\rho] \cap \mathcal{C}_m = \emptyset$. Informally, for a Π^0_1 class, being nowhere dense is effective by nature.)

We now obtain 4 + 4 highness properties according to the relations specified in Subsection 2.1.

3.1.1. *Effectively meager sets*

$$
\begin{array}{ccc}
\mathcal{B}(\in_{\mathcal{M}}) & \xrightarrow{(3)} & \mathcal{D}(\subseteq_{\mathcal{M}}) \\
{\scriptstyle (2)}\big\uparrow & & \big\uparrow{\scriptstyle (4)} \\
\mathcal{B}(\subseteq_{\mathcal{M}}) & \xrightarrow{(1)} & \mathcal{D}(\in_{\mathcal{M}})
\end{array}
$$

We clarify the meaning of these properties of an oracle A, and introduce some terminology or match them with known notions in computability theory.

- $\mathcal{B}(\subseteq_{\mathcal{M}})$: there is an A-effectively meager set $\mathcal{S}$ that includes all effectively meager sets. Such an oracle A will be called *meager engulfing.*
- $\mathcal{B}(\in_{\mathcal{M}})$: there is an A-effectively meager set that contains all computable reals. Such an oracle A will be called *weakly meager engulfing.* Note that the notion of (weakly) meager engulfing is the same in the Cantor space and in the Baire space.
- $\mathcal{D}(\subseteq_{\mathcal{M}})$: there is an A-effectively meager set not included in any effectively meager set. It is easy to see that this is the same as saying that A is not low for weak 1-genericity (see Ref. 11 Theorem 3.1).
- $\mathcal{D}(\in_{\mathcal{M}})$: there is $f \leq_T A$ such that $f \notin \mathcal{F}$ for each effectively meager $\mathcal{F}$. This says that A computes a weakly 1-generic.

An arrow now means containment. The various arrows can be checked easily.

(1) Given an A-effectively meager set $\mathcal{S} \subseteq {}^{\omega}2$, by finite extensions build $f \in {}^{\omega}2$, $f \leq_T A$ such that $f \notin \mathcal{S}$. If $\mathcal{S}$ includes all effectively meager sets, then f is not in any nowhere dense Π^0_1 class, so f is weakly 1-generic.
(2) Trivial.
(3) Let $\mathcal{F}$ be an A-effectively meager set containing all computable reals. If $\mathcal{F}$ is included in an effectively meager set $\mathcal{G}$, then choose a computable $P \notin \mathcal{G}$ for a contradiction.
(4) Trivial.

3.1.2. *Schnorr null sets*

In order to join diagrams later on, for measure we work with the equivalent flipped diagram from Subsection 2.1 where the left upper and right lower corner have been exchanged.

$$\begin{array}{ccc} \mathcal{D}(\in_{\mathcal{N}}) & \longrightarrow & \mathcal{D}(\subseteq_{\mathcal{N}}) \\ \uparrow & & \uparrow \\ \mathcal{B}(\subseteq_{\mathcal{N}}) & \longrightarrow & \mathcal{B}(\in_{\mathcal{N}}) \end{array}$$

- $\mathcal{B}(\subseteq_{\mathcal{N}})$: there is a Schnorr null in A set that includes all Schnorr null sets. Such an oracle A will be called *Schnorr engulfing.* (This was called "Schnorr covering" by Rupprecht[7,8]. We have changed his terminology because the computability theoretic class is not the analog of a cardinal characteristic of type cover($\mathcal{C}$).)

- $\mathcal{B}(\in_{\mathcal{N}})$: there is a Schnorr null in A set that contains all computable reals. Such an oracle A will be called *weakly Schnorr engulfing*.
- $\mathcal{D}(\subseteq_{\mathcal{N}})$: there is a Schnorr null in A set not contained in any Schnorr null sets. One says that A is not low for Schnorr tests following Ref. 2.
- $\mathcal{D}(\in_{\mathcal{N}})$: there is $x \leq_T A$ such that $x \notin \mathcal{F}$ for each Schnorr null $\mathcal{F}$. This says that A computes a Schnorr random.

As before, the arrows are easily verified. One uses the well-known fact that each Schnorr null set fails to contain some computable real; see for example Ref. 6 Ex. 1.9.21 and the solution. This was already observed by Rupprecht[8].

3.2. *Combinatorial relations*

Let us see which highness properties we obtain for the three relations in Subsection 2.2.

- If R is $\leq^*$, then $\mathcal{B}(R)$ is highness, and $\mathcal{D}(R)$ says that an oracle A is of hyperimmune degree.
- Let R be $\neq^*$. The property $\mathcal{B}(\neq^*)$ says that there is $f \leq_T A$ such that f eventually disagrees with each computable function. Recall that a set A is called *DNR* if it computes a function g that is *diagonally nonrecursive*, i.e., there is no e such that the eth partial computable function converges on input e with output $g(e)$ (this is also referred to as diagonally noncomputable or d.n.c., for example in Ref. 6). A is called *PA* if it computes a $\{0, 1\}$-valued diagonally nonrecursive function. The property $\mathcal{B}(\neq^*)$ is equivalent to "high or DNR" by Theorem 5.1 of Ref. 12.
  The property $\mathcal{D}(\neq^*)$ says that there is $f \leq_T A$ such that f agrees infinitely often with each computable function.
- Let R be $\in^*$. Slaloms are usually called *traces* in computability theory. Recall that D_n is the n-th finite set. We say that a trace σ is *computable in* A if there is a function $p \leq_T A$ such that $\sigma(n) = D_{p(n)}$. Now $\mathcal{B}(\in^*)$ says that A computes a trace that traces all computable functions. By Theorem 6 of Ref. 8 this is equivalent to highness.
  The property $\mathcal{D}(\in^*)$ says that there is a function $f \leq_T A$ such that,

for each computable trace σ, f is not traced by σ. This means that A is not computably traceable in the sense of Ref. 2.

4. The full diagram in computability theory

We now present the full analog of Cichoń's diagram. Note that in Theorem IV.7 of his thesis[7] Rupprecht also gave this diagram for the standard part of Cichoń's diagram, without the analogs of $\mathfrak{b}(\in^*)$, $\mathfrak{d}(\in^*)$, $\mathfrak{b}(\neq^*)$, and $\mathfrak{d}(\neq^*)$ explicitly mentioned. Most of the equivalences between the analog of the standard Cichoń diagram and its full form are implicit in the literature; see Fig. 5.

The bijection between concepts in set theory and in computability theory is obtained as follows. In Section 3 we have already specified effective versions of each of the relations R introduced in Section 2. In Cichoń's diagram of Fig. 4, express each characteristic in the original form $\mathfrak{b}(R)$ or $\mathfrak{d}(R)$, and replace it by $\mathcal{B}(R)$ or $\mathcal{D}(R)$, respectively. Replacing most of these notations by their meanings defined and explained in Section 3, we obtain the diagram of Fig. 5.

Note that there is a lot of collapsing: instead of ten distinct nodes, we have only seven. Recall that we rely on a specific way of effectivizing the relations R. T. Kihara has raised the question whether, if one instead chooses an effectivization via higher computability theory, there is less collapsing. He has announced that in the hyperarithmetical case, the analogs of $\mathfrak{d}$ and $\mathrm{cover}(\mathcal{M})$ do not coincide.

As the analog of $\mathrm{cofin}(\mathcal{M}) = \max(\mathfrak{d}, \mathrm{non}(\mathcal{M}))$, we have that

$$\text{not low for weak 1-genericity} = \text{weakly meager engulfing} \cup \text{h.i. degree}.$$

This is so by the degree theoretic characterizations and because highness implies being of hyperimmune degree. The analog of the dual ternary relation $\mathrm{add}(\mathcal{M}) = \min(\mathfrak{b}, \mathrm{cover}(\mathcal{M}))$ is trivial because of the collapsing.

4.1. *Implications*

We verify the arrows and equalities in the computability version of Cichoń's diagram Fig. 5 in case they are not the trivial arrows from Subsections 3.1.1 and 3.1.2, and they have not been referenced in the diagram or have only appeared in Rupprecht's thesis[7]. This only leaves highness properties relating to meagerness.

As indicated in the diagram, Kurtz[15] has shown that the weakly

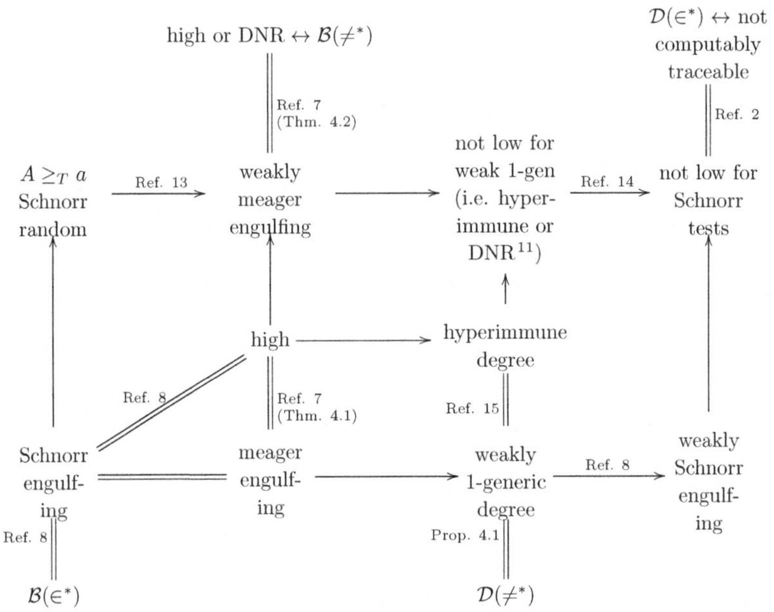

Fig. 5. The analog of Cichoń's diagram in computability.

1-generic and hyperimmune degrees coincide (also see Ref. 16 Corollary 2.24.19).

Proposition 4.1. *A is in* $\mathcal{D}(\neq^*)$ $\Leftrightarrow$ *A has weakly 1-generic degree.*

Proof. ($\Rightarrow$): Suppose $f \leq_T A$ infinitely often agrees with each computable function. Then $f+1$ is not dominated by any computable function, so A is of hyperimmune degree.

($\Leftarrow$): Suppose that A is weakly 1-generic. Let $f(n) =$ least $i \geq 0$ such that $n+i \in A$. Let h be a total computable function. It is enough to argue that $f(x) = h(x)$ for some x. We let $V = \{\sigma * 0^{h(|\sigma|)} * 1 \mid \sigma \in 2^{<\omega}, |\sigma| > 0\}$. Clearly V is a dense computable set of strings. Let σ be such that $A \supset \sigma * 0^{h(|\sigma|)} * 1$. Then $f(|\sigma|) = h(|\sigma|)$. □

Rupprecht (Ref. 7 Corollary V.46) showed that meager engulfing is equivalent to high. Our proof below goes by way of a further intermediate characterization.

Theorem 4.1. *The following are equivalent for an oracle A.*

(i) *A is high.*
(ii) *There is an effective relative to A sequence* $\{G_k\}$ *of nowhere dense* $\Pi^0_1(A)$*-classes such that each nowhere dense* Π^0_1 *class equals some* G_k*.*
(iii) *A is meager engulfing.*

Proof. (i) $\Rightarrow$ (ii): One says that $f \in {}^{\omega}\omega$ is *dominant* if $g \leq^* f$ for each computable function g. Suppose $A \geq_T f$ for some dominant function f. Fix an effective list $P_0, P_1, \cdots$ of all Π^0_1-classes, and for each i let $h_i(n)$ be the least stage $s > n$ such that for every $\sigma \in 2^n$ there exists some $\tau \succeq \sigma$ such that $[\tau] \cap P_{i,s} = \emptyset$. Then each h_i is partial computable, and h_i is total iff P_i is nowhere dense.

Define the closed set

$$F_{i,n} = \begin{cases} \emptyset, & \text{if } (\exists x > n)\ (h_i(x)\uparrow \ \vee \ h_i(x) > f(x)), \\ P_i, & \text{if otherwise.} \end{cases}$$

Note that the predicate "$(\exists x > n)\ (h_i(x)\uparrow \ \vee \ h_i(x) > f(x))$" is $\Sigma^0_1(A)$ and so $\{F_{i,n}\}$ is an A-computable sequence of $\Pi^0_1(A)$-classes, which are all nowhere dense. Now fix i such that P_i is nowhere dense, i.e. h_i is total. Since f is dominant let n be such that $f(x) > h_i(x)$ for every $x > n$. We have $F_{i,n} = P_i$.

(ii) $\Rightarrow$ (iii): For this easy direction, note that an oracle A is meager engulfing iff there exists an A-effectively meager $F_\sigma(A)$-class containing all nowhere dense Π^0_1-classes (i.e., we may replace being effectively meager by being nowhere dense).

(iii) $\Rightarrow$ (i): Suppose A is meager engulfing. Let $\bigcup_i G_i$ be an A-effectively meager $F_\sigma(A)$-class in the sense of Subsection 3.1 containing all nowhere dense Π^0_1-classes. Let $f(n)$ be defined by

$$f(n) = \max_{i<n, \sigma \in {}^n 2} \{|\tau| \mid \tau \supseteq \sigma \text{ is the first found such that } [\tau] \cap G_i = \emptyset\}.$$

Then f is total (as each G_i is nowhere dense) and $f \leq_T A$.

We claim that the function f is dominant. Suppose not. Let h be a computable function and let the increasing sequence $\{x_n\}$ be such that $h(x_n) > f(x_n)$ for every n. We will define a nowhere dense Π^0_1-class P such that $P \not\subseteq \bigcup_i G_i$. Given string τ and $n \in \omega$ we say that τ is n-good if τ is of the form $0^k * 1 * \tau' * 1$ where $k \in \omega$ and τ' can be any binary string of length $h(n + k + 1)$. An infinite binary string X is good if there are strings $\sigma_0, \sigma_1, \cdots$ such that $X = \sigma_0 * \sigma_1 * \cdots$, σ_0 is 0-good, and for each i, σ_{i+1} is $(|\sigma_0| + |\sigma_1| + \cdots + |\sigma_i|)$-good. Now let P be the set of all infinite binary strings X which are good. Clearly the complement of P is open and is generated by a computable set of basic neighbourhoods. It is also clear that P is nowhere dense, since if $\sigma_0 * \cdots * \sigma_i$ is an initial segment of a good path, then the string $\sigma_0 * \cdots * \sigma_i * 1 * \tau' * 0$ is not extendible in P for any τ' of length $h(|\sigma_0| + |\sigma_1| + \cdots + |\sigma_i| + 1)$.

Now we use the sequence $\{x_n\}$ to build a path $X \in P$ such that $X \notin G_i$ for every i. (We remark that since the sequence $\{x_n\}$ is A-computable, the construction below will produce an A-computable path X). We inductively define strings $\eta_0, \eta_1, \cdots$ such that η_0 is 0-good and η_{i+1} is $(|\eta_0|+|\eta_1|+\cdots+|\eta_i|)$-good for every i. At the end we take $X = \eta_0 * \eta_1 * \cdots$ and so $X \in P$. We will also explicitly ensure that $X \notin G_i$ for any i.

Construction of X. Suppose that $\eta_0, \cdots, \eta_i$ have been defined satisfying the above, so that $\eta_0 * \cdots * \eta_i$ is extendible in P. Find the least j such that $x_j > |\eta_0| + \cdots + |\eta_i|$, and take $k = x_j - |\eta_0 * \cdots * \eta_i| - 1 \geq 0$. Let τ' be any string of length strictly equal to $h(x_j)$ such that $[\eta_0 * \cdots * \eta_i * 0^k * 1 * \tau'] \cap G_{i+1} = \emptyset$. This τ' exists because $h(x_j) > f(x_j)$ and can be found A-computably. Now take $\eta_{i+1} = 0^k * 1 * \tau' * 1$, which will be $(|\eta_0 * \cdots * \eta_i|)$-good.

It is straightforward to check the construction that $X = \eta_0 * \eta_1 * \cdots$ is good and that for every i, $[\eta_0 * \cdots * \eta_i] \cap G_i = \emptyset$. Hence P is a nowhere dense set not contained in $\bigcup_i G_i$, a contradiction. □

Theorem 4.2 (Rupprecht, Ref. 7 Corollary VI.12). *The following are equivalent for an oracle A.*

- *(i) A is in $\mathcal{B}(\neq^*)$, that is, there is some $f \leq_T A$ that eventually disagrees with each computable function.*
- *(ii) A is weakly meager engulfing.*

Proof. (i) $\Rightarrow$ (ii): Let $f \leq_T A$ be given. Then the classes $\mathcal{C}_m = \{x : (\forall n \geq m)(x(n) \neq f(n))\}$ are uniformly $\Pi^0_1(A)$ (in the Baire space) and all $\mathcal{C}_m$ are nowhere dense. Thus the set $D = \bigcup_m \mathcal{C}_m$ is effectively meager and contains all computable functions in ${}^\omega\omega$.

(ii) $\Rightarrow$ (i): We now work in the Cantor space. Suppose that (i) fails. Let $V = \bigcup_e V_e$ be a meager set relative to A. For each n we let σ_n be the first string found such that $V_i \cap [\tau * \sigma_n] = \emptyset$ for every $\tau \in 2^n$ and $i \leq n$. Since the degree of A is not high, let p be a strictly increasing computable function not dominated by the function $n + |\sigma_n| \leq_T A$, with $p(0) > 0$. Call a pair (m, τ) good with respect to n if $p^n(0) \leq m < p^{n+1}(0)$ and $m + |\tau| < p^{n+2}(0)$. Here $p^0(0) = 0$ and $p^{n+1}(0) = p(p^n(0))$. Call two good pairs (m_0, τ_0) and (m_1, τ_1) disjoint if (m_i, τ_i) is good with respect to

n_i and $n_0 + 2 \leq n_1$ or $n_1 + 2 \leq n_0$. By the choice of p there are infinitely many numbers m such that (m, σ_m) is good and pairwise disjoint from each other.

Define $f(n)$ to code the natural sequence $(m_0^n, \sigma_{m_0^n}), \cdots, (m_{3n}^n, \sigma_{m_{3n}^n})$ such that each pair in the sequence is good and the pairs are pairwise disjoint (from each other and from all previous pairs coded by $f(0), \cdots, f(n-1)$). Let h be a computable function infinitely often equal to f. We may assume that each $h(n)$ codes a sequence of the form $(t_0^n, \tau_0^n), \cdots, (t_{3n}^n, \tau_{3n}^n)$ where each pair in the sequence is good for some number larger than n, and that the pairs in $h(n)$ are pairwise disjoint. (Unfortunately we cannot assume that (t_i^n, τ_i^n) and (t_j^m, τ_j^m) are disjoint if $n \neq m$.) This can be checked computably and if $h(n)$ is not of the correct form then certainly $h(n) \neq f(n)$ and in this case we can redefine it in any way we want.

We define the computable real α by the following. We first pick the pair (t_0^0, τ_0^0). Assume that we have picked a pair from $h(i)$ for each $i < n$, and assume that the n pairs we picked are pairwise disjoint. From the sequence coded by $h(n)$ there are $3n + 1$ pairs to pick from, so we can always find a pair from $h(n)$ which is disjoint from the n pairs previously picked. Now define α to consist of all the pairs we picked, i.e. if (t, τ) is picked then we define $\alpha \supset (\alpha \upharpoonright t) * \tau$, and fill in 1 in all the other positions. This α is computable because for each i, $h(i)$ must code pairs which are good for some $n > i$. It is easily checked that α is not in $\bigcup_e V_e$. $\square$

4.2. *Allowed cuts of the diagram*

A *cut* of Cichoń's diagram is a partition of the set of nodes into two nonempty sets L, R such that edges leaving L go to the right or upward. For any cut not contradicting the ternary relationships $\text{add}(\mathcal{M}) = \min(\mathfrak{b}, \text{cover}(\mathcal{M}))$ and $\text{cofin}(\mathcal{M}) = \max(\mathfrak{d}, \text{non}(\mathcal{M}))$, it is consistent with ZFC to assign the cardinal $\aleph_1$ to all nodes in L, and $\aleph_2$ to all nodes in R. See Sections 7.5 and 7.6 of Ref. 5 for all of the models.

We expect the same to be true for the computability-theoretic diagram: for any allowed cut L, R, there is a degree satisfying all the properties in R, and none in L. There are still several open questions. Of course, since there are more equivalences on the computability theoretic side, there are fewer possible combinations.

Here is a list of possible combinations.

(1) *There is a set $A \geq_T$ a Schnorr random which is not high yet of hyperimmune degree.* (This corresponds to the cut where L only

contains the property of highness.) The fact follows by considering a low random real, whose existence is guaranteed by the low basis theorem — see for example Ref. 6 Theorem 1.8.37.

(2) *There is a set $A \geq_T$ a Schnorr random which is weakly Schnorr engulfing and of hyperimmune-free degree.* (This corresponds to the cut where L contains the properties of highness and of having hyperimmune degree.) The fact follows by taking a set A of hyperimmune-free PA degree (see e.g. 1.8.32 and 1.8.42 of Ref. 6 for the existence of such A). This set is weakly Schnorr engulfing (by Theorem 5.3 below) and also computes a Schnorr random (by the Scott basis theorem, see e.g. Ref. 16: 2.21.2 or Ref. 6: 4.3.2). Note that Rupprecht shows in Corollary 27 of Ref. 8 that if B is a hyperimmune-free Schnorr random, then B is not weakly Schnorr engulfing; this shows the example A itself cannot be Schnorr random. Intuitively, the example A is "larger than the Schnorr random". One way to view this (from the forcing-theoretic point of view) is as a two-step iteration: first add a Schnorr random B of hyperimmune-free degree and then a hyperimmune-free weakly Schnorr engulfing A, see e.g. Ref. 8 Proposition 18 or Theorem 19.

(3) *There is a set A which is not weakly Schnorr engulfing and computes a Schnorr random.* This follows by taking A to be Schnorr random of hyperimmune-free degree (which is possible by the basis theorem for computably dominated sets, 1.8.42 of Ref. 6), see Ref. 8 Corollary 27.

(4) *There is a weakly meager engulfing set A of hyperimmune degree which computes no Schnorr random.* Miller[17] (see also Ref. 16, 13.8) proved that there is a Δ^0_2 set A which has effective Hausdorff dimension $\frac{1}{2}$ and does not compute a real of higher dimension. Such A necessarily is DNR (Ref. 16, 13.7.6) and does not compute a Martin-Löf random. Rupprecht in Theorem VI.19 of Ref. 7 showed that A is low_2. In particular, A is not high and thus does not compute a Schnorr random either (Ref. 6, 3.5.13). Since A is Δ^0_2, it is of hyperimmune degree.

(5) *There is a set of hyperimmune degree which is not weakly meager engulfing.* Any non recursive low r.e. set will suffice. By Arslanov's completeness criterion (Ref. 6, 4.1.11), such a set cannot be DNR. Rupprecht provides another source of examples in Theorem VI.4 of his thesis[7], showing that no 2-generic real is weakly meager engulfing.

(6) *There is a weakly Schnorr engulfing set which is low for weak 1-genericity.* This is Ref. 8 Theorem 19.

Question 4.1.

(7) Is there a weakly meager engulfing set A which does not compute a Schnorr random, is of hyperimmune-free degree, and is weakly Schnorr engulfing? (L consists of highness, computing a Schnorr random, and being of hyperimmune degree.)
(8) Is there a weakly meager engulfing set which neither computes a Schnorr random nor is weakly Schnorr engulfing? (R consists of weakly meager engulfing, not low for weak 1 genericity, and not low for Schnorr tests.)
(9) Is there a set which is not low for Schnorr tests, is low for weak 1-genericity and not weakly Schnorr engulfing? (R only contains the property of being not low for Schnorr tests.)

Kumabe and Lewis[18] show that there is a set A of minimal Turing degree which is DNR. It was observed in Theorems 1.1 and 3.1 of Ref. 19 that this construction also makes A of hyperimmune-free degree. Since A cannot compute a Schnorr random set, there is a positive answer to (7) or (8) in Question 4.1. We note that a positive answer to (9) would refute Conjecture 31 of Ref. 8. Kjos-Hanssen and Stephan have announced an affirmative answer to further questions stated above.

5. Other cardinal characteristics and their analogs

5.1. *Kurtz randomness and closed measure zero sets*

A closed measure zero set necessarily is nowhere dense. Thus the σ-ideal $\mathcal{E}$ generated by closed measure zero sets is contained in both $\mathcal{M}$ and $\mathcal{N}$; in fact, it is properly contained in $\mathcal{M} \cap \mathcal{N}$. In a combinatorially intricate work[20], Bartoszyński and Shelah computed the cardinal characteristics of $\mathcal{E}$ (see alternatively Section 2.6 of Ref. 5). Main results are:

(A) $\text{add}(\mathcal{E}) = \text{add}(\mathcal{M})$ and, dually, $\text{cofin}(\mathcal{E}) = \text{cofin}(\mathcal{M})$;
(B) $\text{add}(\mathcal{E}, \mathcal{N}) = \text{cover}(\mathcal{M})$ and, dually, $\text{cofin}(\mathcal{E}, \mathcal{N}) = \text{non}(\mathcal{M})$;
(C) if $\text{cover}(\mathcal{M}) = \mathfrak{d}$, then $\text{cover}(\mathcal{E}) = \max\{\text{cover}(\mathcal{M}), \text{cover}(\mathcal{N})\}$; dually, if $\text{non}(\mathcal{M}) = \mathfrak{b}$, then $\text{non}(\mathcal{E}) = \min\{\text{non}(\mathcal{M}), \text{non}(\mathcal{N})\}$.

Here, for two ideals $\mathcal{I} \subseteq \mathcal{J}$, $\text{add}(\mathcal{I}, \mathcal{J})$ denotes the least size of a family of sets in $\mathcal{I}$ whose union does not belong to $\mathcal{J}$. Similarly, $\text{cofin}(\mathcal{I}, \mathcal{J})$ is the smallest cardinality of a subfamily $\mathcal{F}$ of $\mathcal{J}$ such that all members of $\mathcal{I}$ are contained in a set from $\mathcal{F}$.

The notion corresponding to $\mathcal{E}$ and its characteristics on the computability theory side is *Kurtz randomness*: a *Kurtz test* is an effective sequence (G_m) of clopen sets such that each G_m has measure at most 2^{-m}. The corresponding null Π^0_1 class $\bigcap_m G_m$ is called a *Kurtz null set.* It is well-known and easy to see that the definition of Kurtz null set is unchanged if we additionally assume $G_{m+1} \subseteq G_m$ for all m. A real A is *Kurtz random* if it passes all Kurtz tests, i.e., A avoids $\bigcap_m G_m$ for all Kurtz tests (G_m). An oracle A is *low for Kurtz tests* if for every Kurtz test (G_m) relative to A, there is a Kurtz test (L_k) such that $\bigcap_m G_m \subseteq \bigcap_k L_k$. A is *low for Kurtz randomness* if every Kurtz random is Kurtz random relative to A. Finally, A is *low for Schnorr-Kurtz* if every Schnorr random is Kurtz random relative to A.

Greenberg and Miller (Ref. 21, Theorem 1.1) proved that a set is low for Kurtz tests iff it is low for weak 1-genericity. This is the computability theoretic analogue of the dual form cofin($\mathcal{E}$) = cofin($\mathcal{M}$) of (A) above. (They also observed that low for Kurtz randomness is the same as low for Kurtz tests.) Furthermore, they showed (Ref. 21, Corollary 1.3) that a set is low for Schnorr-Kurtz iff it is neither DNR nor high. Thus, by Ref. 12 Theorem 5.1 and our Theorem 4.2, a set is not low for Schnorr-Kurtz iff it is weakly meager engulfing. This corresponds to cofin($\mathcal{E}, \mathcal{N}$) = non($\mathcal{M}$) in (B) above. Finally, it is well-known (see e.g. Ref. 6 Proposition 3.6.4) that a Kurtz random either is of hyperimmune degree (and thus contains a weakly 1-generic) or is already Schnorr random, an analogue of the first part of (C) above. (Note that the antecedent cover($\mathcal{M}$) = $\mathfrak{d}$ is true in computability theory: up to Turing degree, weakly 1-generic = hyperimmune.)

We now look into the computability theoretic aspect of the dual results of the Bartoszyński-Shelah theorems. To this end, say that an oracle A is *Kurtz engulfing* if there is an A-computable sequence $\{G^i\} = \{(G^i_m)\}$ of Kurtz tests relative to A such that each Kurtz null set $\bigcap_m L_m$ is contained in some $\bigcap_m G^i_m$. A is *weakly Kurtz engulfing* if there is such $\{G^i\} = \{(G^i_m)\}$ such that $\bigcup_i \bigcap_m G^i_m$ contains all computable reals. Finally, A is *Schnorr-Kurtz engulfing* if the union of all Kurtz null sets is a Schnorr null set relative to A. Then we obtain:

Theorem 5.1. *The following are equivalent for an oracle A.*

(i) A is high.
(ii) There is a Kurtz test (G_m) *relative to A such that for all Kurtz tests* (L_m) *and almost all m,* $L_{2m} \cup L_{2m+1} \subseteq G_m$.

(iii) *There is a Kurtz test* (G_m) *relative to* A *such that for all Kurtz tests* (L_m) *there is an* m_0 *with* $\bigcap_{m \geq m_0} L_m \subseteq \bigcap_{m \geq m_0} G_m$.

(iv) A *is Kurtz engulfing.*

Proof. (i) $\Rightarrow$ (ii): We use Theorem 6 of Ref. 8: since A is high, there is a trace $\sigma \leq_T A$ tracing all computable functions. Fix a computable coding $c(m,i)$ of all basic clopen sets in 2^ω such that for fixed m, $c(m,\cdot)$ lists all basic clopen sets of measure $\leq 2^{-m}$. Let $G_m = \bigcup\{c(2m,i) : i \in \sigma(2m)\} \cup \bigcup\{c(2m+1,i) : i \in \sigma(2m+1)\}$. Then $\lambda G_m \leq 2m \cdot 2^{-2m} + (2m+1) \cdot 2^{-(2m+1)} < 2^{-m}$ for $m \geq 4$. So, changing finitely many G_m if necessary, we may think of (G_m) as a Kurtz test relative to A. Now, given a Kurtz test $L = (L_m)$, define a function $f = f_L$ by $f(m) = \min\{i : L_m = c(m,i)\}$. Clearly f is computable and $L_m = c(m, f(m))$. Thus $f \in^* \sigma$. This means that $L_{2m} \cup L_{2m+1} = c(2m, f(2m)) \cup c(2m+1, f(2m+1)) \subseteq G_m$ for almost all m, as required.

(ii) $\Rightarrow$ (iii) $\Rightarrow$ (iv): Trivial.

(iv) $\Rightarrow$ (i): Assume A is not high and $\{G^i\} = \{(G^i_m)\}$ is an A-computable sequence of Kurtz tests relative to A. This means that there are sequences (k^i_m) and $(\sigma^i_{m,j})$ computable in A such that each G^i_m is of the form $G^i_m = \bigcup_{j<k^i_m} [\sigma^i_{m,j}]$. It is easy to see that each $\sigma^i_{m,j}$ belongs to 2^k for some $k \geq m$. It suffices to find a Kurtz test (L_m) and a set $X \leq_T A$ such that $X \in \bigcap_m L_m$ yet $X \notin \bigcup_i \bigcap_m G^i_m$, that is, for each i there is m with $X \notin G^i_m$.

Define a function $f \leq_T A$ as follows.

$$f(0) = 0$$
$$f(n+1) = \min\{k : \forall m \leq f(n) + (n+1)^2 \; \forall i < n+1 \; \forall j < k^i_m \; (\sigma^i_{m,j} \in 2^{\leq k})\}$$

Then clearly $f(n+1) \geq f(n) + (n+1)^2$; in particular, f is strictly increasing. Since A is not high, there is a computable function h with $h \not\leq^* f$. We may assume that h is strictly increasing as well. Let J_m be the interval $[h(m), h(m)+m)$ of length m. Put $L_m = \{X : X \upharpoonright J_m \equiv 0\}$. Then $\lambda L_m = 2^{-m}$ and therefore (L_m) is a Kurtz test. We produce the required $X \in \bigcap_m L_m \setminus \bigcup_i \bigcap_m G^i_m$ by recursively defining $X \upharpoonright f(n)$ such that $X \upharpoonright (f(n) \cap J_m) \equiv 0$ for all m.

$X \upharpoonright f(0) = X \upharpoonright 0$ is the trivial sequence. Assume $X \upharpoonright f(n)$ has been defined. If $h(n+1) \leq f(n+1)$ simply extend $X \upharpoonright f(n)$ to $X \upharpoonright f(n+1)$ such that $X \upharpoonright (f(n+1) \cap J_m) \equiv 0$ for all m. So suppose $h(n+1) > f(n+1)$. Note that, when extending X from $f(n)$ to $f(n+1)$, there are at most $|\bigcup_{m \leq n} J_m| \leq n^2$ many places where X necessarily has to assume the value

0. Hence the measure of the set of possible extensions of X to $f(n+1)$ is $\geq 2^{-(f(n)+n^2)}$. On the other hand,

$$\lambda\left(\bigcup\{G^i_{f(n)+(n+1)^2} : i \leq n\}\right) \leq (n+1)2^{-(f(n)+(n+1)^2)} < 2^{-(f(n)+n^2)}$$

This means that we can extend $X \upharpoonright f(n)$ to $X \upharpoonright f(n+1)$ such that $X \upharpoonright (f(n+1) \cap J_m) \equiv 0$ for all m and $[X \upharpoonright f(n+1)] \cap G^i_{f(n)+(n+1)^2} = \emptyset$ for all $i \leq n$. Since $h(n+1) > f(n+1)$ for infinitely many n, X is as required. □

Theorem 5.2. *The following are equivalent for an oracle A.*

(i) A is of hyperimmune degree.
(ii) A is Schnorr-Kurtz engulfing.

Proof. (i) ⇒ (ii): By Proposition 4.1, we know that there is a function $f \leq_T A$ infinitely often equal to all computable reals. As in the previous proof, let $c(n, \cdot)$ be a computable coding of all basic clopen sets of measure $\leq 2^{-n}$. Then $h(n) = c(n, f(n))$ is a sequence of clopen sets computable in A with $\lambda(h(n)) \leq 2^{-n}$. Thus, by Ref. 8 Proposition 3, $N = \bigcap_m \bigcup_{n \geq m} h(n)$ is a Schnorr null set relative to A. We need to show it contains all Kurtz null sets. Let (G_n) be a Kurtz test. Then $G_n = c(n, k(n))$ for some computable function k. Hence $k(n) = f(n)$ for infinitely many n. Now assume $X \in \bigcap_n G_n$. Fix any m. There is $n \geq m$ with $k(n) = f(n)$. Thus $X \in G_n = c(n, k(n)) = c(n, f(n)) = h(n) \subseteq \bigcup_{\ell \geq m} h(\ell)$. Unfixing m we see that X belongs to N, as required.

(ii) ⇒ (i): Assume A is of hyperimmune-free degree and N is a Schnorr null set relative to A. By Ref. 8 Proposition 3, we may assume that $N = \bigcap_n \bigcup_{m \geq n} E_m$ where E_m is a sequence of clopen sets computable in A with $\lambda E_m \leq 2^{-m}$. There are sequences (k_m) and $(\sigma_{m,j})$ computable in A such that each E_m is of the form $E_m = \bigcup_{j<k_m} [\sigma_{m,j}]$. It suffices to find a Kurtz test (L_m) and a set $X \leq_T A$ such that $X \in \bigcap_m L_m$ yet $X \notin N$, that is, there is some m_0 with $X \notin \bigcup_{m \geq m_0} E_m$.

We proceed as in the proof of (iv) ⇒ (i) of Theorem 5.1. Define $f \leq_T A$ by:

$$f(0) = 0$$
$$f(n+1) = \min\{k : \forall m \leq f(n) + (n+1)^3 \; \forall j < k_m \; (\sigma_{m,j} \in 2^{\leq k})\}$$

Since A is of hyperimmune-free degree, there is a computable function h with $h \geq^* f$. We may assume that $h \geq f$ everywhere and that h is strictly increasing. As in the previous proof, let J_m be the interval $[h(m), h(m)+m)$ of length m and define a Kurtz test (L_m) by $L_m = \{X : X \upharpoonright J_m \equiv 0\}$. Let

$m_0 = f(2) + 28$. We produce the required $X \in \bigcap_m L_m \setminus \bigcup_{m \geq m_0} E_m$ by recursively defining $X \restriction f(n)$ such that

(a) $X \restriction (f(n) \cap J_m) \equiv 0$ for all $m(< n)$,
(b) $[X \restriction f(n)] \cap \bigcup\{E_\ell : f(2) + 27 < \ell \leq f(n-1) + n^3\} = \emptyset$,
(c) $\lambda([X \restriction f(n)] \cap \bigcup\{E_\ell : f(n-1) + n^3 < \ell \leq f(n) + (n+1)^3\}) < 2^{-(f(n)+(n-1)n^2)}$ for $n \geq 3$.

Clearly, an X satisfying the first two properties for all n is as required. The third property is used to guarantee the second property along the recursive construction.

$X \restriction f(2)$ is arbitrary satisfying (a). When defining $X \restriction f(3)$, (b) vacuously holds. Between $f(2)$ and $f(3)$, there are at most $|J_0 \cup J_1 \cup J_2| = 3$ many places where X necessarily has to assume the value 0. Hence the set of extensions of X to $f(3)$ satisfying (a) for $n = 3$ has measure at least $2^{-(f(2)+3)}$. On the other hand,

$$\lambda\left(\bigcup\{E_\ell : f(2) + 27 < \ell \leq f(3) + 64\}\right) < 2^{-(f(2)+27)}$$

This means that the relative measure of the latter set in the set of possible extensions is smaller than $\frac{2^3}{2^{3^3}} < 2^{3^2-3^3} = 2^{-2 \cdot 3^2}$. Hence there must be one such extension $X \restriction f(3)$ satisfying

$$\lambda\left([X \restriction f(3)] \cap \bigcup\{E_\ell : f(2) + 27 < \ell \leq f(3) + 64\}\right) < 2^{-(f(3)+2 \cdot 3^2)}$$

Thus (c) holds for $n = 3$.

More generally, suppose $X \restriction f(n)$ has been defined for $n \geq 3$. Between $f(n)$ and $f(n+1)$, there are at most $|\bigcup_{m \leq n} J_m| \leq n^2$ many places where X necessarily has to assume the value 0. Hence the set of extensions of X to $f(n+1)$ satisfying (a) for $n+1$ has measure at least $2^{-(f(n)+n^2)}$. By (c) for n we see that the set of extensions satisfying both (a) and (b) for $n+1$ has measure at least $2^{-(f(n)+n^2+1)}$. On the other hand,

$$\lambda\left(\bigcup\{E_\ell : f(n) + (n+1)^3 < \ell \leq f(n+1) + (n+2)^3\}\right) < 2^{-(f(n)+(n+1)^3)}$$

This means that the relative measure of the latter set in the set of possible extensions is smaller than $\frac{2^{n^2+1}}{2^{(n+1)^3}} < 2^{(n+1)^2-(n+1)^3} = 2^{-n(n+1)^2}$. Hence there must be one such extension $X \restriction f(n+1)$ satisfying

$$\lambda\left([X \restriction f(n+1)] \cap \bigcup\{E_\ell : f(n) + (n+1)^3 < \ell \leq f(n+1) + (n+2)^3\}\right) < 2^{-(f(n+1)+n(n+1)^2)}.$$

Again, this gives (c) for $n+1$. □

Theorem 5.3. *Each PA set is weakly Kurtz engulfing.*

Proof. Fix an oracle A computing a $\{0,1\}$-valued function g such that for all e, if $J(e) := \varphi_e(e) \downarrow$, then we have $J(e) \neq g(e)$. If φ_e is $\{0,1\}$-valued and total then it gives rise naturally to a computable real X where $X(n) = \varphi_e(n)$. Furthermore every computable real can be identified with a total φ_e for some e.

There is a computable sequence $\{R_e\}$ of pairwise disjoint computable sets such that for every e and n, if $\varphi_e(n) \downarrow$ then $J(r_e(n)) \downarrow = \varphi_e(n)$, where $r_e(n)$ is the nth element of R_e. Now define the A-Kurtz test $\{G_k\}$ by

$$G_k = \bigcup_{i<k} [Z_i \upharpoonright 2k],$$

where for every i, Z_i is the infinite binary sequence defined by $Z_i(j) = 1 - g(r_i(j))$. It is then easy to see that every computable real X belongs to $\bigcup_e \bigcap_{k \geq e} G_k$. □

Corollary 5.1. *There is a hyperimmune-free weakly Kurtz engulfing degree.*

Proof. It is well-known that there is a hyperimmune-free PA degree. (Use the fact that the PA degrees form a Π^0_1 class and the basis theorem for computably dominated sets. See Ref. 6 1.8.32 and 1.8.42.) See also item (2) in Section 4.2 above. □

We have no characterization of "weakly Kurtz engulfing" in terms of the other properties and conjecture there is none. More specifically, we conjecture there is a set both weakly meager engulfing and weakly Schnorr engulfing that is not weakly Kurtz engulfing. (Note that the antecedent of the second part of (C), $\text{non}(\mathcal{M}) = \mathfrak{b}$, is false in computability theory: high is strictly stronger than weakly meager engulfing.)

5.2. *Specker-Eda number and its dual*

The *Specker-Eda number* $\mathfrak{se}$ is a cardinal characteristic introduced by Blass[22] in the context of homomorphisms of abelian groups. Whilst the original definition would take us too far afield, there is an equivalent formulation due to Brendle and Shelah[23] that fits in well with the cardinal characteristics we have already considered.

Definition 5.1. A *partial g-slalom* is a function $\varphi : D \to [\omega]^{<\omega}$ with domain D an infinite subset of ω, satisfying $|\varphi(n)| \leq g(n)$ for all $n \in D$.

In defining (total) slaloms in Section 2.2, we implicitly took g to be the identity function. In the set-theoretic context, the specific choice of g is in fact irrelevant for our purposes so long as it goes to infinity; we have given the definition in this way for the sake of the analogy to come. As such, for Definition 5.2 we think of g as being fixed.

Definition 5.2. The Specker-Eda number $\mathfrak{se}$ is the unbounding number for the relation of being traced by a partial slalom:

$$\mathfrak{se} = \mathfrak{b}(\in_p^*) = \min\{|\mathcal{F}| : \mathcal{F} \subseteq \omega^\omega \wedge \forall \text{ partial slalom } \sigma \exists f \in \mathcal{F} \\ \exists^\infty n \in \mathrm{dom}(\sigma)(f(n) \notin \sigma(n))\}.$$

We denote its dual by $\mathfrak{d}(\in_p^*)$:

$$\mathfrak{d}(\in_p^*) = \min\{|\Phi| : \Phi \text{ is a set of partial slaloms } \wedge \\ \forall f \in \omega^\omega \exists \sigma \in \Phi \forall^\infty n \in \mathrm{dom}(\sigma)(f(n) \in \sigma(n))\}.$$

The cardinal $\mathfrak{se}$ sits in the curious part of the diagram in which cardinals are different set-theoretically but their computability-theoretic analogues are equivalent notions: $\mathrm{add}(\mathcal{N}) \leq \mathfrak{se} \leq \mathrm{add}(\mathcal{M})$, and each of these inequalities may be strict — see Ref. 23 Corollary (b).

The fact that we are considering partial slaloms raises the consideration of partial computable rather than just computable traces. For the analogue of $\mathfrak{se}$, we shall see that the choice is immaterial.

Definition 5.3. Given an infinite computably enumerable subset D of ω and a partial computable function g dominating the identity with $D \subseteq \mathrm{dom}(g)$, a (D, g)*-trace* is a partial computable function σ from D to $[\omega]^{<\omega}$ such that for all $n \in D$, $|\sigma(n)| < g(n)$. A *partial trace* is a (D, g)-trace for some D and g. We say that a (D, g)-trace σ *traces* $f : \omega \to \omega$ if $f(n) \in \sigma(n)$ for all but finitely many $n \in D$.

The computability-theoretic analogue of $\mathfrak{se}$ is the property that A computes a partial trace tracing every computable function. This property coincides with being high.

Theorem 5.4. *The following are equivalent for any oracle A.*

(i) A computes a trace tracing every computable function (i.e. A is high).
(ii) A computes a partial trace tracing every computable function.

Proof. (i) $\Rightarrow$ (ii) is trivial. For the reverse direction, it suffices by Theorem 6 of Ref. 8 to show that (ii) implies A is high. So suppose σ is a (D, g)-trace computed by A, that is, D is c.e. in A and both g and σ are partially computable in A, tracing every computable function; we wish to show that A computes a function eventually dominating every computable function. Fix a computable enumeration of D, and for each $m \in \omega$ let n_m be the first natural number greater than or equal to m that appears in this enumeration of D. We define $h : \omega \to \omega$ by $h(m) = \max(\sigma(n_m))$. Now, let f be a computable function from ω to ω; without loss of generality we may assume that f is non-decreasing. There is some $m_0 \in \omega$ such that for all $n \in D \setminus m_0$, $f(n) \in \sigma(n)$. In particular, for all $m \geq m_0$, $f(m) \leq f(n_m) \in \sigma(n_m)$, and so $f(m) \leq h(m)$. $\square$

We turn now to the analogue of $\mathfrak{d}(\in^*_p)$: the property that A computes a function not traceable by any partial trace. With "partial trace" defined as in Definition 5.3, this property coincides with being of hyperimmune degree, and the proof is straightforward.

Theorem 5.5. *The following are equivalent for any oracle A.*

(i) A is of hyperimmune degree.
(ii) A computes a function not traceable by any partial trace.
(iii) For every partial computable function g, there exists $f \leq_T A$ such that f cannot be traced by a (D, g)-trace for any c.e. D.

Proof. (i) $\Rightarrow$ (ii): Let f be a function computable from A which is dominated by no computable function. We may assume that f is increasing. If f is traced by a (D, g)-trace σ then we let $h(n) = \max(\sigma(x_n))$ where x_n is the first number greater than or equal to n that is enumerated in D. Then $h + 1$ is a computable function dominating f, which is impossible.

(ii) $\Rightarrow$ (iii): Trivial.

(iii) $\Rightarrow$ (i): Let $g(\langle e, n\rangle) = \varphi_e(\langle e, n\rangle) + 1$, where φ_e is the eth partial computable function. Let f be an A-computable function that cannot be traced by any (D, g)-trace. Then we claim that f cannot be dominated by a computable function. If φ_e dominates f then we take $D = \{\langle e, n\rangle : n \in \omega\}$ and take $\sigma(x) = \{0, \cdots, \varphi_e(x)\}$. Clearly σ is a (D, g) trace tracing f, contradiction. $\square$

So the notion of "partial computably traceable" coincides with being hyperimmune-free if the bound g is allowed to be partial. If g is required to be total, however, we obtain a different notion. This latter notion is obviously still weaker than being computably traceable, but is now strictly stronger than being hyperimmune-free. In fact:

Proposition 5.1. *Suppose that there is a total computable function g such that for every $f \leq_T A$ there is a (D, g)-trace tracing f. Then A is of hyperimmune-free degree and not DNR.*

Proof. By Theorem 5.5 we get that A is of hyperimmune-free degree. To show that A is not DNR, let f be A-computable. Let σ be a (D, g)-trace tracing f; we may assume that in fact $f(n) \in \sigma(n)$ for all $n \in D$. Viewing σ as a c.e. trace then allows us by Ref. 12 Theorem 6.2 to deduce that A is not DNR. □

In other words, if we consider the property that for every total computable g, there is an $f \leq_T A$ such that no (D, g)-trace traces f, then this property lies between "not low for weak 1-genericity" and "not low for Schnorr tests" in Figure 5. This is analogous to the fact in the set-theoretic setting that $\text{cofin}(\mathcal{M}) \leq \mathfrak{d}(\in_p^*) \leq \text{cofin}(\mathcal{N})$.

5.3. *Final comments*

The *splitting number* $\mathfrak{s}$ is the least size of a subset $\mathcal{S}$ of $\mathcal{P}(\omega)$ such that every infinite set is split by a set in $\mathcal{S}$ into two infinite parts. The analog in computability theory is *r-cohesiveness*: an infinite set $A \subseteq \omega$ is r-cohesive if it cannot be split into two infinite parts by a computable set; that is, if B is computable, then either $A \subseteq^* B$ or $A \cap B$ is finite. A is *cohesive* if it cannot be split into two infinite parts by a computably enumerable set. Clearly, cohesive implies r-cohesive. ZFC proves that $\mathfrak{s} \leq \mathfrak{d}, \text{non}(\mathcal{E})$ (Ref. 24 Theorems 3.3 and 5.19). On the computability side, r-cohesive implies both being of hyperimmune degree, and weakly Kurtz engulfing. On the other hand, $\mathfrak{s} < \text{add}(\mathcal{N})$[25] (see also Ref. 5 Theorem 3.6.21), $\mathfrak{s} > \text{cover}(\mathcal{E})$, and $\mathfrak{s} > \mathfrak{b}$ are known to be consistent (the latter two follow from Ref. 26, see the next paragraph). The first has no analog in recursion theory for high implies cohesive[27], while the last does by a result of Jockusch and Stephan[28], who showed that a cohesive set can be non-high. We do not know whether every (r-)cohesive degree computes a Schnorr random.

The dual of the splitting number is the *unreaping number* $\mathfrak{r}$, the least size of a subset $\mathcal{S}$ of $[\omega]^\omega$ (the infinite subsets of ω) such that every subset of ω is either almost disjoint from, or almost contains, a member of $\mathcal{S}$. To see the duality, consider the relation $R \subseteq [\omega]^\omega \times [\omega]^\omega$ defined by $\langle x, y \rangle \in R$ iff y splits x iff both $x \cap y$ and $x \setminus y$ are infinite. Then $\mathfrak{s} = \mathfrak{d}(R)$ and $\mathfrak{r} =$

$\mathfrak{b}(R)$. The analog of $\mathfrak{r}$ is being of *bi-immune degree*, a property introduced by Jockusch[29]: $A \subseteq \omega$ is *bi-immune* if it splits every infinite computable set or, equivalently, if it splits every infinite computably enumerable set, that is, if neither A nor its complement contain an infinite computable (or computably enumerable) set. In ZFC $\mathfrak{r} \geq \mathfrak{b}, \text{cover}(\mathcal{E})$ holds (Ref. 24 Theorems 3.8 and 5.19). Similarly, Kurtz random (and thus also being of hyperimmune degree) implies bi-immune. Jockusch and Lewis[19] recently showed that DNR implies having bi-immune degree, so that, in fact, not low for weak 1-generic implies bi-immune. This is very different from the situation in set theory, where $\mathfrak{r} < \mathfrak{s}$ (and thus also $\mathfrak{r} < \text{non}(\mathcal{E})$ and $\mathfrak{r} < \mathfrak{d}$) is consistent[26]. We do not know whether there is a weakly Schnorr-engulfing degree that is not bi-immune.

Rupprecht also briefly discusses the analogy between splitting/unreaping and r-cohesive/bi-immune in his thesis[7] Theorems V.41, 42, 43; his treatment is less comprehensive than ours above.

For further open problems see Question 4.1 above.

References

1. D. A. Martin, Completeness, the Recursion Theorem, and effectively simple sets, *Proc. Amer. Math. Soc.* **17**, 838 (1966).
2. S. Terwijn and D. Zambella, Computational randomness and lowness, *J. Symbolic Logic* **66**, 1199 (2001).
3. T. Bartoszyński, Additivity of measure implies additivity of category, *Trans. Amer. Math. Soc.* **281**, 209 (1984).
4. J. Raisonnier, A mathematical proof of S. Shelah's theorem on the measure problem and related results, *Israel J. Math.* **48**, 48 (1984).
5. T. Bartoszyński and H. Judah, *Set Theory. On the structure of the real line* (A K Peters, Wellesley, MA, 1995), 546 pages.
6. A. Nies, *Computability and randomness*, Oxford Logic Guides, Vol. 51 (Oxford University Press, Oxford, 2009).
7. N. Rupprecht, Effective correspondences to cardinal characteristics in Cichoń's diagram, Ph.D. Dissertation, University of Michigan, (University of Michigan, 2010).
8. N. Rupprecht, Relativized Schnorr tests with universal behavior, *Arch. Math. Logic* **49**, 555 (2010).
9. T. Bartoszyński, Combinatorial aspects of measure and category, *Fund. Math.* **127**, 225 (1987).
10. A. Miller, A characterization of the least cardinal for which the Baire

category theorem fails, *Proc. Amer. Math. Soc.* **86**, 498 (1982).
11. F. Stephan and L. Yu, Lowness for weakly 1-generic and Kurtz-random, in *Theory and applications of models of computation*, eds. J.-Y. Cai, S. B. Cooper and A. Li, Lecture Notes in Comput. Sci., Vol. 3959 (Springer, Berlin, 2006) pp. 756–764.
12. B. Kjos-Hanssen, W. Merkle and F. Stephan, Kolmogorov complexity and the Recursion Theorem, *Trans. Amer. Math. Soc.* **363**, 5465 (2011).
13. A. Nies, F. Stephan and S. Terwijn, Randomness, relativization and Turing degrees, *J. Symbolic Logic* **70**, 515 (2005).
14. B. Kjos-Hanssen, W. Merkle and F. Stephan, Kolmogorov complexity and the Recursion Theorem, in *STACS 2006*, ed. B. Durand, Lecture Notes in Comput. Sci., Vol. 3884 (Springer, Berlin, 2006) pp. 149–161.
15. S. Kurtz, Randomness and genericity in the degrees of unsolvability, ph.D. Dissertation, University of Illinois, Urbana, (University of Illinois at Urbana-Champaign, 1981).
16. R. Downey and D. Hirschfeldt, *Algorithmic randomness and complexity* (Springer-Verlag, Berlin, 2010), 855 pages.
17. J. S. Miller, Extracting information is hard: A Turing degree of non-integral effective Hausdorff dimension, *Adv. Math.* **226**, 373 (2011).
18. M. Kumabe and A. E. M. Lewis, A fixed point free minimal degree, *J. London Math. Soc.* **80**, 785 (2009).
19. C. G. Jockusch, Jr. and A. E. M. Lewis, Diagonally non-computable functions and bi-immunity, *J. Symbolic Logic* **78**, 977 (2013).
20. T. Bartoszyński and S. Shelah, Closed measure zero sets, *Ann. Pure Appl. Logic* **58**, 93 (1992).
21. N. Greenberg and J. S. Miller, Lowness for Kurtz randomness, *J. Symbolic Logic* **74**, 665 (2009).
22. A. Blass, Cardinal characteristics and the product of countably many infinite cyclic groups, *J. Algebra* **169**, 512 (1994).
23. J. Brendle and S. Shelah, Evasion and prediction. II, *J. London Math. Soc. (2)* **53**, 19 (1996).
24. A. Blass, Combinatorial cardinal characteristics of the continuum, in *Handbook of Set Theory*, eds. M. Foreman and A. Kanamori (Springer, Dordrecht Heidelberg London New York, 2010) pp. 395–489.
25. H. Judah and S. Shelah, Suslin forcing, *J. Symbolic Logic* **53**, 1188 (1988).
26. A. Blass and S. Shelah, There may be simple $P_{\aleph_1}$- and $P_{\aleph_2}$-points and the Rudin-Keisler ordering may be downward directed, *Ann. Pure*

Appl. Logic **33**, 213 (1987).
27. C. G. Jockusch, Jr., Upward closure and cohesive degrees, *Israel J. Math.* **15**, 332 (1973).
28. C. G. Jockusch, Jr. and F. Stephan, A cohesive set which is not high, *Math. Log. Quart.* **39**, 515 (1993).
29. C. G. Jockusch, Jr., The degrees of bi-immune sets, *Z. Math. Logik Grundlagen Math.* **15**, 135 (1969).

A Non-Uniformly C-Productive Sequence & Non-Constructive Disjunctions

John Case* and Michael Ralston

Computer & Information Sciences, University of Delaware,
Newark, DE 19716 USA
**E-mail: case@udel.edu*
www.cis.udel.edu

Yohji Akama

Department of Mathematics, Graduate School of Science, Tohoku University
Sendai, Japan
E-mail: akama@m.tohoku.ac.jp

Let φ be an acceptable programming system/numbering of the partial computable functions: $\mathbb{N} = \{0, 1, 2, \ldots\} \to \mathbb{N}$, where, for $p \in \mathbb{N}$, φ_p is the partial computable function computed by program p of the φ-system and $W_p =$ domain(φ_p). W_p is, then, the computably enumerable (c.e.) set accepted by p.

The first author taught in a class a recursion theorem proof *employing a pair of cases* that, for *each* q, $\{x \mid \varphi_x = \varphi_q\}$ is not computably enumerable. The particular pair of cases employed was: domain(φ_q) is infinite vs. finite — incidentally, by Rice's Theorem, a *non*-constructive disjunction. Some student asked why the proof involved an analysis by cases. The answer given straightaway to the student was that his teacher didn't know how else to do it.

The present paper provides, among other things, *a better answer*: any proof that, for each q, $\{x \mid \varphi_x = \varphi_q\}$ is not c.e. *provably must* involve some such *non*-constructiveness. Furthermore, *completely characterized* are all the possible such pairs of (index set based) cases that will work.

Along the way, relevantly explored are *uniform* lifts of the concept completely productive (abbr: *c-productive*) sets to *sequences* of sets.

The c-productive sets are the *effectively* non-c.e. sets.

A set *sequence* $S_q, q \in \mathbb{N}$, is said to be *uniformly c-productive* iff there is a computable f so that, for all q, x, $f(q, x)$ is a counterexample to $S_q = W_x$.

Relevance: a completely constructive proof that each S_q is not c.e. would *entail* the S_qs forming a uniformly c-productive sequence.

Relevantly shown, then, is that the *sequence* $\{x \mid \varphi_x = \varphi_q\}, q \in \mathbb{N}$, is *not* uniformly c-productive — and this even though, of course, the *set* $\{\langle q, x\rangle \mid \varphi_x = \varphi_q\}$ *is* c-productive.

For some results we provide upper and/or lower bounds for them as to position in the Arithmetical Hierarchy of the Law of Excluded Middle needed in addition to Heyting Arithmetic.

Results are also obtained for similar problems, including for run-time bounded programming systems and other subrecursive systems.

Proof methods include recursion theorems.

Keywords: Complete Productivity; Uniformly Computable; Index Sets; Constructivity; Arithmetically Limited Law of Excluded Middle; Clocked Programming Systems; Recursion Theorems.

1. Motivation

Let φ be an *acceptable programming system/numbering* of the partial computable functions: $\mathbb{N} = \{0, 1, 2, \ldots\} \to \mathbb{N}$, where, for $p \in \mathbb{N}$, φ_p is the partial computable function computed by *program* p of the φ-system; such numberings are characterized as being intercompilable with naturally occurring general purpose programming formalisms such as a full Turing machine formalism; let $W_p = \text{domain}(\varphi_p)$ [1–4]. W_p is, then, the computably enumerable (c.e.) set $\subseteq \mathbb{N}$ accepted by φ-program p.

The first author taught a recursion theorem proof by employing a pair of cases that, for *each* q, $\{x \mid \varphi_x = \varphi_q\}$ is *not* c.e. The disjunction of cases used was:

$$\text{domain}(\varphi_q) \text{ is } \infty \text{ vs. is finite.} \tag{1}$$

A student asked *why* the proof involved an analysis by cases. The answer given straightaway to the student was that his teacher didn't know how else to do it.

The present paper provides, *among other things*, a *better answer*. Of course, by Rice's Theorem [2], the cases (1) above are *non*-constructive.[a]

The better answer is: *any* proof that, for *each* q, $\{x \mid \varphi_x = \varphi_q\}$ is *not* c.e. *provably must* involve *some* such *non-constructivity*.

We assume the reader is familiar with the definitions and names of the levels of the (classical) Arithmetical Hierarchy as in Rogers' book.[2] The classically valid Law of Excluded Middle (LEM) is: either a sentence holds or its negation holds. The idea of constructive mathematical proof, beginning with the intuitionist (constructivist) Brouwer [b], from about 1908 (see Ref. 5), is that mathematical proofs at every layer and step should permit the explicit presentation/*computation* of *examples proved to exist.* Herein, we need this concept only for parts of mathematics expressible

[a]More about constructivity below.

[b]Technically, Intuitionism is a brand of Constructivism somewhat more subjective in focus than general constructivism. We will not and need not explore herein the subjectivism of Brouwer's Intuitionism.

in the language of First Order Peano Arithmetic (**PA**)[2,6], and we again assume reader familiarity. This "proof-mineability" places some limitations (compared to classical mathematics) on the logical operators $\vee$ and $\exists$ (think of $\vee$ as a finitary version of $\exists$). Classical mathematics has no such absolute constraints on existence proofs. A nice example contrasting non-constructive and constructive proofs is found in Ref. 7 where it is made clear that *un*restricted LEM is a source of *non*-constructivity: in constructively permissible proofs by cases, it must be decidable as to which case/disjunct holds.

Brouwer's student, Heyting, formulated a variant of first order logic[6] called *intuitionistic logic* which is just like first order logic except it is is missing unrestricted LEM; it captures thereby Brouwer's constructivist ideas in the context of logic.[8] *Heyting Arithmetic* (abbreviated: **HA** is just **PA** but underpinned instead by intuitionistic logic (see also Refs. 9,10). It captures Brouwer's ideas about the mathematics *expressible in it.*

There has been recent interest in adding some *arithmetically limited version* of LEM (and other principles) to **HA**. One, then, gets theories of *strictly* intermediate strength between **HA** and **PA**[11–13]. For example, Π^0_2-LEM is the set of all instances of LEM, where, the instances are all and only those of the form

$$(\forall u)(\exists v)R(u,v,w) \ \vee \ \neg(\forall u)(\exists v)R(u,v,w), \tag{2}$$

for some computable predicate R.[c]

If we replace in (1) above 'is finite' by 'is not infinite', then the result above can be easily shown to be provable in (**HA** + Π^0_2-LEM).[d]

If, instead, we replace in (1) above, 'is infinite' by 'is not finite', the result is in Σ^0_2-LEM. Since, from Ref. 12, (**HA** + Π^0_2-LEM) is *strictly* less non-constructive than (**HA** + Σ^0_2-LEM), we choose to employ the indicated replacement for (1) above which is in Π^0_2-LEM:

$$\text{domain}(\varphi_q) \text{ is } \infty \text{ vs. is not } \infty. \tag{3}$$

Note that, while one can put any formula of **PA** into a *classically* equivalent normal form variant with all its quantifiers in front and alternating[2,6], this equivalence, in some cases, is *not* provable in **HA**. *However*, from

[c]It turns out, *in the cases there are quantifiers in front of R*, we can take such Rs to be primitive recursive[2]. *Then*, for example, Π^0_2-LEM becomes a computably decidable set of sentences. In this way, proof-checking in, for example, (**HA** + Π^0_2-LEM) is algorithmic.

[d]See the proofs of Proposition 1 and 2 below in Section 3.1.

Ref. 12, for $\mathcal{A}$, a standard level (such as Π_2^0) in the arithmetical hierarchy, (**HA** + $\mathcal{A}$-LEM) *is* strong enough to *prove* the equivalence of the normal form variant of formulas classically equivalent to formulas defining the members of $\mathcal{A}$.

2. Basic Definition & Relevant Theorem

Recall that the *completely productive* (abbr: *c-productive*) sets ($\subseteq \mathbb{N}$)[2,14–16] are the *effectively non-c.e. sets*, i.e., the sets S such that

$$(\exists \text{ computable } f)(\forall y)[f(y) \in ((S - W_y) \cup (W_y - S))]. \tag{4}$$

Idea: in (4) just above, $f(y)$ is a *counterexample to* $S = W_y$.[e]

Definition 1. A *set sequence* $S_q, q \in \mathbb{N}$, is *uniformly c-productive* iff there is a *computable* f so that, for *all* q, y, $f(q, y)$ is a counterexample to $S_q = W_y$.[f]

Relevance: a completely *constructive* proof that *each* S_q is *not* c.e. would *entail the* S_q*s forming a uniformly c-productive sequence.*[g]

Let $E_q = \{x \mid \varphi_x = \varphi_q\}$. Then:

Theorem 1. *The* set sequence $E_q, q \in \mathbb{N}$, *is* not *uniformly c-productive.*

We'll include the following short, sweet proof of Theorem 1 just above. Theorem 1 is generalized by Corollary 1 below.

Proof. Suppose for contradiction that f is a computable function so that, for *all* q, y, $f(q, y)$ is a counterexample to E_q (i.e., $\{x \mid \varphi_x = \varphi_q\}) = W_y$.

By the *Double Recursion Theorem*[2] there are programs q_0, y_0 each of which *creates a copy of itself and the other* (outside themselves[h]) and each *uses its copies together with an algorithm for* f *to compute* $f(q_0, y_0)$ with:

$$\varphi_{q_0} = \varphi_{f(q_0,y_0)} \; \& \tag{5}$$

$$W_{y_0} = \{f(q_0, y_0)\}. \tag{6}$$

---

[e]Below we'll write $((S - W_y) \cup (W_y - S))$ as $(S \,\triangle\, W_y)$, the symmetric difference of S, W_y.
[f]These set sequences are, in motivation and mathematically, reasonably unrelated to Cleave's creative sequences.[17]
This defined notion will be generalized a bit in a formal defintion below (Definition 2 in Section 3).
[g]That's *entail* or *imply*, *not* hold-iff.
[h]There is no need for infinite regress.

From just above,

$$f(q_0, y_0) \in (\{x \mid \varphi_x = \varphi_{q_0}\} \cap W_{y_0}). \tag{7}$$

Clearly, then, $f(q_0, y_0)$ *fails* to be a counterexample to $E_{q_0} = W_{y_0}$, and we get a contradiction. □

Herein, we will use the linear-time computable and invertible pairing function $\langle\cdot,\cdot\rangle$ from the Royer-Case monograph[18]. This function maps all the pairs of elements of $\mathbb{N}$ 1-1, onto $\mathbb{N}$. We also employ this notation, based on iterating, $\langle\cdot,\cdot\rangle$, as in the Royer-Case monograph[18], to code also triples, quadruples, ... of elements of $\mathbb{N}$ 1-1, onto $\mathbb{N}$. Technically, each φ_p takes one argument $\in \mathbb{N}$. For some p and some $n > 1$, and some $x_1, \ldots, x_n$, we will sometimes write $\varphi_p(x_1, \ldots, x_n)$ as an abbreviation for $\varphi_p(\langle x_1, \ldots, x_n\rangle)$. We'll sometimes so abbreviate for other functions which technically take a single argument $\in \mathbb{N}$.

In the next paragraph we assume reader familiarity with employed notions and relevant characterizations thereof and treated in Rogers' book.[2]

By *contrast*, $\{\langle q, x\rangle \mid \varphi_x = \varphi_q\}$ *is*, of course, c-productive, *Why? Insightful answer:* Let q_1 be such that, say, $\varphi_{q_1} = \lambda y.(1)$. Then, *easily* $\overline{K} \leq_1 \{x \mid \varphi_x = \varphi_{q_1}\} \leq_1 \{\langle q, x\rangle \mid \varphi_x = \varphi_q\}$ — and this entails[2] the c-productivity of $\{x \mid \varphi_x = \varphi_{q_1}\}$ as well as of $\{\langle q, x\rangle \mid \varphi_x = \varphi_q\}$. We see, then, that the c-productivity of $\{\langle q, x\rangle \mid \varphi_x = \varphi_q\}$ can be shown exercising only one q-value, q_1; whereas *uniform* c-productivity must exercise all q-values.

3. Characterizing the Index Set Cases

3.1. *Uniform C-Productivity of S_q, $q \in M$*

An *index set*[2] is a set of φ-programs M so that, for some class of (1-argument) partial computable functions $\mathfrak{S}$, $M = \{p \mid \varphi_p \in \mathfrak{S}\}$.

Example (complementary) index sets include those implicit in the non-constructive disjunction of cases mentioned early on above (in Section 1): $M_{\text{inf}} = \{q \mid \text{domain}(\varphi_q)\ \infty\}$ vs. $M_{\text{fin}} = \{q \mid \text{domain}(\varphi_q) \text{ not } \infty\}$.

Next we give the promised generalization of the notion of uniformly c-productive. For computations $\downarrow$ means *converges* and $\uparrow$ means *diverges*, English terms used as in Rogers' book.[2]

Definition 2. For any *index set* M, we define the sequence of sets S_q, $q \in M$, to be *uniformly c-productive* iff, *for some partial computable* η, for *any* $q \in M$ & *any* y, $\eta(q, y)\downarrow$ to a counterexample to $S_q = W_y$.

It can be seen from the proof of the Characterization Theorem (Theorem 2 in Section 3.2 below) that, for $S_q = E_q$, *when* η just above exists, it can be taken to be *total*.

Below we write δ for domain and ρ for range.

By a pair of Kleene Parametric Recursion Theorem (PKRT)[2] arguments, for each $M \in \{M_{\text{inf}}, M_{\text{fin}}\}$, the corresponding sequence E_q, $q \in M$, *is* uniformly c-productive:

Proposition 1. *The sequence* E_q, $q \in M_{\text{inf}}$, is *uniformly c-productive.*

Proof. By KPRT, there is a computable function f st, for each q, y, z,

$$\varphi_{f(q,y)}(z) = \begin{cases} \uparrow, & \text{if } f(q,y) \in W_y \text{ in } \leq z \text{ steps;} \\ \varphi_q(z), & \text{otherwise.} \end{cases} \tag{8}$$

Suppose $q \in M_{\text{inf}}$.

Case one: $f(q,y) \in W_y$. Then by (8) just above, $\delta\varphi_{f(q,y)}$ not ∞, so, $\varphi_{f(q,y)} \neq \varphi_q$. Hence, $\varphi_{f(q,y)} \notin E_q$.

Case two: $f(q,y) \notin W_y$. Then by (8) just above, $\varphi_{f(q,y)} = \varphi_q$. Hence, $\varphi_{f(q,y)} \in E_q$.

Therefore, in any case, $f(q,y) \in (E_q \,\triangle\, W_W)$. □

Proposition 2. *The sequence* E_q, $q \in M_{\text{fin}}$, is *uniformly c-productive.*

Proof. By KPRT, there is a computable function f st, for each q, y, z,

$$\varphi_{f(q,y)}(z) = \begin{cases} 1, & \text{if } f(q,y) \in W_y \text{ in } \leq z \text{ steps;} \\ \varphi_q(z), & \text{otherwise.} \end{cases} \tag{9}$$

Suppose $q \in M_{\text{fin}}$.

Case one: $f(q,y) \in W_y$. Then by (8) just above, $\delta\varphi_{f(q,y)}$ is ∞, so, $\varphi_{f(q,y)} \neq \varphi_q$. Hence, $\varphi_{f(q,y)} \notin E_q$.

Case two: $f(q,y) \notin W_y$. Then by (9) just above, $\varphi_{f(q,y)} = \varphi_q$. Hence, $\varphi_{f(q,y)} \in E_q$.

Therefore, in any case, $f(q,y) \in (E_q \,\triangle\, W_W)$. □

These two just prior proofs each involve Σ^0_1-LEM, already strictly subsumed in Π^0_2-LEM.[12] Here is why. Each of these two proofs employ for its only non-constructivity, for some computable f, the *disjunction* of cases

$$f(q,y) \in W_y \;\vee\; \neg[f(q,y) \in W_y], \tag{10}$$

clearly in Σ^0_1-LEM.

This provides a now-known-to-be *necessarily non*-constructive proof that, for *each* $q \in \mathbb{N}$, $E_q = \{x \mid \varphi_x = \varphi_q\}$ is *not* c.e. It's non-constructivities (plural) can all be handled in (**HA** + Π^0_2-LEM).

3.2. *The Characterization*

There are divisions into non-constructive (top level) disjunctions *besides* the above example for proving that, for each $q \in \mathbb{N}$, $E_q = \{x \mid \varphi_x = \varphi_q\}$ is not c.e.

For example, the division into $\{q \mid \textit{domain}(\varphi_q) \textit{ is not recursive}\}$ vs. its complement also works, *but* it's Π^0_3-LEM, *more non*-constructive than Π^0_2-LEM above.[12]

Let F_x, $x \in \mathbb{N}$, be a *canonical indexing*[2,3] of all/only the *finite functions*: $\mathbb{N} \to \mathbb{N}$. Below we identify partial functions with their graphs (as sets of pairs). We have for our characterization:

Theorem 2. *For any index set M and corresponding sequence of sets $E_q(= \{x \mid \varphi_x = \varphi_q\})$, $q \in M$, the sequence is uniformly c-productive iff*

$$(\exists \text{ c.e. } A \subseteq \overline{M})(\forall x)(\exists y \in A)[\varphi_y \supseteq F_x]. \tag{11}$$

Wolog*: in each direction y can be taken to be* algorithmic *in x.*

The disjoint index sets partitioning $\mathbb{N}$, $\{q \mid \text{domain}(\varphi_q) \neq \emptyset\}$ (c.e.) vs. its complement, $\{q \mid \text{domain}(\varphi_q) = \emptyset\}$, do *not* work — since, by our just above characterization and the Rice-Shapiro-Myhill-McNaughton Theorem.[2]

Corollary 1. *For* any c.e. index set *M, E_q, $q \in M$, is* not *uniformly c-productive.*

Since $\mathbb{N}$ is trivially a c.e. index set, Theorem 1 above in Section 1 follows from Corollary 1 just above. Since[2] the c.e. sets are the Σ^0_1 sets and their complements are the Π^0_1 sets, the problem of proving, for each $q \in \mathbb{N}$, E_q is not c.e. is upper-bounded (can be done) with (**HA** + Π^0_2-LEM) (by Propositions 1 and 2 in Section 3.1 above) and lower-bounded (can't be done) with (**HA** + Σ^0_1-LEM)[i] — the latter at least for employing a division into cases involving *index sets*. We conjecture that (**HA** + Δ^0_2-LEM) is a *maximal* lower-bound (cannot be done) — at least for employing a division into cases involving *index sets*. In general, we don't know about divisions

[i] From Ref. 12, (**HA** + Π^0_1-LEM) is strictly less non-constructive than (**HA** + Σ^0_1-LEM).

into cases *not* involving index sets and we don't know whether (**HA** + Π^0_2-LEM) is a *minimum* required with respect to the Arithmetically-Limited-LEMs from Ref. 12.

Proof. PROOF OF THE CHARACTERIZATION THEOREM 2. Fix index set M.

Claim 1. *If $E_q, q \in M$, is uniformly c-productive, then*

$$(\exists\ C.E.\ A \subseteq \overline{M})(\forall x)(\exists y \in A)[\varphi_y \supseteq F_x]. \tag{12}$$

PROOF OF CLAIM 1. Let partial computable η be a witness to the uniform c-productivity of $E_q, q \in M$; that is, for each q, y st $q \in M$, $\eta(q,y)\downarrow \in (E_q \triangle W_y)$.

By Kleene's S-m-n theorem[2] there is a computable function t st, for each x,

$$W_{t(x)} = \{z \mid \varphi_z \supseteq F_x\}. \tag{13}$$

By Kleene's Parametric Recursion Theorem[2] there is a computable function s st, for each x, z,

$$\varphi_{s(x)}(z) = \begin{cases} F_x(z), & \text{if } z \in \delta F_x; \\ \varphi_{\eta(s(x),t(x))}(z), & \text{otherwise.} \end{cases} \tag{14}$$

Subclaim 1. $(\forall x)[s(x) \in \overline{M}]$

PROOF OF SUBCLAIM 1. Suppose by way of contradiction there exists *some* x such that that $s(x) \in M$. Fix that x. By the assumption that $s(x) \in M$,

$$\eta(s(x), t(x))\downarrow \in (E_{s(x)} \triangle W_{t(x)}). \tag{15}$$

Case one: $\varphi_{\eta(s(x),t(x))} \in W_{t(x)}$. Then, for all z, one of two subcases holds.

Subcase one: $z \in \delta F_x$. In this subcase, as $\varphi_{\eta(s(x),t(x))}$ extends F_x, $\varphi_{\eta(s(x),t(x))}(z) = F_x(z)$, and, by the first clause of (14), $\varphi_{s(x)}(z) = F_x(z)$; therefore, $\varphi_{s(x)}(z) = \varphi_{\eta(s(x),t(x))}(z)$.

Subcase two: $z \notin \delta F_x$. In this subcase, by the second clause of (14), $\varphi_{s(x)}(z) = \varphi_{\eta(s(x),t(x))}(z)$.

Thus, $(\forall z)[\varphi_{s(x)}(z) = \varphi_{\eta(s(x),t(x))}(z)]$. This means $\eta(s(x),t(x)) \in E_{s(x)} \wedge \eta(s(x),t(x)) \in W_{t(x)}$; a contradiction to (15).

Case two: $\varphi_{\eta(s(x),t(x))} \notin W_{t(x)}$. Then there exists *some* $z \in \delta F_x$ such that $\varphi_{\eta(s(x),t(x))}(z) \neq F_x(z)$, thus, by Case one of (14), $\varphi_{\eta(s(x),t(x))} \neq \varphi_{s(x)}$. This means $\eta(s(x),t(x)) \notin E_{s(x)} \wedge \eta(s(x),t(x)) \notin W_{t(x)}$; a contradiction to (15).

Both cases lead to a contradiction; the subclaim immediately follows.

□ **Subclaim 1**

Let $A = \rho s$. The claim then follows from Subclaim 1, the fact that A is the range of a computable function, and (14).

□ **Claim 1**

Claim 2. *If there exists a c.e. set A that is a subset of $\overline{M}$ such that, for any finite function F, there is a program in A which computes a function that extends F, then the sequence of sets $E_q, q \in M$, is uniformly c-productive.*

Proof of Claim 2. Assume the existence of such an A.

Let s be a code number for the following program on inputs q, y, e.

> Determine how many steps it takes to get an output from program y on input e; if it never terminates, never terminate from this point. Compute the canonical index of the finite function F whose graph is the set of $(z, \varphi_q(z))$ such that $\Phi_y(e) > z$ and $\Phi_y(e) > \Phi_q(z)$. Dovetail an algorithmic enumeration of A, until, if ever, an element a is found such that φ_a extends F. If such an a is found, output that a. Otherwise, $\uparrow$.

By Kleene's Parametric Recursion Theorem[2] there is a computable function f st, for each q, y, z,

$$\varphi_{f(q,y)}(z) = \begin{cases} \varphi_q(z), & \text{if } \Phi_y(f(q,y)) > z \\ & \quad \wedge\ \Phi_y(f(q,y)) > \Phi_q(z); \\ \varphi_{\varphi_s(q,y,f(q,y))}(z), & \text{if } \varphi_y(f(q,y))\downarrow \\ & \quad \wedge\ \varphi_s(q,y,f(q,y))\downarrow \\ & \quad \wedge\ [\Phi_y(f(q,y)) \leq z \\ & \qquad \vee\ \Phi_y(f(q,y)) \leq \Phi_q(z)]; \\ \uparrow, & \text{otherwise.} \end{cases} \tag{16}$$

Subclaim 2. f is a witness to the uniform c-productivity of $E_q, q \in M$.

Proof of Subclaim 2. Fix q and y.

Case one: $f(q,y) \in W_y$. Thus, $\Phi_y(f(q,y))$ converges. Because of this, we can compute the canonical index of the finite function F whose graph is $\{(z, \varphi_q(z)) \mid \Phi_y(f(q,y)) > z \ \wedge \ \Phi_y(f(q,y)) > \Phi_q(z)\}$; from that and the assumption about A, it follows that $\varphi_s(q,y,f(q,y))\downarrow$ to an $a \in A$ such that φ_a extends F. By careful examination of the clauses in (16) and the behavior of program s on $q, y, f(q,y)$, it follows that $\varphi_{f(q,y)} = \varphi_a$. By assumption about A, $a \in \overline{M}$, from which it follows that $f(q,y) \in \overline{M}$, thus $f(q,y) \notin E_q$.

Case two: $f(q,y) \notin W_y$. Then, for all z such that $\varphi_q(z)\downarrow$, $\varphi_{f(q,y)}(z) = \varphi_q(z)$ by the first clause of (16), and for all z such that $\varphi_q(z)\uparrow$, $\varphi_{f(q,y)}(z)\uparrow$ by the third clause of (16); thus, $\varphi_{f(q,y)} = \varphi_q$. Therefore, $f(q,y) \in E_q$.

Thus, by Case one and Case two, $(\forall q, y)[f(q,y) \in E_q \,\triangle\, W_y]$: f is a witness to the uniform c-productivity of the sequence $E_q, q \in M$.

□ **Subclaim 2**

The claim follows from Subclaim 2 and the computability of f.

□ **Claim 2**

The theorem follows immediately from Claims 1 and 2. □

3.3. *Another Corollary of the Characterization*

Above, we've looked at proving *each* E_q is *not* c.e., with the proof's top level *non*-constructivity confined to disjunctions with *two* disjuncts as to which of *two* partitioning index sets contains q. How about such disjunctions but with numbers of such irreducible disjuncts *more than* 2? Thanks to our characterization:

Corollary 2. *For each $n \geq 2$, there are n pairwise disjoint, non-trivial index sets $M_0, \ldots, M_{n-1}$ unioning to $\mathbb{N}$ such that:*

- *For each $i < n$, the sequence $E_q = \{x \mid \varphi_x = \varphi_q\}$, $q \in M_i$,* is *uniformly c-productive,* but
- *For each $i, j < n$ such that $i \neq j$, the sequence E_q, $q \in (M_i \cup M_j)$, is* not *uniformly c-productive.*

A simple extension of the just above Corollary's proof (just below) yields the $n = \omega$, infinitary, irreducible disjunctions case too. We omit the details of that extension.

Proof. Assume $n \geq 2$.

Let

$$O = \{\langle i, j\rangle \mid i \neq j \ \wedge \ i < n \ \wedge \ j < n\}. \tag{17}$$

Let

$$O_0 = \{q \mid \varphi_q(0)\uparrow \ \vee \ [\varphi_q(0)\downarrow \notin O \ \wedge \ W_q \text{ infinite}]\}. \tag{18}$$

Let

$$O_1 = \{q \mid \varphi_q(0)\downarrow \notin O \ \wedge \ W_q \text{ finite}\}. \tag{19}$$

For each $i < n$, let

$$S_i = \begin{cases} \{q \mid [\varphi_q(0)\downarrow \in \{\langle x, i\rangle \mid \langle x, i\rangle \in O\} \ \wedge \ W_q \text{ finite}] \ \vee \\ \qquad [\varphi_q(0)\downarrow \in \{\langle i, x\rangle \mid \langle i, x\rangle \in O\} \ \wedge \ W_q \text{ infinite}]\}. \end{cases} \tag{20}$$

Let

$$M_0 = (S_0 \cup O_0). \tag{21}$$

Let

$$M_1 = (S_1 \cup O_1). \tag{22}$$

For each $i \geq 2$ st $i < n$, let

$$M_i = S_i. \tag{23}$$

Claim 3.1. *The union of all M_i is N.*

Proof of Claim 3.1. Suppose by way of contradiction there exists some q such that q is not in the union of all the M_i.

Case one: $\varphi_q(0)\uparrow$. Then, $q \in M_0$ by (18); a contradiction.

Case two: $\varphi_q(0)\downarrow \notin O$, and W_q infinite. Then, $q \in M_0$ by (18); a contradiction.

Case three: $\varphi_q(0)\downarrow \notin O$, and W_q finite. Then, $q \in M_1$ by (19); a contradiction.

Case four: $\varphi_q(0)\downarrow \in O$. Then, by (17), there exists some i, j such that $i \neq j, i < n, j < n, \langle i, j\rangle \in O$, and $\varphi_q(0) = \langle i, j\rangle$.

Subcase one: W_q finite. Then, by (20), q is in M_j; a contradiction.

Subcase two: W_q infinite. Then, by (20), q is in M_i; a contradiction.

Hence, in any case we have a contradiction. Therefore, q is in the union of all M_i, and the claim follows.

□ **Claim 3.1**

Claim 3.2. *For any $i < n$, and any $j < n$, st $i \neq j$, M_i and M_j are disjoint.*

Proof of Claim 3.2. Suppose by way of contradiction there exists some $i, j < n$, and there exists q such that $i \neq j$ and $q \in (M_i \cap M_j)$.

Case one: $\varphi_q(0)\uparrow$. Then, $q \in M_0$ by (18). By (19) and (20), q is not in any other set; a contradiction.

Case two: $\varphi_q(0)\downarrow \notin O$, and W_q infinite. Then, $q \in M_0$ by (18). By (19) and (20), q is not in any other set; a contradiction.

Case three: $\varphi_q(0)\downarrow \notin O$, and W_q finite. Then, $q \in M_1$ by (19); By (18) and (20), q is not in any other set; a contradiction.

Case four: $\varphi_q(0)\downarrow \in O$, and W_q infinite. Then, by (17), $\varphi_q(0) = \langle i', j' \rangle, i' < n, j' < n$, and $i' \neq j'$. q is not in O_0 nor O_1 by (18) and (19). By (20), q is in $S_{i'}$, and not in $S_{j'}$. Then, by (21), (22), and (23), q is in exactly and only $M_{i'}$, a contradiction.

Case five: $\varphi_q(0)\downarrow \in O$, and W_q finite. Then, by (17), $\varphi_q(0) = \langle i', j' \rangle, i' < n, j' < n$, and $i' \neq j'$. q is not in O_0 nor O_1 by (18) and (19). By (20), q is in $S_{j'}$, and not in $S_{i'}$. Then, by (21), (22), and (23), q is in exactly and only $M_{j'}$, a contradiction.

All of the cases lead to a contradiction. Therefore, there are no such i, j, q; the claim immediately follows.

□ **Claim 3.2**

Claim 3.3. *For any $i, j < n$ st $i \neq j$, the sequence $E_q, q \in (M_i \cup M_j)$ is* not *uniformly c-productive.*

Proof of Claim 3.3. By Theorem 2, $(M_i \cup M_j)$ is uniformly c-productive if and only if $\overline{M_i \cup M_j}$ contains a c.e. set A st A has a program for some extension of *each* finite function. Suppose for contradiction the claim fails; hence, there is such a c.e. A.

By (20), all finite function extensions of $(0, \langle j, i \rangle)$ have all their programs in M_i, and all infinite partial computable extensions of $(0, \langle j, i \rangle)$ have all their programs in M_j; thus, all partial computable extensions of $(0, \langle j, i \rangle)$ have all their programs in $(M_i \cup M_j)$; hence, the complement of that set (which includes A) has no programs for such extensions, a contradiction.

□ **Claim 3.3**

Claim 3.4. *For any $i < n$, the set M_i is uniformly c-productive.*

PROOF OF CLAIM 3.4. By Kleeene's S-m-n Theorem[2], there exist computable *functions* q_1, q_2, q_3 such that, for each x, z,

$$\varphi_{q_1(x)}(z) = \begin{cases} F_x(z), & \text{if } z \in \delta F_x; \\ 0, & \text{otherwise.} \end{cases} \tag{24}$$

$$\varphi_{q_2(x)}(z) = F_x(z). \tag{25}$$

$$\varphi_{q_3(x)}(z) = \begin{cases} \langle i, i \rangle, & \text{if } z = 0; \\ F_x(z), & \text{otherwise.} \end{cases} \tag{26}$$

Case one: $i \geq 1$ st $i < n$. Let p be the code number of a program as follows:

$$\varphi_p(x) = \begin{cases} q_1(x), & \text{if } (\exists y)[F_x(0)\!\downarrow = \langle y, i \rangle \ \wedge \ F_x(0) \in O]; \\ q_2(x), & \text{otherwise.} \end{cases} \tag{27}$$

Subclaim 3. φ_p is total.

PROOF OF SUBCLAIM 3. Determining if $F_x(0)$ is defined is computable, treating it as an ordered pair and extracting the second component to compare to i is also computable, and determining if something is a member of O is computable in a similar fashion. Thus, the subclaim follows.

□ **Subclaim 3**

Subclaim 4. The range of φ_p is a subset of $\overline{M_i}$

PROOF OF SUBCLAIM 4. For all x, one of the following two cases holds:

Case (a): $(\exists y)[F_x(0)\!\downarrow = \langle y, i \rangle \ \wedge \ F_x(0) \in O]$. Then, by (27) $(\exists y)[\varphi_{\varphi_p(x)}(0)\!\downarrow = \langle y, i \rangle \ \wedge \ \langle y, i \rangle \in O]$, and $W_{\varphi_p(x)}$ infinite. Therefore, $\varphi_p(x)$ is *not* in M_i.

Case (b): Otherwise. Thus, $\varphi_{\varphi_p(x)} = F_x$, and therefore $W_{\varphi_p(x)}$ is finite. By (20), all programs which are in M_i and compute finite functions are such that the value of their computed function on input 0 is $\langle y, i \rangle$ for some $y < n$. If F_x were such a function, Case (a) would hold instead. Therefore, $\varphi_p(x)$ is *not* in M_i.

□ **Subclaim 4**

Subclaim 5. For all x, there is a program in the range of φ_p which computes a partial function that extends F_x.

Proof of Subclaim 5. This follows directly from (27) and (3).

□ **Subclaim 5**

By (3), (4), and (4), the range of φ_p is a recursively-enumerable subset of $\overline{M_i}$ such that for any finite function F_x, there is a program in that recursively-enumerable subset which extends F_x. By Theorem 2, M_i is uniformly c-productive in this case.

Case two: $i = 1$. Let p be the code number of a program as follows:

$$\varphi_p(x) = \begin{cases} q_1(x), & \text{if } (\exists y)[F_x(0){\downarrow} = \langle y, i \rangle \ \wedge \ F_x(0) \in O] \ \vee \ F_x(0){\downarrow} \notin O; \\ q_2(x), & \text{otherwise.} \end{cases} \tag{28}$$

The proof of Subclaim 3 is applicable to an identical subclaim *mutatis mutandis.* The proof of Subclaim 4 works if Case (a) is replaced with two subcases; Subcase i is identical to Case (a), while Subcase ii is the case where $F_x(0){\downarrow} \notin O$; in that instance, $\varphi_p(x)$ is in M_0 by 18, and thus by (3.2) is not in M_1. The proof of Subclaim 5 holds *mutatis mutandis.* Thus, just as per Case one, M_1 is uniformly c-productive.

Case three: $i = 0$. Let p be the code number of a program as follows:

$$\varphi_p(x) = \begin{cases} q_3(x), & \text{if } 0 \notin F_x; \\ q_1(x), & \text{if } (\exists y)[F_x(0){\downarrow} = \langle y, i \rangle \ \wedge \ F_x(0) \in O]; \\ q_2(x), & \text{otherwise.} \end{cases} \tag{29}$$

The proof of Subclaim 3 is applicable to an identical subclaim *mutatis mutandis.* The proof of Subclaim 4 works if Case (b) is renamed to Case (c) and a new Case (b) is inserted between Case (a) and Case (c); the new Case (b) is: 0 is not in F_x. In Case (b), φ_q is finite and $\varphi_q(0){\downarrow} \notin O$; thus, by (19), q in M_1. Thus, by (3.2), q is not in M_0. The proof of Subclaim 5 holds *mutatis mutandis.* Thus, just as per Cases one and two, M_0 is uniformly c-productive.

As M_i is uniformly c-productive for all $i < n$, the claim holds.

□ **Claim 3.4**

Claims 3.1, 3.2, 3.3, 3.4 together provide exactly the statement of the theorem. □

4. Further Examples and Future Work

We see that $\overline{E_q} = \{x \mid \varphi_x \neq \varphi_q\}$. Clearly, for *each* q such that domain$(\varphi_q) = \emptyset$, $\overline{E_q}$ *is c.e.*

Let $M_{\text{ne}} = \{q \mid \text{domain}(\varphi_q) \neq \emptyset\}$.

Then, by an omitted Σ_1^0-LEM Kleene Parametric recursion theorem argument, the sequence, $\overline{E_q}$, $q \in M_{\text{ne}}$, *is* uniformly c-productive — as witnessed by a *total* computable function. Again, our division into cases is of the form (10), and it is open whether this is minimally non-constructive.

The sequence $\{x \mid W_x = W_q\}$, $q \in \mathbb{N}$, and its complementary sequence satisfy results like the E_qs and $\overline{E_q}$s *mutatis mutandis*, e.g., for the Characterization Theorem, F_x is replaced by Rogers'[2] D_x, the finite *set* with canonical index x. We omit the straightforward details.

The next subsection, Section 4.1, presents an aside: the first author initially suspected that each index set is either c.e. or c-productive, but couldn't prove it; we subsequently proved the opposite from very basic principles, i.e., that *some* index set S (perhaps a possible S_q) is *neither* c.e. *nor* c-productive.[j] An anonymous referee of the just prior draft of this chapter nicely called our attention to Dekker and Myhill's[19] vastly earlier construction of such an index set. Their proof employed the Friedberg-Muchnik solution to Post's Problem[2,20,21], and Hay in Ref. 22 calculates the Turing degree of their index set. Since our construction employs simpler principles, we've retained it herein.

4.1. *A Simple Index Set Neither C.E. Nor C-Productive*

We'll prove the existence of an index set neither c.e. nor productive, directly without using the Friedberg-Muchnik theorem. Specifically, for any *immune* set (i.e., an infinite set which has no infinite c.e. subset[2]) S, we'll show below in the proof of Corollary 3 that the index set

$$S' = \{p \mid \varphi_p(0) \in S\}$$

is neither c.e. nor productive.

Theorem 3. *Given any set S, there exists an index set S' such that S' is c.e if and only if S is c.e., and, if S' is c-productive, then so is S.*

Proof. Let $S' = \{p \mid \varphi_p(0)\downarrow \in S\}$.

Clearly S' is an index set.

Clearly too, if S is c.e., then S' is c.e..

Claim 3. *If S' is c.e., then S is c.e..*

---

[j]Curiously, then, by *a* proof of Rice's Theorem[2], $\overline{S}$ *must be* $\geq_1 \overline{K}$, hence *c-productive*.

Proof of Claim 3. Assume S' is c.e.. Then there exists s' such that $W_{s'} = S'$. Let s be a program such that $W_s = \{x \mid (\exists p \in W_{s'})[\varphi_p(0) = x]\}$. Clearly, $S = W_s$, therefore S is c.e.

□ **Claim 3**

Claim 4. *If S' is c-productive, then S is c-productive.*

Proof of Claim 4. Assume S' is c-productive as witnessed by computable function f'.

Let p be such that $\varphi_p(x, y) = 1$ if $\varphi_y(0) \in W_x$; and undefined otherwise.

By S-m-n there is a computable function $\lambda x.(p_x)$ st, for each x, y, $\varphi_{p_x}(y) = \varphi_p(x, y)$.

Let θ be the partial computable function st, for each x, $\theta(x) = \varphi_{f'(p_x)}(0)$.

Suppose by way of contradiction that θ is not total. Let x be least st $\theta(x)\uparrow$. Then, by the assumption that f' witnesses the c-productivity of S', there is some $w = f'(p_x)$ st $\varphi_w(0)\uparrow$. Then, $w \notin W_{p_x}$. Furthermore, by the definition of S', $w \notin S'$. But this means that f' does not witness the c-productivity of S' on input p_x, a contradiction. Therefore, $f = \theta$ is total. So, f is computable st, for each x, $f(x) = \varphi_{f'(p_x)}(0)$.

By the assumption that f' witnesses the c-productivity of S', it follows that, for each x, exactly one of the following two cases holds:

Case one: $f'(p_x) \in (W_{p_x} - S')$. By the construction of p_x and the definition of S', this implies that $\varphi_{f'(p_x)}(0) \in (W_x - S)$. Thus, $f(x) \in (W_x - S)$.

Case two: $f'(p_x) \in (S' - W_{p_x})$. Thus, $f(x) \in (S - W_x)$.

Therefore, by Cases one and two, f is a witness to the c-productivity of S.

□ **Claim 4**

The theorem follows from the claims. □

The desired result for the present subsection follows.

Corollary 3. *There exists an index set which is neither c.e. nor c-productive.*

Proof. Immune sets are neither c.e. nor c-productive, thus, by Theorem 3 above, there exists an index set S', corresponding to an immune set S, which is neither c.e. nor c-productive. □

4.2. *Some Subrecursive Examples*

In this section we consider a wide variety of "natural" *sub*recursive programming systems ψ, including for such subrecursive classes as: for $k > 0$, the functions computable in time bounded by a k-degree polynomial; the polynomial time computable functions; the elementary recursive functions; and other levels of the Meyer-Ritchie[23] loop hierarchy.

Let $C_q = \{x \mid \psi_x = \psi_q\}$. Trivially, each $\overline{C_q}$ *is* c.e.

In this section our main goal is to prove Theorem 5 below (in Section 4.2.2 below) that the sequence $C_q, q \in \mathbb{N}$, *is* uniformly c-productive, and, interestingly, as witnessed by a two-argument function f *computable in linear-time in the lengths of its inputs.* We had had an f linear-time in the length of its second input, and Jim Royer supplied a suggestion which enabled getting f linear-time in the lengths of both its inputs.

Before we prove that theorem, we describe a bit informally (but with references to more detail) nice clocked programming systems for subrecursive classes such as the above. We spell out two Assumptions as to for which subrecursive classes and corresponding programming systems we can prove this theorem. We prove (mostly with a few citations and hints as to underlying ideas from the Royer-Case monograph[18]), *as an example*, Theorem 4 (in Section 4.2.2) below that, for $k > 0$, the nice clocked systems for the class of functions computable in k-degree polynomial time *do* satisfy our two Assumptions.

Let θ^k be such a programming system. For this system, let f be the *associated* function mentioned above. For each *fixed* q, y, $f(q, y)$ *is* a θ^k-program, which, *then*, on any input z, runs within time $\mathcal{O}(|z|^k)$. For *these particular* θ^k*-systems*, the *lower*-bound run time cost of running corresponding *universal (simulator)* programs is known to be very high. We examine the best known (exponential) *upper*-bound cost of simulating, for the associated f, the run time cost, as a function of *all of* q, y, z, of computing $\theta^k_{f(q,y)}(z)$. Lastly, Theorem 6 (in Section 4.2.2) below lays out an explicit $\mathcal{O}$-formula for this exponential cost.

4.2.1. *Preliminaries*

Residually unexplained notation or terminology is from Rogers' book and/or the Royer-Case monograph[2,18].

As above we choose our multi-argument "pairing" $\langle \cdot, \ldots, \cdot \rangle$ (and unpairing) function(s) to be in $\mathcal{L}$time.[k] For one-argument functions α we write $\alpha(z_1, \ldots, z_m)$ to mean $\alpha(\langle z_1, \ldots, z_m \rangle)$ [18].

For *this* section we suppose $\mathfrak{S}$ is a c.e. class of one-argument computable functions [2] and that ψ is a "natural" *sub*recursive programming system (numbering) for this $\mathfrak{S}$, with ψ's universal function $\lambda\langle q, z \rangle \centerdot (\psi_q(z))$ being (at least) computable[l], and *where* $\mathfrak{S}$ and ψ satisfy the "closure" properties Assumptions 1 and 2 spelled out further below.

First it is pedagogically useful to provide typical examples of *such* $\mathfrak{S}$ and ψ.

We let φ^{TM} be the acceptable programming system based on *efficiently numbered*, deterministic, base 2 I/O, multi-tape Turing machines (TMs) — all from Ref. 18 Chapter 3 & Errata. Φ^{TM} is the corresponding TM step-counting Blum Complexity Measure. [24]

Suppose $k > 0$.

We let $\mathcal{P}\mathrm{time}_k$ (Ref. 18 Definition 3.3) be the class of functions: $\mathbb{N} \to \mathbb{N}$ each computed by *some* φ^{TM}-program which, on each input z, has Φ^{TM}-complexity in $\mathcal{O}(|z|^k)$.

$\mathcal{L}$time denotes $\mathcal{P}\mathrm{time}_1$, the class of linear-time computable functions. Of course $\mathcal{P}$time$= \bigcup_{k>0} \mathcal{P}\mathrm{time}_k$ is the class of polynomial time computable functions.

We let θ^k be a very natural, so-called $\mathcal{L}$time-effective *clocked* programming system (numbering) with respect to $(\varphi^{\mathrm{TM}}, \Phi^{\mathrm{TM}})$ and *for* the class $\mathcal{P}\mathrm{time}_k$ (see especially Ref. 18 Sections 3.2.2 & 4.2 & Chapter 6). Associated with θ^k are functions trans and positive-valued bound, each nicely $\in \mathcal{L}$time. Each θ^k-program q *directly/efficiently codes both* a φ^{TM}-program p *and* a positive coefficient a for the run time (upper) bound $(a|z|)^k$.[m]

Here is how to think about the running of θ^k-program q on input z.

---

[k] Of course, as in Rogers' book [2], for each $m > 1$, $\lambda y_1, \ldots, y_m \centerdot (\langle y_1, \ldots, y_m \rangle) : \mathbb{N}^m \to \mathbb{N}$ is 1-1 and onto. Herein we won't need names for the unpairing functions.

[l] Of course, since all the ψ_qs are total, $\lambda\langle q, z \rangle \centerdot (\psi_q(z))$ is not in $\mathfrak{S}$.
In Ref. 18 there are many calculations of complexity upper-bounds (and some lower-bounds) for universal functions. In the proof of the last theorem below in Section 4.2.2 (Theorem 6), we employ one such upper-bound.

[m] As in Ref. 18 we are using $(a|z|)^k$ instead of the more natural $a(|z|)^k$. Curiously, the latter does not provide $\mathcal{L}$time-effective full composition for all of $\mathcal{P}$time, but the former (with an extra $+1$) does. $\mathcal{L}$time-effective inner and outer composition with $\mathcal{L}$time *for* $\mathcal{P}\mathrm{time}_k$ *does* work for the latter, but with somewhat larger coefficients for the compostions and a little smaller exponent for Ref. 18 Theorem 6.4. We employ this latter theorem in the proof of our Theorem 6 below, so we'll stick with $(a|z|)^k$ for $\mathcal{P}\mathrm{time}_k$ for our Section 4.2.2 below.

The φ^{TM}-program trans(q) does the work: trans(q) on z computes $(a|z|)^k$ and directly runs p on z for no more than $(a|z|)^k$ steps. If p halts by then, trans(q)'s (as well as q's) output is p's output; else, trans(q) outputs some default value (which is, then, also q's output value). The working of trans(q), of course, may take a bit more than time on z than $(a|z|)^k$, e.g., because it spends time on computing this run time bound coded in q; however, in any case, trans(q) on z runs within time $(\mathrm{bound}(q)|z|)^k$. In symbols:

$$(\forall q, z)[\theta_q^k(z) = \varphi^{\mathrm{TM}}_{\mathrm{trans}(q)}(z) \;\;\&\;\; \Phi^{\mathrm{TM}}_{\mathrm{trans}(q)}(z) \leq (\mathrm{bound}(q)|z|)^k]. \tag{30}$$

Both general and specific technical details about various clocked systems are in Ref. 18 Section 3.2.2 & Chapters 4–6.

Technical details for many $\mathcal{L}$time-effective clocked systems, including *also* for the polynomial time computable functions, the elementary recursive functions, other levels of the Meyer-Ritchie [23] loop hierarchy, ..., can be found in Ref. 18 Section 4.2 & Chapters 5 & 6.

As we will see in Theorem 4 further below, $(\mathfrak{S}, \psi) = (\mathcal{P}\mathrm{time}_k, \theta^k)$ satisfies our above mentioned Assumptions 1 and 2 which are presented shortly below. Actually, the other complexity classes and corresponding clocked systems mentioned at the end of the just prior paragraph also satisfy Assumptions 1 and 2, but we omit herein verification details regarding that.

Assumption 1. *$\mathfrak{S}$ contains the identity function and is closed under inner and outer composition with $\mathcal{L}$time; hence, in particular, $\mathfrak{S}$ contains $\mathcal{L}$time and is closed under finite variants.* Moreover, *the closure of $\mathfrak{S}$ under* outer *composition with $\mathcal{L}$time is $\mathcal{L}$time-effective for ψ in the following somewhat weak sense.*[n] *For each $g \in \mathcal{L}$time and each $m > 0$, there is a function comp $\in \mathcal{L}$time st, for each $q_1, \ldots, q_m, z$,*

$$\psi_{comp(q_1,\ldots,q_m)}(z) = g(\psi_{q_1}(z), \ldots, \psi_{q_m}(z)). \tag{31}$$

Assumption 2. *ψ satisfies the* Constructive $\mathcal{L}$time-effective Parametric Recursion Theorem, *i.e., for each $m > 0$, there is a function $r \in \mathcal{L}$time st, for each $p, y_1, \ldots, y_m, z$,*

$$\psi_{r(p,y_1,\ldots,y_m)}(z) = \psi_p(r(p, y_1, \ldots, y_m), y_1, \ldots, y_m, z). \tag{32}$$

Intuitively, (32) says that ψ-program $r(p, \vec{y})$ has $p, \vec{y}$ stored inside, and, on input z: it creates a self-copy (seen on the right-hand side of (32)); it pulls

[n] This weak sense is a little more than enough to prove our Theorem 5 below.

$p, \vec{y}$ out of storage; and it runs ψ-task/program p on its self-copy, $\vec{y}, z$. Task p represents the use *which $r(p, \vec{y})$ makes of its self-copy/self-knowledge (and of its stored $p, \vec{y}$ and input z).*[o]

We employ the convenient discrete log function from Ref. 18 Page 22: for any $x \in \mathbb{N}$, $\log x \stackrel{\text{def}}{=} (\lfloor \log_2 x \rfloor$, if $x > 1; 1$, if $x \leq 1)$.

4.2.2. *Results*

Theorem 4. *$(\mathfrak{S}, \psi) = (\mathcal{P}time_k, \theta^k)$ satisfies our Assumptions 1 and 2 just above.*

Proof. The essential idea behind the proof of this theorem (and of similar results for the other nicely clocked programming systems mentioned above) is that clocked systems *inherit effective* closure properties (which can be thought of as instances of control structures, e.g., see Ref. 18 Section 4.2.3& Ref. 4) from the system, e.g., φ^{TM}, out which they are built.[p]

One gets Assumption 1 *and much more*, for $(\mathcal{P}\text{time}_k, \theta^k)$, from Ref. 18 Theorem 6.3(a, b).

Assumption 2 *and more*, for $(\mathcal{P}\text{time}_k, \theta^k)$, follows from Ref. 18 Theorem 6.3(d). □

As above, let $C_q = \{x \mid \psi_x = \psi_q\}$. As noted above: trivially, each $\overline{C_q}$ *is* c.e., and, as indicated above, the next theorem's proof employs help from Jim Royer. Its proof is constructive *relative to* the limited non-constructive Σ_1^0-LEM from Ref. 12.

Theorem 5. *The sequence $C_q, q = 0, 1, 2, \ldots$* is *uniformly c-productive — as witnessed by a two-argument function $f \in \mathcal{L}time$.*

---

[o]Our Assumptions 1 and 2 imply the analogous pair of assumptions in Ref. 25, but, thanks to the weak $\mathcal{L}$time-effectivity part of Assumption 1 above, the assumptions herein are apparently stronger.
Actually, all the the classes and corresponding systems mentioned above satisfy much stronger $\mathcal{L}$time-effective closure under compositions (including inner ones too) with $\mathcal{L}$time. For $k \geq 2$, $\mathcal{P}\text{time}_k$ is *not* closed under compositions with $\mathcal{P}\text{time}_k$, but, each of the other subrecursive classes mentioned just above together with corresponding clocked system satisfies full $\mathcal{L}$time-effective closure under arbitrary compositions of functions *within the class.*
Moreover, each class above and corresponding clocked system satisfies the Constructive, $\mathcal{L}$time-efffective *Multiple* Parametric Recursion theorem. For some details, see the above mentioned parts of Ref. 18.

[p]They also inherit corresponding complexity e.g., Φ^{TM}, properties from the control structures.

Proof. We let $\Phi^{\text{SlowedDownTM}}$ be the special, *slowed-down/delayed* step-counting measure associated with the acceptable φ^{TM}-system from Ref. 18 Theorem 3.20. In the proof of Ref. 18 Theorem 3.20, for the case of $(\varphi^{\text{TM}}, \Phi^{\text{TM}})$, $\Phi^{\text{SlowedDownTM}}$ is obtained from the standard Φ^{TM} measure associated with φ^{TM}, by a $\log\log$-delay trick. $\Phi^{\text{SlowedDownTM}}$ has the nice property (among others) that the *predicate* $T \stackrel{\text{def}}{=}$

$$\lambda w, y, z\boldsymbol{.}(1, \text{ if } \Phi_y^{\text{SlowedDownTM}}(w) \leq z; 0, \text{ otherwise}) \in \mathcal{L}\text{time!} \tag{33}$$

□

It is an immediate φ^{TM}-programming exercise to show

$$\mathcal{L}\text{time is closed under definition by } \textit{if-then-else}. \tag{34}$$

Hence, by (33, 34), g just below is clearly $\in \mathcal{L}$time.

$$g(w, x, y, z) = \begin{cases} 1 + x, & \text{if } T(w, y, z); \\ x, & \text{otherwise.} \end{cases} \tag{35}$$

Therefore, by Assumption 1, there is a function cmp $\in \mathcal{L}$time st,

$$(\forall w, q, y, z)[\psi_{\text{cmp}(q)}(w, y, z) = g(w, \psi_q(z), y, z)]. \tag{36}$$

Hence, by (35, 36), for all w, q, y, z,

$$\psi_{\text{cmp}(q)}(w, y, z) = \begin{cases} 1 + \psi_q(z), & \text{if } T(w, y, z); \\ \psi_q(z), & \text{otherwise.} \end{cases} \tag{37}$$

By the $m = 2$ case of Assumption 2, we have that there is a function $r \in$ **LinearTime** st, for all p, q, y, z,

$$\psi_{r(p,q,y)}(z) = \psi_p(r(p, q, y), q, y). \tag{38}$$

From the $p = \text{cmp}(q)$ case of (38) and the $w = r(\text{cmp}(q), q, y)$ case of (37), for all, q, y, z,

$$\psi_{r(\text{cmp}(q),q,y)}(z) = \begin{cases} 1 + \psi_q(z), & \text{if } T(r(\text{cmp}(q), q, y), y, z); \\ \psi_q(z), & \text{otherwise.} \end{cases} \tag{39}$$

Let $f(q, y) = r(\text{cmp}(q), q, y)$. f is a composition of functions $\in \mathcal{L}$time; hence, f is also $\in \mathcal{L}$time.

It is, then, straightforward to verify that, for each $q, y \in \mathbb{N}$,

$$f(q, y) \in (C_q \bigtriangleup W_y); \tag{40}$$

however, this verification involves consideration of the cases $f(q, y) \in W_y$ vs. not, and this is what employs Σ_1^0-LEM.

Future work: *minimize* non-constructivity here & elsewhere, a Reverse Mathematics project.

From here until the end of the present section we will work primarily with $(\mathfrak{S}, \psi) = (\mathcal{P}\text{time}_k, \theta^k)$. To make this clear below, we will mostly write θ^k in place of ψ.

Let f be from Theorem 5 *for the case of* $(\mathfrak{S}, \psi) = (\mathcal{P}\text{time}_k, \theta^k)$. We know already that $f(q, y)$ can be computed in linear-time in the lengths of q, y. Of course, for *each fixed* q, y, $f(q, y)$ is a θ^k-program, and, so, $\theta^k_{f(q,y)}$ runs within time some k-degree polynomial evaluated at the lengths of its arguments.

We are interested next in how hard it is to compute $\theta^k_{f(q,y)}(z)$ *as a function of all of* q, y, z. We, then, run up against the high cost of *universality* for the θ^k-system. From Ref. 18 Theorem 6.5, the general cost of universality even for θ^1 is worse than $\mathcal{P}$time. Kozen[26] Theorem 7.4 showed the general cost of universality for sensible systems for all of $\mathcal{P}$time is worse than $\mathcal{P}$space.

Theorem 6. *For some constants* $b, c > 0$, $\theta^k_{f(q,y)}(z)$ *is computable in*

$$\mathcal{O}(2^{b(ck+1)|(|q|+|y|)|}|z|^k \log |z|) \textit{ time.} \tag{41}$$

Proof. Since $f \in \mathcal{L}$time,

$$(\exists \text{ a constant } b > 0)(\forall q, y)[|f(q, y)| \leq b(|q| + |y|)]. \tag{42}$$

Mostly from Ref. 18 Theorem 6.4 and parts of its proof, we have: for some constant $c > 0$, $\theta^k_p(z)$ is computable in

$$\mathcal{O}(2^{(ck+1)|p|}|z|^k \log |z|) \text{ time.} \tag{43}$$

In particular, we see that the run time just above is exponential in the size of the program p to be simulated.

The proof of Ref. 18 Theorem 6.4 makes use of the inequalities (30) above re trans and bound, that they are each in $\mathcal{L}$time, the sentence *just before* Ref. 18 Theorem 4.16, and, also, Ref. 18 Corollary 3.7 which provides an upper-bound on the general time cost of *time-bounded* $(\varphi^{\text{TM}}, \Phi^{\text{TM}})$-universal simulation. *That* cost is *exponential in the length of the time-bound's value.*

Therefore, from (42, 43), we have the desired theorem. □

References

1. H. Rogers, Gödel numberings of partial recursive functions, *Journal of Symbolic Logic* **23(3)**, 331 (1958).

2. H. Rogers, *Theory of Recursive Functions and Effective Computability* (McGraw Hill, New York, 1967). Reprinted, MIT Press, 1987.
3. M. Machtey and P. Young, *An Introduction to the General Theory of Algorithms* (North Holland, New York, 1978).
4. J. Royer, *A Connotational Theory of Program Structure* (Lecture Notes in Computer Science 273. Springer-Verlag, 1987).
5. L. Brouwer, in *Brouwer's Cambridge Lectures on Intuitionism*, ed. D. van Dalen (Cambridge University Press, 1981)
6. E. Mendelson, *Introduction to Mathematical Logic*, fifth edn. (Chapman & Hall, London, 2009).
7. A. Bauer, Constructive gem: irrational to the power of irrational that is rational (2009), See `math.andrej.com/2009/12/28/` for the article.
8. A. Heyting, *Intuitionism: An Introduction* (North-Holland, Amsterdam, 1971). Third edition.
9. A. Troelstra (ed.), *Metamathematical Investigation of Intuitionistic Arithmetic and Analysis*, Lecture Notes in Mathematics, Vol. 344 (Springer, 1973).
10. A. Troelstra and D. van Dalen, *Constructivism in Mathematics: An Introduction, Volume I* (Elsevier, 1988).
11. M. Nakata and S. Hayashi, A limiting first order realizability interpretation, *Scientiae Mathematicae Japonicae Online* **5**, 421 (2001).
12. Y. Akama, S. Berardi, S. Hayashi and U. Kohlenbach, An arithmetical hierachy of the laws of excluded middle and related principles, *Proceedings of 19th Annual Symposium on Logic in Computer Science (LICS'04)* , 192 (2004).
13. S. Hayashi, Mathematics based on incremental learning, excluded middle and inductive inference, *Theoretical Computer Science* **350**, 125 (2006).
14. E. Post, Recursively enumerable sets of positive integers and their decision problems, *Bulletin of the American Mathematical Society* **50**, 284 (1944).
15. J. Myhill, Creative sets, *Z. Math. Logik Grundlagen Math* **1**, 97 (1955).
16. J. Dekker, Productive sets, *Trans. of AMS* **78**, 129 (1955).
17. J. Cleave, Creative Functions, *Zeitschrift für Mathematische Logik und Grundlagen der Mathematik* **7**, 205 (1961).
18. J. Royer and J. Case, *Subrecursive Programming Systems: Complexity and Succinctness* (Birkhäuser Boston, 1994). See `www.eecis.udel.edu/{\textasciitilde{}}case/RC94Errata.pdf` for corrections.
19. J. Dekker and J. Myhill, Some theorems on classes of recursively enu-

merable sets, *Trans. of AMS* **89**, 25 (1958).

20. R. Friedberg, Two recursively enumerable sets of incomparable degrees of unsolvability (solution of Post's Problem, 1944), *Proceedings of the National Academy of Sciences* **43**, 236 (1957).
21. A. Muchnik, On the unsolvability of the problem of reducibility in the theory of algorithms, *Doklady Akademii Nauk SSSR* **108**, 194 (1956).
22. L. Hay, The class of recursively enumerable subsets of a recursively enumerable set, *Pacific Journal of Mathematics* **46**, 167 (1973).
23. A. Meyer and D. Ritchie, The complexity of loop programs, in *Proceedings of the 22nd National ACM Conference*, (Thomas Book Co., 1967) pp. 465–469.
24. M. Blum, A Machine-Independent Theory of the Complexity of Recursive Functions, *Journal of the ACM* **14**, 322 (1967).
25. J. Case, *Zeitschrift für Mathematische Logik und Grundlagen der Mathematik* **37**, 97 (1991), `http://www.eecis.udel.edu/{\textasciitilde{}}case/papers/mkdelta.pdf` corrects missing set complement signs in definitions in the journal version.
26. D. Kozen, Indexings of subrecursive classes, *Theoretical Computer Science* **11**, 277 (1980).

Minimal Pairs in the C.E. Truth-table Degrees *

Rod Downey

School of Mathematics, Statistics and Operations Research
Victoria University of Wellington
PO Box 600, Wellington, New Zealand
Rod.Downey@vuw.ac.nz

Keng Meng Ng

Division of Mathematical Sciences
School of Physical & Mathematical Sciences
Nanyang Technological University
21 Nanyang Link, Singapore
kmng@ntu.edu.sg

We construct a pair of c.e. sets $A \equiv_{wtt} \emptyset'$ and $B \equiv_T \emptyset'$ whose truth table degrees form a minimal pair.

Keywords: Turing degrees, truth table degrees, algorithmic randomness

1. Introduction

Strong reducibilities such as the m-reducibility have been around implicitly, if not explicitly, since the dawn of computability theory. The explicit recognition of the existence of differing kinds of oracle access mechanisms began with the seminal work of Post[12]. Of interest to us from Post's paper are the so-called tabular reducibilities $\leq_{tt}$, truth table reducibility, and $\leq_{wtt}$, *weak* truth table reducibility. Following the work of Nerode[11], we know that $\leq_{tt}$ can be characterized by saying $A \leq_{tt} B$ iff there is a Turing procedure Φ with $\Phi^B = A$ *and* Φ^X *is total for all oracles* X. Weak truth table reducibility is defined via $X \leq_{wtt} Y$, iff there is a procedure Ψ with the use $\psi(x)$ is a computable function.

Quite aside from their intrinsic interest, reducibilities stronger than Turing reducibility crop up everywhere in effective mathematics. For exam-

*The first author is partially supported by the Marsden Fund of New Zealand, while the second author is partially supported by the MOE grants MOE2011-T2-1-071 and MOE-RG26/13.

ple $\leq_{wtt}$ can be used for classifying the degrees of bases of vector spaces (Downey and Remmel[5]), it can be used to study the presentations of reals (Downey and Terwijn[6]), and are also applied to concepts used for the classifications of approximations to effective processes. For example, X is of hyperimmune-free Turing degree iff for all $Y \leq_T X$, $Y \leq_{tt} X$.

One reason for recent interest in truth table reducibility is due to its intrinsic relationship with algorithmic randomness. This relationship seems to occur because the totality of truth table procedures allows measure to be transformed from one space to another. For example, long ago, Demuth[4] showed that if $X \leq_{tt} Y$ is noncomputable and Y is random, then there is a random $Z \equiv_T X$. The reason being that the uniform measure for the space Y can be effectively transformed into another measure relative to which X is random, and this, in turn, can be re-translated into a uniform measure for Z. Other examples include Franklin and Stephan's use of tt-reducibility for classifying concepts related to Schnorr randomness (Franklin and Stephan[7]) and similar investigations of Greenberg, Franklin, Stephan and Wu[8].

The present paper was motivated by the consideration of the reals tt-below the collection of nonrandom strings. For example, let $N_C = \{x \mid C(x) < |x|\}$, where C is plain Kolmovorov complexity. We know that N_C is wtt-complete amongst c.e. sets. Kummer[10] showed that in fact N_C is also tt-complete. The argument was particularly interesting in that the process of constructing the tt-reduction was *nonuniform*.

On the other hand, Kummer showed that whether N_K, the analogous set for prefix-free complexity, is tt-complete depends on the choice of universal machine. That is, there are universal machines U_1, U_2 such that $N_{K_{U_1}}$ is tt-complete, and such that $N_{K_{U_2}}$ is not. The proof of the latter fact was rather novel and has seen applications as far as complexity, such as Allender, Buhrman, Friedman and Loff[1] where it is used for results relating randomness to PSPACE.

In Ref. 2, Cai, Downey, Epstein, Lempp and Miller looked at the question of the tt-computational power of sets of random strings for various universal machines computing the prefix-free complexity K. This group showed that a set A is tt- below all such sets of random strings iff A is computable. On the other hand, these authors also showed that for any two universal machines U_1, U_2, there is a noncomputable set $X \leq_{tt} N_{K_{U_1}}, N_{K_{U_2}}$.

Since all such sets of non-random strings are necessarily wtt-complete, the question arises whether there exist wtt-complete c.e. sets A, B which form a tt-minimal pair. If this is false then the second result of Cai et al. would be a simple corollary. In Ref. 2, Cai et al. showed that there are c.e.

Turing complete sets forming a tt-minimal pair, but left the wtt-question open.

This question seems hard. Earlier, Jockusch and Mohrherr[9] showed that the diamond lattice can be embedded into the c.e. tt-degrees preserving the greatest and the least element. That is, there are c.e. sets A and B such that $A \oplus B \equiv_{tt} \emptyset'$ with $X \leq_{tt} A, B$ implies X computable. Also Degtev[3] had shown the existence of c.e. Turing complete sets forming a minimal pair in the tt-degrees of c.e. sets. It is unknown whether $A \equiv_{wtt} B \equiv_{wtt} \emptyset'$ is possible.

In the present paper, we will make modest progress towards the resolution of the question of a wtt-complete tt-minimal pair by proving that there can be $A \equiv_{wtt} \emptyset'$ and $B \equiv_T \emptyset'$ forming a tt-minimal pair.

The reader might guess that this result can be proved by a straightforward modification of the proof in the Cai et al. paper. This is unfortunately far from the case; the method used in this paper is quite different. Our argument uses the determinacy of finite games and their computable winning strategies in a way hitherto unseen. We hope that this might lead to progress on the full questions. Also, in itself, the technique is very pretty and might well have other applications.

We prove the following Theorem:

Theorem 1.1. *There is a pair of c.e. sets $A \equiv_{wtt} \emptyset'$ and $B \equiv_T \emptyset'$ whose tt-degrees form a minimal pair, i.e. $X \leq_{tt} A, B \Rightarrow X$ is computable.*

2. Proof of Theorem 1.1: Informal description

We will use the well-known fact that every finite game is computably determined. Moreover there is a uniform procedure to pass from a finite game to a winning strategy for one of the two players in the finite game. We shall build c.e. sets A and B. To ensure that $B \geq_T \emptyset'$ we specify computable approximations to the markers $\{\gamma(n)\}$, satisfying the usual marker rules. For each n:

- $\gamma(n)$ is moved monotonically.
- $\gamma(n)$ is moved at a stage s only if $B \upharpoonright \gamma(n)[s]$ has changed.
- $\gamma(n)$ is only moved finitely often.
- If n enters $\emptyset'$ at a stage s then $\gamma(n)$ is moved at or after stage s.

It is straightforward to check that these rules ensure that $B \geq_T \emptyset'$. To ensure $A \geq_{wtt} \emptyset'$, we will reserve intervals $\{I_n\}$ where each I_n is used for coding $\emptyset'(n)$. We ensure the following hold:

- The intervals $\{I_n\}$ and their lengths $|I_n|$ are chosen and effectively determined in advance.
- If n enters $\emptyset'$ at stage s then A changes on interval I_n at or after stage s.

If these two points are ensured during the construction then clearly $A \geq_{wtt} \emptyset'$. As we will see, during the construction we will be enumerating various numbers (for the sake of tt-minimal pair requirements) into $A \upharpoonright I_n$, even when n has not yet entered $\emptyset'$. For the coding $A \geq_{wtt} \emptyset'$ to work, we must ensure that the interval I_n in A is not filled up before n enters $\emptyset'$. Otherwise if n enters $\emptyset'$ later on then there is no way to record this in A.

The requirements to meet are

$$R_e : \left(\Phi_e^A = \Phi_e^B = X\right) \Rightarrow X \text{ is recursive,}$$

where Φ_e is the eth potential tt-functional. For each n such that we observe $\Phi_e^A = \Phi_e^B = r$, we will begin a procedure to *certify* the computations. This certification procedure will be concluded when A or B changes below a small number, or when we obtain the desired certification. In the former case we will wait for both sides to agree again before beginning another certification procedure on the new computations. In the latter case we will *believe in the common value* r, and define the value of $X(n) = r$.

2.1. *The strategy to satisfy R while keeping $B \geq_T \emptyset'$*

We briefly describe the strategy to make C and B a tt-minimal pair while keeping $C \equiv_T B \equiv_T \emptyset'$; for more details we refer the reader to Cai et al.[2] The strategy for processing a pair (e, n) is as follows. We let $c_{e,n}$ be the least number on the C-side below the $\varphi_e(n)$-use such that the entry of any combination of $\gamma(i) > \gamma(c_{e,n})$ will not change the value of $\Phi_e^C(n)$. If $c_{e,n} \leq e$ then we say that $c_{e,n} \uparrow$; otherwise $c_{e,n}$ is always picked to be larger than e. We can similarly define the number $b_{e,n} > e$ for $\Phi_e^B(n)$. We then pick the larger of the two, say $b_{e,n}$, and enumerate $\gamma(b_{e,n})$ into C and B (and lifting all larger γ-markers). If $b_{e,n} = c_{e,n}$ we can then enumerate $\gamma(b_{e,n})$ into B and $\gamma(c_{e,n} + 1)$ into C.

We then get a disagreement $\Phi_e^C(n) = 0 \neq 1 = \Phi_e^B(n)$ which will be preserved unless $\gamma(x)$ is enumerated into $C \cup B$ for some $x \leq b_{e,n}$. This can be used to argue show that each marker $\gamma(x)$ is moved only finitely often, and so $B, C \geq_T \emptyset'$. We will only believe in the pair (e, n) when we find that one of $b_{e,n} \uparrow$ or $c_{e,n} \uparrow$. In this case at least one of the two sides of a certified agreement can be forever preserved at the original value (unless, of course $\gamma(e)$ is moved, where e is a small number relative to the eth requirement).

Note that herein lies the fundamental difference between making the atoms Turing complete versus *wtt*-complete. In the case of *wtt*-complete this naive plan cannot work, because the uses for $\gamma(i)$, $i > b_{e,n}$ are not lifted by this action and so this disagreement *can be removed* even by a large $\gamma(i)$ change. When considering $A \geq_{wtt} \emptyset'$ we need to consider the possible effects of enumerating any combination of markers below the use. To help us organize this process we will need to consider games.

2.2. *The game $G(e,n,s)$*

We will define the finite game $G(e,n,s)$. The idea is that we will only believe a pair (e,n) when we find that no appropriate value of $b_{e,n}$ on the B-side is found. In this case we can ensure that the computation on the B-side can be preserved at the original value.

Definition 2.1. For $e,n,s \in \omega$, the game $G(e,n,s)$ is defined as follows. We assume the game $G(e,n,s)$ starts at stage s of the construction with both sides $\Phi_e^A(n)[s] = \Phi_e^B(n)[s]$ agreeing. For convenience we always assume that the starting common value is $r = 0$. The game is a finite game of perfect information played between two players. The first player, called the SPOILER, always starts first and the game alternates between a SPOILER move and a PRESERVER move. The PRESERVER wants to keep the starting value $\Phi_e^A(n) = 0$ (by enumerating elements in A). The SPOILER wants to flip the value of $\Phi_e^A(n)$ to 1 (also via enumeration into A). By artificially increasing the use by a recursive amount we may assume that $I_0, \cdots, I_N \subseteq \varphi_e(n)$ for some N.

At any point in the game, the SPOILER's next legal move consists of a finite set $\{j_{n_0}, j_{n_1}, \cdots\}$ where $j_{n_k} \in I_{n_k}$ and n_ks are distinct numbers $\leq N$, such that the SPOILER has not previously played in any of the intervals $I_{n_0}, \cdots$, and such that $\Phi_e^A(n) = 1$ after playing these elements into A. In other words, in the entire game, the SPOILER can play at most one element in each interval I_n, $n \leq N$, and once he has played in an interval, he cannot later play another element in the same interval. A single move of the SPOILER can contain finitely many elements in different intervals, as long as the $\Phi_e^A(n)$-computation takes value 1 after playing these elements into A.

A legal move of the PRESERVER is similar. It consists of a finite set $\{j_{n_0}, j_{n_1}, \cdots\}$ where $j_{n_k} \in I_{n_k}$ and n_ks are distinct numbers $\leq N$, such that the PRESERVER has not previously played in any of the intervals $I_{n_0}, \cdots$, and such that $\Phi_e^A(n) = 0$ after playing these elements into A. Furthermore we require that each j_{n_k} is larger than some element (not necessarily in the same interval) already played by the SPOILER. So the PRESERVER plays like the SPOILER, with the extra restriction that it must play elements which are larger than something previously played by the SPOILER.

Note that "passing" is not allowed by either player since on a SPOILER turn the value of $\Phi_e^A(n)$ must be 0. The game ends when one of the two players has no more moves; in that case the player who made the last move wins.

In the game $G(e, n, s)$ either the PRESERVER or the SPOILER has a winning strategy.

2.3. *Making $A \geq_{wtt} \emptyset'$ and $B \geq_T \emptyset'$*

Let's consider a single pair (e, n) in isolation. How do we certify the pair (e, n)? If $b_{e,n}$ does not exist, then we have a certification of the pair (e, n); since in this case the computation on the B-side can be preserved forever. Otherwise we will begin a diagonalization procedure by considering the following cases:

2.3.1. *Case 1: The PRESERVER has a winning strategy on the A-side*

In this case we try and diagonalize by enumerating $\gamma(b_{e,n})$ into B and we will obtain $\Phi_e^A(n) = 0 \neq 1 = \Phi_e^B(n)$. We play the PRESERVER strategy on the A-side. What this means is that we enumerate the smallest element of $\omega - A$ in I_x whenever we find that x enters $\emptyset'$. If ever we discover that $\Phi_e^A(n)$ flips to 1 due to a (sequence of) coding actions, we will consider this to be a SPOILER move in the game $G(e, n, s)$. During the construction we immediately respond with a PRESERVER-move according to PRESERVER's winning strategy in $G(e, n, s)$; this immediately restores the value of $\Phi_e^A(n)$ back to 0.

Since the SPOILER moves first in the game, and every SPOILER-move is due to coding, also every PRESERVER-move is to respond to a previous SPOILER-move, and the PRESERVER only plays larger numbers, every enumeration into each interval can be accounted for against the coding of $\emptyset'$. At the end of the game when the SPOILER has no more legal moves, we end up in a situation where the truth table forever oputputs 0, regardless of any future coding, since the SPOILER has no more moves in the game. So in this situation, when the PRESERVER has a winning strategy, we can ensure that the diagonalization succeed by keeping the A-side output 0. The only time a diagonalization set up like this can be unsuccessful is if a number $x \leq b_{e,n}$ enters $\emptyset'$. We stop the process if this happens.

There are two outcomes to this diagonalization procedure:

- *The procedure is halted due to coding of some number $x \leq b_{e,n}$.* In this case all the moves on the A-side can be accounted for against coding, while the lifing of $\gamma(b_{e,n})$ on the B-side is accounted for against the coding of x.
- *The procdure is never halted.* We successfully preserve the disagreement.

Note that since the PRESERVER has a winning strategy on the A-side, *what we could have done*, was to have immediately believed in the computations and promise to always play the PRESERVER strategy to keep $\Phi_e^A(n) = 0$. This is fine when considering a single pair (e, n), but not desirable in full construction, because the interactions between different pairs get too complicated. So we want to keep attempting to diagonalize (if we can), until the B-side can change no more, i.e. $b_{e,n} \uparrow$, and only then do we believe in the computations. Note that this "permanent" computation is, of course permanent, assuming that no marker $\gamma(k), k \leq e$ enters B. Believing in such a permanent computation is fine because $\gamma(k)$ can change only finitely often for $k \leq e$, and the entire strategy for e can restart every time this happens.

Now we assume that the SPOILER has a winning strategy on the A-side. For a number $M \leq N$ we say that SPOILER has a M-winning strategy if there is a winning strategy for SPOILER such that in every possible run of the game the SPOILER has a response that plays only numbers from intervals with index $\geq M$. Since the SPOILER has a winning strategy in the game $G(e, n, s)$, there is a largest number $M_A \leq N$ such that SPOILER has an M_A-winning strategy in $G(e, n, s)$. (At worst we can take $M_A = 0$ which certainly works; any winning strategy is a 0-winning strategy). There are now two further cases.

2.3.2. *Case 2: The SPOILER has a winning strategy on the A-side and $M_A \geq b_{e,n}$*

We play the M_A-winning strategy for the SPOILER on the A-side to keep $\Phi_e^A(n) = 1$. We halt the procedure if a number $\leq b_{e,n}$ enters $\emptyset'$. While the diagonalization procedure is running, no number $\leq b_{e,n}$ enters $\emptyset'$, and so we know that $\Phi_e^B(n) = 0 \neq 1 = \Phi_e^A(n)$.

On the A-side we use the PRESERVER to code $\emptyset'$ into A. We explain what this means. We make the first move of the M_A-winning strategy for the SPOILER and flip the A-side to 1. We then wait for coding to flip the truth table back to 0. When this happens (if ever) we consider this as the next move of the PRESERVER. We then respond with a SPOILER move to flip it to 1, and so on. The only time we cannot run this strategy is when some small number, say $x \leq N$, enters $\emptyset'$ (smaller than any of the SPOILER's moves so far). Then we may have no more legal next move for PRESERVER and so the SPOILER's M_A-winning strategy does not know how to respond. In this case we halt the procedure, and wait for the next recovery $\Phi_e^A(n) = \Phi_e^B(n)$, and then do this certification over again.

There are three possible outcomes of this diagonalization procedure:

- *A number $x \leq b_{e,n}$ enters $\emptyset'$.* In this case we must stop the diagonalization procedure because the B-side may no longer have output 0. In this case the attempted diagonalization is destroyed, but it is okay because we have not yet believed in the computation, and we have one more number in $\emptyset' \upharpoonright N$. In A we have enumerated a bunch of numbers for the SPOILER (in accordance to the SPOILER's winning strategy), which are now wasted, but these numbers are all larger than $M_A \geq b_{e,n} \geq x$, so we can blame all the moves of the SPOILER on the coding of x.
- *A number x which is smaller than all the SPOILER moves so far has entered $\emptyset'$, so that the PRESERVER has no more legal moves in the game.* The procedure is also halted in this case. In this case all of the SPOILER moves are wasted, but are all involving numbers larger than x, so again we can account these against the coding of x.
- *The procedure is never halted.* We then have a permanent disagreement $\Phi_e^A(n) \neq \Phi_e^B(n)$. In this case on the A-side we have enumerated a bunch of numbers (for the sake of the M_A-winning SPOILER strategy). These numbers may be small and cannot be accounted for against coding (but each enumerated number must be at least as large as M_A). However this is okay because we now have a permanent disagreement, and so we can tolerate the wastages in I_n made by the SPOILER.

2.3.3. *Case 3: The* SPOILER *has a winning strategy on the A-side and* $M_A < b_{e,n}$

Since there are no $(M_A + 1)$-winning strategies for the SPOILER on the A-side, by determinacy the PRESERVER has a winning strategy which allows her to win, provided that SPOILER only plays numbers $\geq M_A + 1$. We then enumerate $b_{e,n}$ into B to make $\Phi_e^B(n) = 1$. (Note that $b_{e,n} \downarrow$ if and only if there is some combination of markers $\gamma(i)$, i in some finite set $F \subseteq \{e + 1, \cdots\}$ so that the entry of these markers will change the computation. In the case F exists we can choose $\min F = b_{e,n}$. So we can in this case enumerate $\gamma(b_{e,n})$, as well as all larger $\gamma(i), i \in F$ into B, to change the computation. For convenience we shall simply say that we enumerate $b_{e,n}$ into B).

We stop the procedure if a number $x < b_{e,n}$ enters $\emptyset'$. On the A-side we play the (M_A+1)-winning strategy for the PRESERVER (as in Case 1 above), and we use the SPOILER moves to code. There are again two outcomes to this procedure.

- *The procedure is halted when some number $x < b_{e,n}$ enters $\emptyset'$.* In this case the lifting of $\gamma(b_{e,n})$ on the B-side is accounted for against the coding of x. On the A-side all SPOILER-moves (used for coding) as well as all PRESERVER response involves numbers $\geq b_{e,n} > x$. Hence all wastage in A can be blamed on the coding of x.
- *The procedure is never halted.* In this case no number less than $b_{e,n}$ enters $\emptyset'$. Hence $\Phi_e^B(n) = 1$ is held permanently. Since $M_A + 1 \leq b_{e,n}$, this means that all SPOILER moves involves numbers larger than M_A, and so the PRESERVER always has a response. Hence we are able to ensure $\Phi_e^A(n) = 0$.

2.4. *Some global considerations*

Notice that if the pair (e, n) has been certified, then $\Phi_e^B(n)$ *must remain at the believed value* 0 *no matter what happens in future.* Therefore, once we believe in the pair (e, n) there is no need for us to actively pursue the PRESERVER-strategy on the A-side. In fact, once the pair (e, n) is certified, we no longer need to look at the game $G(e, n)$ on the A-side, since $\Phi_e^B(n)$ must be 0. The sole purpose of looking at games and winning strategies on the A-side is to allow us to actively preserve a disagreement before certification, and is never used *after* we have obtained certification. Thus, there are *no real interactions* between two certified pairs (e, n) and (e', n'). We may treat this as a finite injury between pairs (e, n). The only interaction are between two pairs each trying to diagonalize, and because we have not yet believed in either (e, n) or (e', n'), it is easy to sort out any conflict between the two pairs; We can simply initialize the lower priority one (the priority is explained in the formal construction). If we need to make both A and B *wtt*-complete, then the winning strategies for different pairs (each of which are already certified) must be made to somehow cohere, and this introduces severe difficulties.

3. The formal construction

At each stage s of the construction, there are three lists:

- The *active list.*
- The *inactive list.*
- The *certified list.*

The certified list contains all pairs (e, n) which have been certified, i.e. $b_{e,n} \uparrow$. As mentioned previously if (e, n) has been certified then it may be removed from all future consideration. This list is ordered by magnitude of $\langle e, n \rangle$. The inactive list contains all pairs (e, n) which is waiting to be placed in the active list, and which the strategy for (e, n) has not begun. This list is also ordered by the magnitude of $\langle e, n \rangle$. Finally the active list contains all pairs (e, n) which we have begun the strategy. Each active pair (e, n) may be in *waiting phase*, or be in *diagonalization phase.* This list is a queue and is ordered by the time where each pair (e, n) is placed in the queue. The priority ordering is static in the sense that elements in the list do not have their priority reversed while they remain in the queue. Of course it can happen that an element leaves the list and later re-joins the end of the queue. In this case the priority of the element is lowered relative to the other elements of the list, but this only happens when it gets kicked out of the list.

At the beginning $s = 0$, place all requirements (e, n) in the inactive list, and do nothing. Suppose we are at stage $s > 0$. Suppose $k \in \emptyset'_s - \emptyset'_{s-1}$. We enumerate $\gamma(k)$ into B and change I_k on the A-side. For every requirement on the active list which needs to respond according to some winning strategy, we do so (in fact, there can be at most one requirement which needs to respond). Initialize all strategies on the active list which are halted by this coding action (i.e. place all these strategies in the inactive list).

Now we find the smallest pair $\langle e, n\rangle$ such that e is not represented in the active list. Place (e, n) from the inactive list into the active list (this new pair joins the active list/queue at the end). This new pair starts off in the waiting phase. Now we check to see if there is a pair (e, n) in the active list currently in waiting phase, which can be moved to the diagonalization phase, we do so for the highest priority pair (this is described below). Otherwise if there is no waiting pair which can begin diagonalizing, end the construction.

A waiting pair (e, n) can begin diagonalization, if $\Phi_e^A(n) = \Phi_e^B(n)$ again (for convenience, we denote the common value 0). If $b_{e,n} \uparrow$ we move this waiting pair into the certified list, and end the stage immediately without initializing anybody. Otherwise assume that $b_{e,n} \downarrow$. Suppose that $\ell = \ell_{e,n}$ is the largest such that I_ℓ is below some $\varphi_i^A(j)$ for some active pair (i, j) of higher priority than (e, n). (Again, remember that the priority ordering amongst active pairs is according to the "queue number", i.e., the time where a pair enters the active list.) Intuitively, any winning strategy which (e, n) follows henceforth should be restrained from modifying I_ℓ and below; so (e, n) does not interfere with any pair of higher priority. During the construction there will only be the following positive actions:

(i) $\gamma(k)$ enters B due to coding.
(ii) $\gamma(b_{e,n})$ enters B due to Case 1 or 3 below.
(iii) I_k is modified due to coding.
(iv) I_k is modified in response to a `PRESERVER`-strategy (in which case there is a smaller number blamed on coding).
(v) I_k is modified in response to a `SPOILER`-strategy.
(vi) I_k is modified by an initial `SPOILER`-move.

(i) and (iii) are called *coding actions.* (ii) and (vi) are called *initial actions.* We also assume that ℓ is larger than the last stage s where the pair (e, n) is initialized not due to any of the above reason (this is the case, if for example, some (e', n') of higher active priority starts diagonalizing but does not put a small number in A or B). Also if (e, n) is initialized due to reason (ii) or (vi) (an initial move) we also increase ℓ. We also assume that $\ell_{e,n}$ is larger than $\ell_{e,i}$ for every $i < n$.

As in Subsection 2.3, there are three possibilities. This time we make minor changes (to take I_ℓ into consideration).

(I) *Case 1: The `PRESERVER` has a winning strategy on the A-side.* In this case we play the `PRESERVER` strategy on the A-side and enumerate

$\gamma(b_{e,n})$ into B. We obtain $\Phi_e^A(n) = 0 \neq 1 = \Phi_e^B(n)$. We stop the procedure if a number $< b_{e,n}$ or $\leq \ell$ enters B. (Note that ℓ may be much larger than $b_{e,n}$).

(II) *Case 2: The* **SPOILER** *has a winning strategy on the A-side and* $M_A \geq \max\{b_{e,n}, \ell\}$. We play the M_A-winning strategy for the **SPOILER** on the A-side. DO NOT LIFT $\gamma(b_{e,n})$. On the A-side we use the **PRESERVER** to code, and on the B-side we code with $\gamma(i)$. We halt the procedure if a number $\leq \max\{b_{e,n}, \ell\}$ enters B, or if the **PRESERVER** has no further move in this game.

(III) *Case 3: The* **SPOILER** *has a winning strategy on the A-side and* $M_A < \max\{b_{e,n}, \ell\}$. On the A-side we play the $(M_A - 1)$-winning strategy for the **PRESERVER**, and use the **SPOILER** to code. We then enumerate $\gamma(b_{e,n})$ into B to make $\Phi_e^B(n) = 1$. We stop the procedure if a number $\leq \max\{b_{e,n}, \ell\}$ enters B.

We apply an initial action for (e, n). If Case 1 or 3 applies we initialize all active requirements which are affected, i.e. whose procedure is halted by the enumeration of $\gamma(b_{e,n})$ into B. (Note that this may cause an active requirement of *higher priority* than (e, n) to be initialized.) We place these requirements back to the inactive list. If Case 2 applies we do not do this, as we did not change B, and the initial action in case 2 on the A-side does not affect any higher priority requirement, because of ℓ. Finally in any case we initialize all lower priority active requirements.

4. Verification

Lemma 4.1. *During the construction, every enumeration into A is accounted for. Hence the weak truth table reduction works, and* $A \geq_{wtt} \emptyset'$.

Proof. Looking at the list of positive actions (i) to (vi), we see that only (iii), (iv), (v) and (vi) are relevant. (iii) happens at most once for each I_k.

Only the pairs (e, n) with $e < k$ can modify I_k via (iv), (v) or (vi). For each e at most one such pair (e, n) will be diagonalizing at any one time. For each such pair (e, n) (that is diagonalizing via a **PRESERVER**- or a **SPOILER**-strategy), it will modify I_k at most one time before it is initialized. We only consider all pairs (e, n) such that $\varphi_e(n) > \max I_k$; otherwise (e, n) has no effect on I_k. Also we must have $k > \max\{\ell_{e,n}, b_{e,n}\}$ (since $k > M_A$ in Case 2, and in in the other two cases we play the **PRESERVER**-strategy so nothing smaller than $\max\{\ell_{e,n}, b_{e,n}\}$ can be played). If (e, n) is never initialized then e will never need to be considered again. If the pair (e, n)

is initialized not due to a positive action (or due to (ii) or (vi)) then $\ell_{e,i}$ is increased large for every $i \geq n$, and so e will cease to be active in I_k again, and so we may neglect this case.

So to find the size of I_k we need to consider an upperbound for the quantity:

$$\sum_{e<k} (\#\text{ times } (e,n) \text{ is initialized for some } n \text{ by a positive action})$$

Assume (e,n) is initialized due to a positive action; then we ask what can this action be? Since (e,n) was initialized by a positive action, we must have some small number less than $\max\{\ell_{e,n}, b_{e,n}\}$ entering A or B: Cases 1 and 3 are clear, and in Case 2 if the PRESERVER runs out of moves, and since k was played by either the SPOILER or the PRESERVER, then we must also have a number smaller than k entering $\emptyset'$.

So in any case we must have a number smaller than k which enters A or B. But this is not enough, because for instance $\gamma(k-1)$ could be lifted many times due to other requirements; so we need to argue that in fact we have a change in $\emptyset' \upharpoonright k$. This number is put in by one of (i), (iii), (iv) or (v). If it is (i) or (iii) or (iv) then we clearly have a change in $\emptyset' \upharpoonright k$. If it is due to (v) and performed by another pair (e', n'), we check if (e', n') is of lower priority than (e,n), at the point of the action. If this is so then this change must be above $\varphi_e(n) > \max I_k$. If (e', n') is of higher priority then there must also have been a recent change in $\emptyset'$, and in fact less than k. In any case we can blame this initialization on a change in $\emptyset' \upharpoonright k$. □

Lemma 4.2. *A requirement (e,n) is initialized only finitely often.*

Proof. Let (e,n) be the least such that (e,n) is initialized infinitely often. Thus there is some stage s_0 large enough such that after s_0,

- A and B are stable up to $\varphi_e(n)$.
- Every smaller pair (i,j) is never again initialized.

Claim 4.1. *There are only finitely many pairs (p,q) such that at some point (p,q) is active and of higher priority than (e,n) and $\langle p,q\rangle > \langle e,n\rangle$.*

To see this, look at the first stage after s_0 where (e,n) is active. At this point there are only finitely many larger (p,q) of higher active priority. No new (p,q) can sneak in after this, because (e,n) is never satisfied, and so the only thing that can remove (e,n) is an initialization to (e,n). But then (e,n) will immediately be placed back into the active list above (p,q), if

$p > e$. (Of course smaller (p, q) can sneak in above (e, n), but not a larger one.)

Thus we may further assume that s_0 is large enough so that

- No new (p, q) with $\langle p, q \rangle > \langle e, n \rangle$ appears in the active list above (e, n).
- All (p, q) in the active list above (e, n) is never again initialized.

Thus after s_0, the set of requirements of higher active priority than (e, n) is fixed and does not change. From this point on, what cam result in an initialization to (e, n)? Since all higher priority requirements are stable, this initialization must be due to a positive action. It cannot be due to coding, by assumption on s_0, so (i) and (iii) are out. Similarly (ii) is not possible (even if it is done be a lower priority requirement). In fact nothing is possible. □

Lemma 4.3. *Each R_e is satisfied, and the tt-degrees of A and B form a minimal pair.*

Proof. Since (e, n) is initialized finitely often, the final state of (e, n) is either permanently diagonalized, or it is certified, or we wait forever in the case of non-convergence. □

Lemma 4.4. *Each $\gamma(k)$ is lifted finitely often, and so $B \geq_T \emptyset'$.*

Proof. Suppose that $\gamma(m)$ is stable for all $m < k$. Note that only requirements R_e for $e < k$ can lift $\gamma(k)$. Coding only happens at most once. So it must be moved infinitely often due to an initial action by some pair.

By Lemma 4.2, for each n, the first n elements of the active list/queue is eventually stable. Furthermore each $\ell_{i,j}$ for each stable pair (i, j) of the active list is also eventually stable. How many stable elements of the list can have $b_{i,j} \leq k$? Only finitely many. Let $M-1$ denote the largest position of the list which is occupied by a stable pair with $b_{i,j} \leq k$. Once everything below M is stable, we claim that $\gamma(k)$ cannot be moved. Suppose it is moved by some pair (e, n). Before the action, the pair (e, n) must occupy a position $> M$ (by the stability of the first M elements of the list). However every requirement between the Mth position and (e, n) must have $b > k$, which means that they would be initialized, and so the pair (e, n) must move to position M itself, a contradiction. □

References

1. Eric Allender, Harry Buhrman, Luke Friedman, and Bruno Loff, Reductions to the set of random strings: The resource-bounded case, in *Proc. 37th International Symposium on Mathematical Foundations of Computer Science (MFCS '12), 2012*, Lecture Notes in Computer Science.
2. Mingzhong Cai, Rod Downey, Rachel Epstein, Steffen Lempp and Joseph Miller, Random strings and truth table degrees of Turing complete c.e. sets, in preparation.
3. Alexander Degtev, *tt*- and *m*-degrees, *Algebra i Logika*, 12 (1973), 143-161. (trans 12 (1973), 78-89)
4. Oswald Demuth. Remarks on the structure of *tt*-degrees based on constructive measure theory. *Commentationes Mathematicae Universitatis Carolinae*, 29:233–247, 1988.
5. Rod Downey and Jeffrey Remmel, Classification of degree classes associated with r.e. subspaces, *Ann. Pure and Appl Logic*, 42 (1989) 105-125
6. Rod Downey and Sebastian Terwijn, Computably Enumerable Reals and Uniformly Presentable Ideals, *Archive for Mathematical Logic* Vol. 48 (2002), 29-40.
7. Johanna Franklin and Frank Stephan, Schnorr trivial sets and truth-table reducibility. *The Journal of Symbolic Logic*, 75:501–521, 2010.
8. Johanna Franklin, Noam Greenberg, Frank Stephan, and Guohua Wu, Anti-complexity, lowness and highness notions, and reducibilities with tiny use, *Journal of Symbolic Logic* 78 (2013), pp. 1307-1327.
9. Carl Jocksuch and Jeanleah Mohrherr, Embedding the diamond lattice in the recursively enumerable truth-table degrees *Proc. Amer. Math. Soc.*, 94 (1985), 123-128.
10. Martin Kummer. On the complexity of random strings. In *13th Annual Symposium on Theoretical Aspects of Computer Science* (*STACS '96*), Lecture Notes in Computer Science 1046, pages 25–36. Springer, 1996.
11. Anil Nerode, General topology and partial recursive functionals, In *Summaries of talks presented at the Summer Institute for Symbolic Logic*, pages 247–251. Cornell University, 1957.
12. Emil Post. Recursively enumerable sets of positive integers and their decision problems. *Bulletin of the American Mathematical Society* Vol. 50 (5) (1944), 284–316.

A Survey on Recent Results on Partial Learning

Ziyuan Gao[a,d], Sanjay Jain[b], Frank Stephan[c] and Sandra Zilles[a,e]

[a]*Department of Computer Science, University of Regina, Regina, SK, Canada S4S 0A2*
[d]*gao257@cs.uregina.ca,* [e]*zilles@cs.uregina.ca*
[b]*School of Computing, National University of Singapore, Block COM1, 13 Computing Drive, Singapore 117417*
[b]*sanjay@comp.nus.edu.sg*
[c]*Department of Mathematics, National University of Singapore, Block S17, 10 Lower Kent Ridge Road, Singapore 119076*
[c]*fstephan@comp.nus.edu.sg*

In the context of recursion-theoretic inductive inference, a partial learner identifies an r.e. language L if and only if it outputs one index e of the language L infinitely often and outputs all other indices only finitely often. Osherson, Stob and Weinstein showed in 1986 that there is a partial learner that succeeds to learn every r.e. language under this learning criterion. Therefore, subsequent investigations refined the question to what happens if the partial learner has to satisfy additional constraints like consistency, conservativeness, confidence and reliability. The present paper gives an overview of recent investigations on this topic.

1. Introduction

Gold[15] introduced a learning paradigm to model the nature of language acquisition. His model featured a learnability criterion known as "identification in the limit", whereby a learner, as it is presented piecewise with information about an unknown target language L, outputs a sequence of conjectures based on a pre-assigned system of naming all the languages to be learnt, and after some finite time must settle on exactly one correct description of L. The present paper deals almost exclusively with recursively enumerable (r.e.) languages; accordingly, languages will be described using a fixed Gödel numbering of all r.e. languages of natural numbers. The underlying model is mainly learning from positive examples, that is, the learner processes an infinite sequence of numbers whose range equals the target language L. Any such sequence is called a text for L.

The main elements of Gold's model were later adapted to a wide variety of alternative learning paradigms[2,7,8]. Notably, many of these paradigm variations, such as behaviourally correct learning[3,4,7,8] and vacillatory learning[5,8], impose some sort of strict convergence criterion on the learner, in the sense that the learner must never make a wrong conjecture after processing a sufficient amount of finite information about the target language. However, as Osherson, Stob and Weinstein[25] observed, if a learning paradigm is viewed as a model of child language acquisition, then such a requirement on the learner cannot be justified merely on *a priori* grounds. It would thus not be unreasonable to consider other plausible learning paradigms in which the convergence requirement is relaxed. In this regard, Osherson, Stob and Weinstein[25] introduced in Exercise 7.5 *partial learning* as an alternative to the paradigm of "identification in the limit". According to this stronger learning model, a learner M is judged to have successfully identified a language L if and only if, on every text for L, M outputs exactly one index e infinitely often, and e is a Gödel number for L. Osherson, Stob and Weinstein observed that partial learning is so powerful that the whole class of r.e. languages is identifiable under this learning criterion. Yet their discovery did not spell the end of partial learning: the notion was later picked up, combined with the constraint of consistency*, and studied comprehensively for the case of learning graphs of recursive functions[18]. This study uncovered quite an interesting connection between partial learning and Gold's more stringent model of learning in the limit, specifically when consistency is imposed in both models. It was shown that if the order of data presentation of a target recursive function is arbitrary, then consistent partial learning and consistent learning in the limit coincide. Moreover, the precise inference degrees of these various learning criteria were determined; a non-trivial fact established by way of these results was that successful consistent partial identification of recursive functions may, under specific conditions, depend on the order of data presentation.

A similar line of research was also extensively pursued in recent work[9,10,12,13]; the object of study in these papers was the class RE^A of all A-r.e. languages, where the partial learnability of RE^A was characterised by the Turing degree of A. Much of this research focussed on the partial learning model imposed with additional learning constraints such as confi-

*Consistency requires the learner, for any finite segment σ of any text, to output an index of a language that contains all the elements listed in σ[2].

dence[†], conservativeness[‡], reliability[§] and consistency[13]; various blends of these constraints were also considered. Apart from revealing a new hierarchy of learnability notions, it was found that several variants of partial learning admit fairly elegant class structural characterisations. In particular, it was shown that the family of all partially conservatively learnable classes of r.e. languages has a characterisation that is analogous to the "tell-tale" characterisation obtained by Angluin[2] for every explanatorily learnable class of r.e. languages. Consistently partially learnable classes of r.e. languages may be described even more concisely: it was shown that a class of r.e. languages is consistently partially learnable if and only if it is a subclass of a uniformly recursive family[13]. This paper reviews much of the work done thus far on partial learning; each of the subsequent sections presents results under a particular learning scenario.

Figure 1 summarises most of the paper's results on partial learning of recursively enumerable languages. The learning criteria are abbreviated as follows: BC for behaviourally correct learning[7], Ex for explanatory learning, Vac for vacillatory learning, Fin for finite learning, $Part$ for partial learning, $Conf$ for confidence, $Consv$ for conservativeness, $ClsCons$ for class-consistency (where the learner is required to be consistent only on valid texts for potential targets), $Cons$ for global consistency (where the learner is required to be consistent on all input texts), $EssClsCons$ and $EssCons$ for the "essential" versions of the two consistency models, $Prud$ for prudence, Rel for reliability, and $ClsPresv$ for the class-preserving versions of a model, requiring the learner to return only hypotheses that represent sets from the target class, on any input. For any learning criterion A, $A[\mathbb{K}]$ means learnability under criterion A relative to the oracle $\mathbb{K}$. A directed arc from criterion A to criterion B means that the collection of classes learnable under model A is contained in that learnable under model B. If there is no path from A to B, then the collection of classes learnable under model A is not contained in that learnable under model B.

[†]A confident partial learner outputs, on any text of any language, exactly one index infinitely often[12].

[‡]In the context of partial learning, conservativeness requires the learner to output an index of a superset of the target language only if that index is the unique one that is output infinitely often[9].

[§]A partial learner is reliable if for every language L and every text for L, the learner only outputs indices of L infinitely often[13].

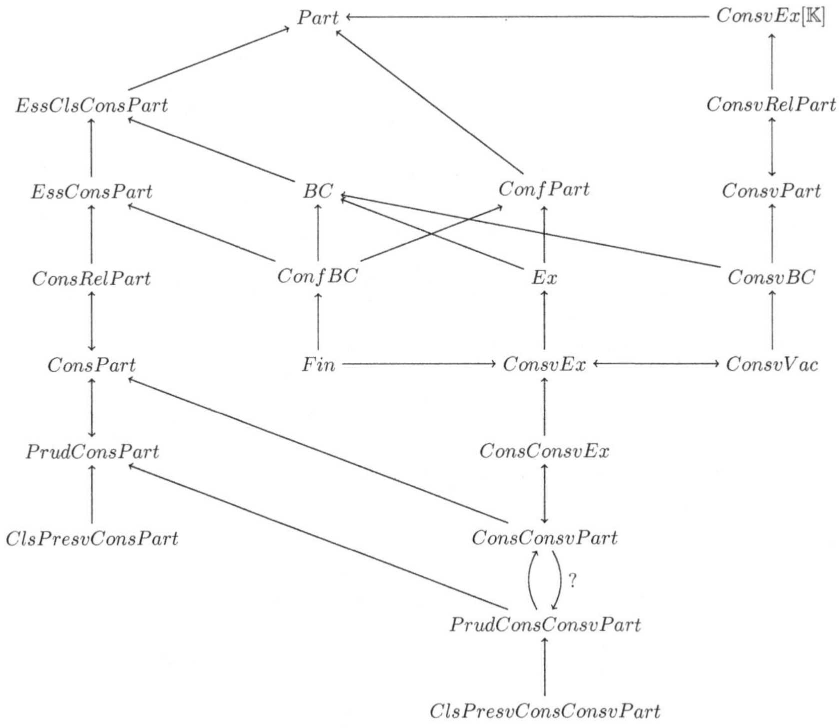

Fig. 1. Learning hierarchy

2. Basic Definitions and a First Result on Partial Learning

Gold's model of *identification in the limit* (also known as *explanatory learning*), as well as the *behaviourally correct learning* and *partial learning* models, are formalised as follows. First, define a text $T = a_0, a_1, a_2, \ldots$ of a language L to be a sequence containing all the members of L in arbitrary order; this sequence may contain repeated elements of L and also pause symbols (denoted by $\#$). The learner, denoted by M, reads a text $a_0, a_1, a_2, \ldots$ of a language L and produces a sequence $e_0, e_1, e_2, \ldots$ of hypotheses; that is to say, if M is modelled as a recursive function mapping $(\mathbb{N} \cup \{\#\})^*$ into $\mathbb{N}$, then for all n, $M(a_0 a_1, \ldots a_n) = e_n$. Each hypothesis e_n describes M's conjecture W_{e_n}, where $W_0, W_1, W_2, \ldots$ is a fixed Gödel numbering of all r.e. languages. Furthermore, $\varphi_0, \varphi_1, \varphi_2, \ldots$ is a Gödel numbering of all partial-recursive functions. The range of a (finite or infinite) sequence σ is denoted by $range(\sigma)$.

Definition 2.1. Assume that a learner M processes a text of a language L and outputs a sequence $e_0, e_1, \ldots$ of hypotheses.

(1) Gold[15] defined that the learner M *explanatorily* learns L from the text iff there is a k such that $e_k = e_h$ for all $h \geq k$ and $W_{e_k} = L$.
(2) Case and Lynes[7] defined that the learner M *behaviourally correctly* learns L from the text iff $W_{e_k} = L$ for almost all k.
(3) Case[5] defined that the learner M *vacillatorially* learns L from the text iff M *behaviourally correctly* learns L and the set $\{e_0, e_1, e_2, \ldots\}$ is finite.
(4) Osherson, Stob and Weinstein[25] defined that M *partially* learns L from the text iff there is exactly one e with $\exists^\infty k\,[e_k = e]$ and this e satisfies $W_e = L$.

Furthermore, a class $\mathcal{S}$ of r.e. languages is learnable under criterion I iff there is a recursive learner that learns every $L \in \mathcal{S}$ from every text with respect to the criterion I.

The class of all finite languages is an example of an explanatorily learnable class; the learner for it just conjectures at each time a canonical index for the set of elements seen so far; it is obvious that this set grows only finitely often and after the last element of the language to be learnt has been seen the learner stabilises on the index of the elements of this language. However, given a non-recursive r.e. language like the halting problem $\mathbb{K}$, the class of all unions $\mathbb{K} \cup D$ with D being finite is in principle learnable by the same algorithm; however, the learner cannot distinguish between irrelevant data (which are members of $\mathbb{K}$) and relevant data (which require an update), therefore one would have to update in each step and conjecture the union of $\mathbb{K}$ and all data-items seen so far. This learner is then behaviourally correct and one can show that there is no explanatory learner for this class. Gold[15] showed that the class consisting of $\mathbb{N}$ and all of its finite subsets is not explanatorily learnable; this proof directly transfers to behaviourally correct learning and thus this is the standard example for a class which is neither explanatorily nor behaviourally correctly learnable.

Osherson, Stob and Weinstein[25] noted that partial learning is more powerful than the stricter notions of explanatory and behaviourally correct learning, for it allows to learn the whole class of r.e. languages and therefore also to learn any class of r.e. languages.

Theorem 2.1. *The class of all r.e. languages is partially learnable.*

Proof. One should first note that partial learners can often be built by not specifying when a hypothesis is output, but only how often it is output and the fine-tuning of these conjectures can then be realised with maintaining a queue. So the basic principle would be that one makes a learner M which outputs the hypothesis e at least n times iff there is a time $t > n$ such that for each $m < n$, m appears in the first t data items iff m is enumerated into the e-th r.e. language W_e within t time steps. If a hypothesis is correct, then it will eventually qualify for each n and be output infinitely often; if a hypothesis is incorrect, then it will be output only a finite number of times: Either there is a data item m which is not in W_e and if n is so large that $n > m$ and m has appeared within the first n data items, then e will not be output n times or more. Or there is an element $m \in W_e$ which does not occur in the input and for every $n > m$ being so large that $m \in W_e$ within n steps, it holds that e will not be output n or more times. The only remaining problem with this approach is that M outputs all correct indices infinitely often.

One way to cure this would be to work with one-one numberings rather than acceptable numberings. Another way to cure this is to use padding. Padding means that there is a recursive one-one function *pad* such that $W_{pad(e,k)} = W_e$ for all e, k; padding can be done in all acceptable numberings. Now one can make a new learner N such that N outputs $pad(e, k)$ at least n times iff there is a time $t > n$ such that the simulated learner M has so far output e at least n times and furthermore output all the indices $0, 1, \ldots, e - 1$ together exactly k times. In the case that the indices below e are all output only finitely often, say altogether k times, and in the case that furthermore M outputs e infinitely often then N will output $pad(e, k)$ infinitely often and all other indices only finitely often. This padding technique is quite useful and has been employed in many results in the field. □

3. Confidence

Gao, Stephan, Wu and Yamamoto[12] introduced *confidence* to moderate the full strength of partial learning. Confidence was originally defined by Osherson, Stob and Weinstein in Section 4.6.2 of their book[25] as the requirement that the learner must, on every text for any (possibly non-r.e.) language, converge to some fixed hypothesis; in the partial learning model, this was relaxed to the condition that the learner must output on every text exactly one index infinitely often. Gao, Stephan, Wu and Yamamoto[12] gave the following definition.

Definition 3.1. A partial learner M is *confident* iff for every text T for any language (including the case where the language is not r.e.) there is exactly one index e that is output infinitely often by M.

A weaker notion of behaviourally correct learning, known as *confident behaviourally correct* learning, was introduced in prior work[11].

Definition 3.2. A class $\mathcal{S}$ of r.e. languages is confidently behaviourally correctly learnable iff there is a recursive learner M that behaviourally correctly learns every $L \in \mathcal{S}$ and has the property that it outputs on every language L, even non-r.e. ones, a sequence of conjectures $e_0, e_1, e_2, \ldots$ that stabilises semantically, that is, $W_{e_{n+1}} = W_{e_n}$ holds for almost all n.

That confidence is indeed a proper restriction on partial learning is witnessed by the class of all cofinite languages, which cannot be confidently partially identified[12]. In addition, behaviourally correct learnability does not in general imply confident partial learnability, as the following example demonstrates. However, confident behaviourally correct learnability is a more restrictive notion which implies confident partial learnability.

Example 3.1. The class $\{\mathbb{K} \cup D : D \text{ is finite}\}$ is behaviourally correctly learnable but not confidently partially learnable.

Conversely, it is known[13] that any class consisting of all finite languages and one infinite language is confidently partially learnable but not behaviourally correctly learnable. Somewhat more surprising, however, is the fact that explanatory learning implies confident partial learnability[10]; the same result holds even when the criterion of explanatory learning is weakened to explanatory learning with at most c anomalies (whereby the learner is permitted to converge to a hypothesis that disagrees with the target language on at most c numbers), as shown by Gao, Stephan and Zilles[13].

Theorem 3.1. *If a class of r.e. languages is explanatorily learnable with at most c anomalies, then it is also confidently partially learnable.*

Proof. Assume that M is a learner which explanatorily learns a class S with up to c anomalies. Without loss of generality, M chooses at every mind change a hypothesis larger than the previous one, so that M never vacillates between two hypotheses forever.

The new learner N maintains in each stage s a sequence $e_{0,s}, e_{1,s}, \ldots, e_{c+1,s}$ of hypotheses with $e_{0,s} = 0$ for all s and $e_{1,s}$ being the current hypothesis of M. Furthermore, $e_{k+1,s}$ is obtained from $e_{k,s}$ by fixing the least conjectured anomaly x. In the case that x is observed in the data so far and $W_{e_{k,s},s}$ does not contain x, then $W_{e_{k+1,s}} = W_{e_{k,s}} \cup \{x\}$; in the case that x has not yet been observed in the data and $x \in W_{e_{k,s},s}$, then $W_{e_{k+1,s}} = W_{e_{k,s}} - \{x\}$. The indices computed by this padding are without loss of generality larger than x. In the case that no such x exists then $e_{k,s} = e_{k+1,s} = \ldots = e_{c+1,s}$.

The hypothesis number s of the new learner N is now $e_{k,s}$ for the maximal k such that $e_{0,s} = e_{0,s+1} \wedge e_{1,s} = e_{1,s+1} \wedge \ldots \wedge e_{k,s} = e_{k,s+1}$. The rationale behind this is that if M does not converge then N outputs $e_{0,s}$ infinitely often, that is, N outputs the fixed value 0 infinitely often. If M converges and the number of anomalies is k then these anomalies are found eventually and the sequence of the $e_{k+1,s}$ converges to a value e_{k+1} which is the index of the language to be learnt. This e_{k+1} is furthermore output infinitely often for the following reason: either there are infinitely many candidates for an anomaly $k+1$ which are each eventually replaced at some stage s by another candidate for that anomaly causing $e_{k+2,s}$ to be different from $e_{k+2,s+1}$ and causing $e_{k+1,s} = e_{k+1}$ to be output at that stage s or there are infinitely many s where $e_{k+1} = e_{k+1,s} = e_{c+1,s}$ and in these cases e_{k+1} is output as well. Thus N does also in this case converge to a right index. Furthermore, there is, independent of the choice of the target language, a maximal k such that the sequences of the $e_{0,s}$, $e_{1,s}$, ..., $e_{k,s}$ converge and the limit of this last sequence is then the only hypothesis which is output infinitely often. □

Interpreted as a model of language learning, the relation between identification in the limit and confident partial learning provides evidence that learners that are powerful enough to explanatorily learn a class of languages are also able, in a quite broad sense, to isolate a single hypothesis infinitely often in the case that they diverge, for example on any text for a nonnatural language. Although confident partial learning may appear at first sight to be a bit restrictive, Gao, Stephan and Zilles[13] showed that it admits the following general characterisation.

Theorem 3.2. *A class $\mathcal{S}$ of r.e. languages is confidently partially learnable iff there is a recursive learner M that outputs on every text exactly one index infinitely often, and whenever $L \in \mathcal{S}$ and T is a text for L, then M on T does not output infinitely often any index smaller than the smallest r.e. index of L.*

As a corollary of this result, it follows that confident partial learning has the mathematically appealing property of being closed under finite unions – a property that is also satisfied by confident vacillatory learning as well as confident behaviourally correct learning. The intuitive explanation for this deduction is that if M_1 and M_2 are two confident partial learners of the classes $\mathcal{S}_1$ and $\mathcal{S}_2$ respectively, then one can define a further learner N that outputs on any text T the pair $\langle d_1, d_2 \rangle$ at least n times iff M_1 outputs d_1 and M_2 outputs d_2 at least n times. Note that N outputs $\langle d_1, d_2 \rangle$ infinitely often iff M_1 outputs d_1 and M_2 outputs d_2 infinitely often. Now, if $L \in \mathcal{S}_1 \cup \mathcal{S}_2$ then there is an index e of L such that $e \leq d_1$ or $e \leq d_2$; it follows by the properties of Cantor's pairing function that $e \leq \langle d_1, d_2 \rangle$ and therefore N outputs an upper bound of e infinitely often. Hence $\mathcal{S}_1 \cup \mathcal{S}_2$ is confidently partially learnable.

The study of confident partial learning relative to oracles illustrates quite an interesting interplay between inductive inference and recursion theory. For example, consider the class $\mathcal{C}$ of graphs of recursive functions f_e satisfying the following conditions:

- $f_e(x) = 0$ for all $x < e$;
- $f_e(x) = 1$ for $x = e$;
- $f_e(x) \downarrow = \varphi_e(x) \downarrow$ for almost all x.

For a given recursive function f, the sequence $(0, f(0)), (1, f(1)), (2, f(2)), \ldots$ is called the *canonical text* for f. While it is not exceedingly difficult to show that $\mathcal{C}$ is behaviourally correctly learnable, the problem of showing that $\mathcal{C}$ cannot be confidently partially learnt from canonical texts is much more involved. Gao and Stephan utilise for their proof of Theorem 4 in their paper[10] a relativised version of the Low Basis Theorem of Jockusch and Soare[19]. They eventually show that every oracle which permits to confidently partially learn $\mathcal{C}$ is high$_2$. Here the classification of low, high, low$_2$ and high$_2$ is included for the reader's convenience.

Definition 3.3. An oracle A is high iff $A' \geq_T \mathbb{K}'$; an oracle A is high$_2$ iff $A'' \geq_T \mathbb{K}''$; an oracle A is low iff $A' \leq_T \mathbb{K}$; an oracle A is low$_2$ iff $A'' \leq_T \mathbb{K}'$.

The proof by Gao and Stephan[10] that $\mathcal{C}$ is not confidently partially learnable with respect to canonical texts establishes a deeper fact: there is a behaviourally correctly learnable class of graphs of recursive functions that is confidently partially learnable only relative to high$_2$ oracles. Moreover, one can show that there do indeed exist high$_2$ oracles that permit the class REC of all graphs of recursive functions to be confidently partially learnt; on the other hand, there are also high$_2$ oracles that are not powerful enough to allow *REC* to be confidently partially learnt. These results complement the characterisation theorem obtained by Adleman and Blum[1] many years earlier for the class of oracles relative to which *REC* is explanatorily learnable.

Theorem 3.3. *An oracle A is high iff there is an A-recursive explanatory learner for the class REC of the graphs of all recursive functions.*

As explanatory learners can be transformed into confident partial learners, one direction of the result of Adleman and Blum transfers to confident partial learning; the other direction is more complicated and Gao and Stephan[10] showed that it needs a weakening to high_2 though there is no characteristation.

Theorem 3.4. *If A permits to confidently partially learn REC then A is high_2. Furthermore, the oracles which permit to confidently partially learn REC are all high ones plus some but not all of those which are high_2 but not high.*

It should be noted that there is even a subclass $G \subseteq REC$ such that G is behaviourally correctly learnable (without any oracle) and such that whenever G is confidently partially learnable relative to A then A is high_2.

4. Reliability

Reliable partial learning[12] is a counterpart of confident partial learning — while a confident learner always converges to a hypothesis (according to the given learning criterion), the reliable learner only converges to a hypothesis iff it is correct. Originally defined by Minicozzi[24] as the requirement that the learner identifies all languages on whose texts it converges syntactically, the partial learning analogue stipulates that the learner must, on every text T for any (possibly non-r.e.) language L, output only indices of L infinitely often. For explanatory learning from texts, this notion is very restrictive and only permits to learn classes of finite languages. Only in the framework of partial learning, the notion of reliability becomes interesting for learning from text.

Definition 4.1. A partial learner M is *reliable* iff for every language L and on every text for L, M only outputs indices of L infinitely often.

The empirical justification for defining such a criterion is based on the observation that children may implement a reliable partial learning function, so that they eventually abandon any initially conjectured hypothesis that is incorrect, while returning infinitely often to exactly one correct hypothesis when presented with a text for a language to be learnt. The usual algorithm to partially learn all r.e. languages is already reliable, that is, an incorrect hypothesis is always output only finitely often. Thus the class of all r.e. languages is already reliably partially learnable and so it makes sense to go into the direction which asks: which classes more comprehensive than the class of all r.e. languages are also partially reliably learnable.

Gao, Stephan, Wu and Yamamoto[12] investigated this question and showed that every uniformly $\mathbb{K}$-recursive class is reliably partially learnable using a padded version of the given indexing as hypothesis space. Furthermore, they considered for an oracle A the class RE^A of all A-r.e. languages from an acceptable numbering as a hypothesis space; here "acceptable" can be understood in the strict sense that for every numbering of A-r.e. languages there is a recursive function f which translates the indices of this numbering into equivalent indices of the given acceptable numbering. Gao, Stephan, Wu and Yamamoto[12] showed the following result.

Theorem 4.1. *The class of all A-r.e. languages is reliably partially learnable using A-r.e. indices iff A is low.*

They also showed[12] the following corresponding result for the class of all A-recursive languages reads as follows.

Theorem 4.2. *The class of all A-recursive languages is reliably partially learnable iff $A \leq_T \mathbb{K}$ and A is low$_2$.*

These results stand in contrast to the case of non-reliable partial learners where there is even an uncountable class of oracles A such that the class of all A-recursive languages can be learnt by a recursive partial learner. So the requirement of reliability is a real restriction in this general setting.

5. Consistency

Angluin[2] proposed *consistency* as a constraint on conjectures for r.e. languages. A learner M is said to be consistent if and only if it always generates at least all the data it has seen so far.

Definition 5.1. A learner M is *consistent* iff $range(\sigma) \subseteq W_{M(\sigma)}$ for all data sequences σ.

Consistency is a fairly intuitive learning constraint, for it is conceivable that no learner will hold on to a hypothesis that is falsified by the available data. Osherson, Stob and Weinstein showed also in Proposition 4.3.4A of their book[25] that only recursive languages can be consistently identified. Going further, one may ask whether the limitations of consistent explanatory learning may be overcome by relaxing the requirement to partial learning. This, however, is not the case: it was proven[13] that consistently partially learnable classes of r.e. languages must be subclasses of uniformly recursive families. It was shown, moreover, that a consistently partially learnable class of r.e. languages can even be *prudently* consistently partially learnt. The definition by Osherson, Stob and Weinstein[25] is the following.

Definition 5.2. A learner M is *prudent* iff it learns the class $\{W_{M(\sigma)} : \sigma \in (\mathbb{N} \cup \{\#\})^*\}$. In other words, M learns every language for which it outputs a conjecture, on some input data.

Gao, Stephan and Zilles[13] investigated prudence for partial learning.

Theorem 5.1. *The following statements are equivalent for a class $\mathcal{C}$ of r.e. languages:*

(I) *$\mathcal{C}$ is a subclass of a uniformly recursive family;*
(II) *$\mathcal{C}$ is consistently partially learnable;*
(III) *$\mathcal{C}$ is prudently consistently partially learnable;*
(IV) *$\mathcal{C}$ is prudently consistently partially learnable using a uniformly recursive hypothesis space.*

The learnability of classes of recursive functions from canonical text versus arbitrary text found quite a lot of attention. For example, Grieser[16] solved an important open problem by showing that there is a class of functions which has a total recursive consistent learner using canonical text while it does not have a class-consistent learner succeeding on arbitrary texts. Jain and Stephan[18] studied consistent partial learning of recursive functions and in particular which classes of recursive functions can be consistently partially learnt from canonical text using the oracle A. Jain and Stephan[18] showed that there are only two cases: If A has hyperimmune Turing degree then these are all classes of recursive functions and if A has hyperimmune-free Turing degree these are those classes which can be learnt using any oracle (or none).

Theorem 5.2. *Let A be an oracle. If A is hyperimmune then some A-recursive learner can consistently partially learn the class REC from canon-*

ical text. If A is not equivalent to any hyperimmune set and some A-recursive learner can consistently partially learn $\mathcal{S}$ from canonical text, then some recursive learner can do the same.

Jain and Stephan[18] also compared the inference degrees of learning in the limit to those of consistent partial identification with respect to arbitrary texts (that is, texts for recursive functions in which the order of the elements can be arbitrary) for the case of learning recursive functions. They showed that in the case of learning recursive functions relative to a given oracle A from arbitrary texts, consistent learning in the limit is just as powerful as consistent partial identification. Moreover, they provided a characterisation of all classes of recursive functions that are consistently partially identifiable relative to a given oracle A.

Theorem 5.3. *For every oracle A and every class $\mathcal{C}$ of recursive functions, the following conditions are equivalent:*

(1) $\mathcal{C}$ is consistently partially learnable relative to A from arbitrary texts;
(2) $\mathcal{C}$ is consistently explanatorily learnable relative to A from arbitrary texts;
(3) There is a strictly increasing A-recursive function g^A such that every $f \in \mathcal{C}$ has an index e which satisfies $\Phi_e(x) \leq g^A(\max(\{x, f(x)\}))$ for almost all x, where Φ denotes a fixed Blum complexity measure such that for all e, $\Phi_e(x)$ is the complexity assigned to the computation of $\varphi_e(x)$.

It is not immediately obvious whether or not the preceding theorem still holds if one relaxes the criteria of consistent partial learning and explanatory learning to *class consistent* partial learning and *class consistent* explanatory learning respectively; in the class consistent versions of these criteria, the learner is only required to be consistent on texts for recursive functions in the given target class.

Grieser[16] had already observed that the order of data presentation is very important for consistent explanatory learning of functions; Jain and Stephan[18] investigated how much these order constraints influence the consistent partial learnability of a class.

Theorem 5.4.

(1) If A is not high, then any class of $\{0,1\}$-valued recursive functions which is consistently partially learnable relative to A from arbitrary

texts is consistently partially learnable (unrelativised) from canonical texts.

(2) *If A is r.e. and not recursive, then there is a class of recursive functions that is consistently partially learnable relative to A from arbitrary texts but not consistently partially learnable (unrelativised) from canonical texts.*

It is well-known that restricting the requirement of consistency only to members of a class makes the criterion more general, for example the class $\{\mathbb{K}\}$ is class-consistently explanatorily learnable but not consistently partially learnable. A further generalisation is linked to the model of partial learning: here the consistency is not required for all hypotheses but only from some time point onwards; note that explanatory and behaviourally correct learners satisfy this constraint of "essential class-consistent learning" by definition. The corresponding criterion is defined for partial learning as below and was studied only recently[10,13].

Definition 5.3. A learner M *essentially class consistently* partially learns $\mathcal{S}$ iff M partially learns $\mathcal{S}$ and for every $L \in \mathcal{S}$ and every text T for L and almost all $\sigma \preceq T$, M outputs on σ an index e with $range(\sigma) \subseteq W_e$. A learner M *essentially globally consistently* partially learns $\mathcal{S}$ iff M partially learns $\mathcal{S}$ and for every L (not necessarily in $\mathcal{S}$) and every text T for L and almost all $\sigma \preceq T$, M outputs on σ an index e with $range(\sigma) \subseteq W_e$.

It was found[13] that essentially class consistent partial learnability is stronger than behaviourally correct learnability, providing yet another link between partial learning and the traditional models of learning in which the learner is required to almost always output correct hypotheses. Moreover, the three constraints of essential class consistency, confidence and conservativeness, are fairly robust when they are compared pairwise; one can show that any two of these notions have incomparable learning strengths. The foregoing results[13] are summed up below.

Theorem 5.5.

(1) *Every behaviourally correctly learnable class is essentially class consistently partially learnable.*

(2) *Every confidently behaviourally correctly learnable class is essentially global consistently partially learnable.*

(3) *The notions of essentially class consistent partial learning and confident partial learning and conservative partial learning are pairwise*

incomparable.

It has been shown[11] Theorem 27 that a confidently behaviourally correctly learnable class of recursive functions is essentially globally consistently partially learnable – a result that is parallel to Theorem 5.5(1). An analogous result holds for classes of r.e. languages.

Theorem 5.6. *Every confidently behaviourally correctly learnable class of r.e. languages is essentially globally consistently partially learnable.*

Proof. Assume that M is a learner that confidently behaviourally correctly learns a class $\mathcal{S}$ of r.e. languages. Theorem 5.5(1) implies that there is a learner N such that N outputs on any text for every language L a sequence of conjectures $d_0, d_1, d_2, \ldots$ satisfying the following conditions.

- There is exactly one index e such that $\exists^\infty m[d_m = e]$ and $\forall^\infty m[W_{d_m} = W_e]$ both hold.
- If $L \in \mathcal{S}$, then $W_e = L$.

One can further define a learner P as follows. P maintains a list I of indices for bookkeeping. I is defined in stages, starting with $I_0 = \mathbb{N}$. On input σ, P checks whether it can compute the least $n \leq |\sigma|$ such that $n \in I_{|\sigma|-1}$ and $\{0, 1, \ldots, n\} \cap range(\sigma) \subseteq W_{d_n,|\sigma|}$ both hold. If such an n exists, then P outputs d_n and sets $I_{|\sigma|} = I_{|\sigma|-1} - \{n\}$. If no such n exists, then P outputs a grammar for range(σ) and sets $I_{|\sigma|} = I_{|\sigma|-1}$. P essentially globally consistently partially learns $\mathcal{S}$ for the following reasons.

- If P is fed with a text for some $L \in \mathcal{S}$, then there is exactly one index e such that $\exists^\infty m[d_m = e]$ and $W_e = L$ both hold. Thus there are infinitely many n such that $d_n = e$ and $\{0, 1, \ldots, n\} \cap range(\sigma) \subseteq W_{d_n}$ hold. On the other hand, for every $f \neq e$, there are only finitely many m for which $d_m = f$, and so P will output f only finitely often. Moreover, since it holds for almost m that $d_m = e$, P must be essentially globally consistent.
- Suppose that P is fed with a text T for a language L such that $\forall^\infty m[W_{d_m} = W_h]$ holds. If $L \not\subseteq W_h$, then there is an n such that $\{0, 1, \ldots, k\} \cap L \not\subseteq W_{d_k}$ holds for all $k \geq n$, and so P will output $range(T[k])$ for almost all k. If $L \subseteq W_h$, then P will almost always output either a superset of L, or the range of the current input. Hence P is essentially globally consistent.

This establishes that P is an essentially globally consistent partial learner of $\mathcal{S}$. □

Like confident partial learning, essentially globally consistent partial learning is closed under finite unions – a property that fails to hold for consistent explanatory learning. Thus partial learning may be quite a liberal criterion; Gao, Stephan and Zilles[13] showed that this holds even when it is combined with various learning constraints.

Theorem 5.7. *The union of essentially globally consistently partially learnable classes is essentially globally consistently partially learnable. This does not apply to essentially class consistent partial learning.*

6. Conservativeness

Another learning constraint often considered in parallel to consistency is *conservativeness*, which was also introduced by Angluin[2]. She defined a learner M to be conservative if and only if, for all sequences σ, τ with $\sigma \prec \tau$, $M(\tau) \neq M(\sigma)$ holds only if $\text{range}(\sigma) \not\subseteq W_{M(\sigma)}$. Loosely interpreted, this means that a conservative learner never revises its hypothesis until it has seen data not generated by its original conjecture. Mazurkewich and White[23] provide evidence that children do indeed overgeneralise, specifically with respect to the English dative alternation, that is, alternations between a prepositional indirect-object construction and a double-object construction; for example, the pair of sentences "John gave the book to Fred" and "John gave Fred the book." Hence there is some empirical justification for assuming that learners may not always be conservative. Moreover, conservativeness is a proper restriction on the learning potential of explanatory learners: Angluin[2] established the existence of a uniformly recursive family that is explanatorily learnable but cannot be conservatively explanatorily learnt.

Gao, Jain and Stephan[9] studied conservativeness in the context of partial learning. The partial learning variant of conservativeness is defined as follows. Again, suppose that M partially learns a class $\mathcal{S}$ of r.e. languages.

Definition 6.1. A partial learner M is *conservative* iff for every language $L \in \mathcal{S}$ and every text T for L the learner outputs exactly one index of L (which is also output infinitely often) and no index of a proper superset of L.

It is quite a curious fact that unrelativised conservative partial learning is weaker than learning in the limit with an oracle for the diagonal halting problem $\mathbb{K}$. More precisely, Gao, Jain and Stephan[9] showed the following theorem.

Theorem 6.1. *If $\mathcal{S}$ is conservatively partially learnable, then $\mathcal{S}$ is explanatorily learnable relative to $\mathbb{K}$. If $\mathcal{S}$ consists of only infinite languages and is explanatorily learnable relative to $\mathbb{K}$, then $\mathcal{S}$ is also conservatively partially learnable.*

The condition in the second part of the above theorem, that the class $\mathcal{S}$ consists of only infinite languages, cannot be dropped; this is implied by an example given by Gao, Jain and Stephan[9].

Example 6.1. If A is not recursive, then there is a class $\mathcal{S}$ that is conservatively explanatorily learnable relative to A but not conservatively partially learnable (unrelativised).

The overall picture of conservativeness in the partial learning model, as well as in models where some sort of hypothesis convergence is enforced – that is, explanatory, vacillatory, and behaviourally correct learning, is summed up in the following theorem of Gao, Jain and Stephan[9].

Theorem 6.2. *Let $\mathcal{S}$ be a class of r.e. languages. Then the following statements hold:*

(1) $\mathcal{S}$ is conservatively explanatorily learnable iff $\mathcal{S}$ is conservatively vacillatorily learnable;
(2) If $\mathcal{S}$ is conservatively vacillatorily learnable, then $\mathcal{S}$ is conservatively behaviourally correctly learnable;
(3) The converse of (II) *is not true in general;*
(4) If $\mathcal{S}$ is conservatively behaviourally correctly learnable, then $\mathcal{S}$ is conservatively partially learnable;
(5) The converse of (IV) *is not true in general.*

Angluin[2] introduced the notion of a *tell-tale* set of a language L to be learnt: it is a finite subset E_L of L such that no language L' in the class of languages to be learnt satisfies $E_L \subseteq L' \subset L$. Based on this notion, she completely characterised all uniformly recursive families that are explanatorily learnable. Formally, Angluin[2] showed the following result.

Theorem 6.3. *A uniformly recursive family $\{L_d : d \in \mathbb{N}\}$ of languages is explanatorily learnable iff there exists a uniformly recursively enumerable family $\{H_d : d \in \mathbb{N}\}$ such that for all e,*

(1) H_e is finite;
(2) $H_e \subseteq L_e$;
(3) for all $d \in \mathbb{N}$, if $H_e \subseteq L_d$ then L_d is not a proper subset of L_e.

Subsequent investigations by Lange, Kapur and Zeugmann in various papers[21,22,29] dealt with several variants of explanatory learning and led to the discovery of a similar tell-tale characterisation of *conservative explanatory learning* in the setting of uniformly recursive families. The characterisation needs the notion of a canonical index e of a finite set D:

$$e \text{ is the canonical index of a finite set } D \Leftrightarrow e = \sum_{x \in D} 2^x.$$

The notion D_e denotes the finite set with canonical index e. The differences to Angluin's tell-tale criterion above are that here the finite tell-tales sets are not enumerated but given by their canonical indices and that in some cases it is furthermore necessary to change the hypothesis space; these differences seem to be essential for characterisations of conservative learning as shown by Lange and Zeugmann[21].

Theorem 6.4. *Let $\mathcal{L} = \{L_d : d \in \mathbb{N}\}$ be a uniformly recursive family. Then $\mathcal{L}$ is conservatively learnable iff there is a hypothesis space $\mathcal{G} = \{G_d : d \in \mathbb{N}\}$ of uniformly recursive languages and a recursive function f such that for all e,*

(1) there is a $d \in \mathbb{N}$ with $G_d = L_e$;
(2) $D_{f(e)} \subseteq G_e$;
(3) for all $d \in \mathbb{N}$, if $D_{f(e)} \subseteq G_d$, then G_d is not a proper subset of G_e.

The preceding characterisation of conservative learning was later generalised by de Jongh and Kanazawa[20] to the case of uniformly r.e. families, although their formulation of the tell-tale condition was somewhat more complicated. A characterisation similar to the one of Lange and Zeugmann was obtained for the notion of conservatively partially learnable classes of r.e. languages[9] – like the criterion of Lange and Zeugmann, it uses canonical indices rather than r.e. indices in Angluin's original characterisation for explanatory learning; furthermore, a hypothesis space is chosen by enumerating the family of the W_{j_e} as explained below. Gao, Jain and Stephan[9] obtained the result as follows.

Theorem 6.5. *A class $\mathcal{S}$ is conservatively partially learnable iff there is a recursive sequence of pairs (i_e, j_e) such that*

(1) For all $L \in \mathcal{S}$ there is an e with $L = W_{j_e}$;
(2) For all e, $D_{i_e} \subseteq W_{j_e}$;
(3) For all d, e, if $D_{i_e} \subseteq W_{j_d} \subset W_{j_e}$, then $W_{j_d} \notin \mathcal{S}$.

7. Combining Reliability with Other Notions

Whilst there are classes of r.e. languages that cannot be partially learnt if both consistency and confidence are enforced, one can nevertheless always impose consistency in tandem with reliability. Central to this observation is the characterisation in Theorem 5.1 which says that a class $\mathcal{S}$ of r.e. languages is consistently partially learnable iff it is a subclass of a uniformly recursive family. The next theorem shows that for consistent partial learning, one can add in reliablility in the same way as one can do for partial learning.

Theorem 7.1. *A class $\mathcal{S}$ of r.e. languages is consistently partially learnable iff it is consistently reliably partially learnable.*

Proof. As was shown in the proof of Theorem 18 by Gao, Stephan and Zilles[13], a consistently partially learnable class $\mathcal{S}$ of r.e. languages is a subclass of some uniformly indexed family $\mathcal{S}' = \{L_1, L_2, L_3, \ldots\}$; without loss of generality, it may be assumed that $\mathcal{S}'$ is one-one and contains all the cofinite languages. Define a learner M as follows. For a text T, M on $T(0)T(1)\ldots T(n)$ finds the least pair $\langle i, j\rangle$ such that $j \leq n$, $|\{m : m < n \wedge M(T(0)T(1)\ldots T(m)) = i\}| = j$ and

$$L_i \cap \{0, 1, \ldots, j\} \subseteq \{T(0), T(1), \ldots, T(n)\} \subseteq L_i \cup \{\#\}. \qquad (*)$$

Having identified the pair $\langle i, j\rangle$, M conjectures L_i. Condition $(*)$ ensures that M is always consistent. Now if range$(T) \neq L_i$, then either $L_i \cap \{0, 1, \ldots, j\} \not\subseteq$ range(T) for some j, or range$(T) \not\subseteq L_i$. Hence there is a least j for which the pair $\langle i, j\rangle$ will never be a candidate at sufficiently large stages; this implies, in particular, that M will not conjecture L_i more than j times, and so it is a reliable learner. Furthermore, if $L_i =$ range(T), then for every j, M will identify the pair $\langle i, j\rangle$ at some stage, since it holds that for all $\langle i', j'\rangle < \langle i, j\rangle$, $\langle i', j'\rangle$ can only qualify at most once. Consequently, M conjectures L_i infinitely often, and so it reliably consistently partially learns $\mathcal{S}'$. □

In contrast to this, it follows from the definition that reliability and confidence cannot be combined; the reason is that on texts for non-r.e. languages, reliability requires the learner not to output any hypothesis infinitely often while confidence requires the learner to output exactly one hypothesis infinitely often. One can, however, make the following combination of confidence and reliability; note that although the learner is no longer confident in the precise sense of the word, it can still be made confident by the usual padding trick.

Remark 7.1. Call a learner N to be confident and weakly reliable iff N satisfies the following conditions:

- N outputs on each text T either one or two hypotheses infinitely often;
- all hypotheses output infinitely often are for the same language L;
- the number of hypotheses output infinitely often is exactly one iff L is the language to be learnt.

Given now a standard confident learner M, one can obtain such an N from M using the following method: N outputs $pad(e, 0)$ at least n times iff M outputs e at least n times; N outputs $pad(e, k+1)$ at least n times iff (i) N outputs $pad(e, 0)$ at least n times and (ii) there is a time $t > n$ such that for all $m < k$, m occurs within the first t data-items iff m is enumerated into W_e within t steps and the same connection fails for $m = k$. So every confident learner can be made confident and weakly reliable.

For conservative learning, the next proposition shows that it can always be combined with reliable learning unless further constraints are postulated.

Proposition 7.1. *Every conservative partial learner M can be replaced by a conservative reliable partial learner N learning the same languages.*

Proof. The main idea is that one can make the translation by just filtering the stream of M's hypotheses with some consistency checks between data and hypotheses. So N outputs a hypothesis e on a text T at least n times iff M outputs e on T at least n times and there exists a $t > n$ such that for all $m < n$, m occurs within the first t data-items of T iff m is enumerated into W_e within t steps. One can easily verify the following facts:

- N outputs only hypotheses conjectured by M and therefore the conservativeness of M transfers to the conservativeness of N; furthermore, N does not output any hypothesis more frequently than M;

- If M outputs on a text T for L a hypothesis e for L infinitely often then so does N, as each repetition of the hypothesis e will eventually pass the filter;
- If M outputs on a text T for L an incorrect hypothesis e infinitely often then it will pass the filtering only finitely often and thus N is reliable.

This completes the proof. □

Osherson, Stob and Weinstein demonstrated via a locking sequence argument in Proposition 4.6.1A of their book[25] that any reliably explanatorily learnable class contains only finite languages. In contrast to this, a consistent, conservative and reliable partial learner can identify infinite languages. One may take, for example, the class $\{\mathbb{N}\}$: On any text, a consistent, conservative and reliable partial learner M outputs a canonical index for $\mathbb{N}$ as many times as a new datum (not previously seen) appears, and conjectures for every initial segment σ of the input text a canonical index for range(σ) whenever $|\sigma| = 1$, or if $|\sigma| > 1$ and range(σ') = range(σ), where σ' is the prefix of σ with length $|\sigma| - 1$. As a further illustration of how partial learning can enrich the classes of languages that are learnable under the stricter criterion of explanatory learning with respect to various learning constraints, the next theorem states that any consistent, conservative, explanatory learner of a class $\mathcal{S}$ can be turned into a consistent, conservative, as well as reliable learner for $\mathcal{S}$. Gao, Stephan and Zilles[14] gave the following result.

Theorem 7.2. *If a class $\mathcal{S}$ of r.e. languages is consistently conservatively (with both constraints fulfilled at once) explanatorily learnable, then it is also consistently conservatively reliably (with all three constraints fulfilled at once) partially learnable.*

8. Conclusion and Open Problems

The partial learning criterion of converging to a hypothesis is to output this hypothesis infinitely often while every other hypothesis is output only finitely often. This learning criterion is quite general and permits to learn even the class of all A-r.e. languages using A-r.e. indices in the case that A is low. Similarly one can even learn for uncountably many oracles A the class of all A-recursive languages using A-recursive indices. In both results, the learner itself is recursive, so the learning power stems from this general convergence criterion. Therefore it is quite natural to ask what happens if one combines this general learning criterion with some restrictive constraint like confidence, conservativeness, consistency and reliability. While, when partially learning classes of r.e. languages, reliability is not an additional restrictive constraint, the other three notions are and the types of restrictions imposed lead to different, incomparable learning notions. The present survey points to a wide variety of results[9,10,12–14,18] about these combined criteria and also provides various characterisations for those classes which can be partially learnt with the additional constraint of confidence, consistency or conservativeness, respectively. However, two important problems remain open. The first open problem relates to the just mentioned field of learning non-recursive objects using a recursive learner.

Problem 8.1. *Are there uncountably many oracles A such that the class of all A-r.e. languages has a recursive partial learner using A-r.e. indices?*

The second question deals with function learning and has already been open for 5 years[18]. It asks whether an equivalence of consistent function learning carries over to class-consistent function learning.

Problem 8.2. *Consider the model of learning recursive functions from texts in arbitrary order: Is every class-consistently partially learnable class of recursive functions also class-consistently explanatorily learnable?*

The study of partial learnability, whilst intrinsically interesting, is not an end in itself: some of the proof techniques illustrate applications of deep recursion-theoretic notions, especially when considering learnability with respect to oracles; moreover, the various connections between partial learning, Gold's model of learning in the limit, and Case and Lynes' model of behaviourally correct learning, show that the traditional learning models, in which the learner is required to almost always output correct hypotheses, are probably not as restrictive as they may seem at first sight. Perhaps the philosophical import of these relations is no less significant.

References

1. Leonard M. Adleman and Manuel Blum. Inductive inference and unsolvability. *The Journal of Symbolic Logic* 56(3) (1991):891–900.
2. Dana Angluin. Inductive inference of formal languages from positive data. *Information and Control* 45(2) (1980):117–135.
3. Ganesh Baliga, John Case and Sanjay Jain. The synthesis of language learners. *Information and Computation* 152 (1999):16–43.

4. Janis Bārzdiņš. Two theorems on the limiting synthesis of functions. In *Theory of Algorithms and Programs, vol. 1*, pages 82–88. Latvian State University, 1974. In Russian.
5. John Case. The power of vacillation in language learning. *SIAM Journal on Computing*, 28 (1999):1941–1969.
6. John Case and Timo Kötzing. Difficulties in forcing fairness of polynomial time inductive inference. *Algorithmic Learning Theory*, Twentieth International Conference, ALT 2009, Porto, Portugal, October 3-5, 2009. Proceedings. Springer LNAI 5809 (2009):263–277.
7. John Case and Chris Lynes. Machine inductive inference and language identification. *Proceedings of the Ninth International Colloquium on Automata, Languages and Programming*, Lecture Notes in Computer Science 140 (1982):107–115.
8. John Case and Carl Smith. Comparison of identification criteria for machine inductive inference. *Theoretical Computer Science* 25 (1983):193–220.
9. Ziyuan Gao, Sanjay Jain and Frank Stephan. On conservative learning of recursively enumerable languages. *Ninth Conference on Computability in Europe,* CiE 2013, Milan, Italy, July 1–5, 2013. Proceedings. Springer LNCS 7921 (2013):181–190.
10. Ziyuan Gao and Frank Stephan. Confident and consistent partial learning of recursive functions. *Algorithmic Learning Theory,* Twenty-third International Conference, ALT 2012, Lyon, France, October 2012, Proceedings. Springer LNAI 7568 (2012):51–65.
11. Ziyuan Gao and Frank Stephan. Confident and consistent partial learning of recursive functions. Technical Report TRB7/14, School of Computing, National University of Singapore, 2014. Long version of[10].
12. Ziyuan Gao, Frank Stephan, Guohua Wu and Akihiro Yamamoto. Learning families of closed sets in matroids. *Computation, Physics and Beyond; International Workshop on Theoretical Computer Science*, WTCS 2012, Springer LNCS 7160 (2012):120–139.
13. Ziyuan Gao, Frank Stephan and Sandra Zilles. Partial Learning of Recursively Enumerable Languages. *Algorithmic Learning Theory,* Twenty-fourth International Conference, ALT 2013, Singapore, October 2013, Proceedings. Springer LNAI 8139 (2013):101–115.
14. Ziyuan Gao, Frank Stephan and Sandra Zilles. Partial Learning of Recursively Enumerable Languages. Manuscript, long version of[13].
15. E. Mark Gold. Language identification in the limit. *Information and Control* 10 (1967):447–474.

16. Gunter Grieser. Reflective inductive inference of recursive functions. *Theoretical Computer Science A* 397 (2008):57–69.
17. Sanjay Jain, Daniel Osherson, James S. Royer and Arun Sharma. 1999. *Systems that learn: an introduction to learning theory.* MIT Press, Cambridge, Massachusetts.
18. Sanjay Jain and Frank Stephan. Consistent partial identification. *Proceedings of the Twenty-second Annual Conference on Computational Learning Theory,* COLT 2009:135–145.
19. Carl G. Jockusch, Jr and Robert I. Soare. Π_1^0 classes and degrees of theories. *Transactions of the American Mathematical Society* 173 (1972):33–56.
20. Dick de Jongh and Makoto Kanazawa. Angluin's theorem for indexed families of r.e. sets and applications. *Proceedings of the Ninth Annual Conference on Computational Learning Theory,* pages 193–204, ACM Press, 1996.
21. Steffen Lange and Thomas Zeugmann. A guided tour across the boundaries of learning recursive languages. In Klaus P. Jantke and Steffen Lange, eds., *Algorithmic Learning for Knowledge Based Systems.* Lecture Notes in Artificial Intelligence 961 (1995):190–258.
22. Steffen Lange, Thomas Zeugmann and Shyam Kapur. Monotonic and dual monotonic language learning. *Theoretical Computer Science* 155 (1996):365–410.
23. Irene Mazurkewich and Lydia White. The acquisition of dative-alternation: unlearning overgeneralizations. *Cognition* 16 (1984):261–283.
24. Eliana Minicozzi. Some natural properties of strong-identification in inductive inference. *Theoretical Computer Science* 2 (1976):345–360.
25. Daniel N. Osherson, Michael Stob and Scott Weinstein. 1986. *Systems that learn: an introduction to learning theory for cognitive and computer scientists.* Cambridge, Massachusetts.: MIT Press.
26. Leonard Pitt. Inductive inference, DFAs, and computational complexity. *Analogical and Inductive Inference, Proceedings of the Second International Workshop,* AII 1989. Springer LNAI 397 (1989):18–44.
27. Hartley Rogers, Jr. 1987. *Theory of recursive functions and effective computability.* Cambridge, Massachusetts: MIT Press.
28. Rolf Wiehagen and Thomas Zeugmann. Learning and consistency. *Algorithmic Learning for Knowledge-Based Systems,* GOSLER Final Report, Springer LNAI 961 (1995):1–24.
29. Thomas Zeugmann, Steffen Lange and Shyam Kapur. Characteriza-

tions of monotonic and dual monotonic language learning. *Information and Computation* 120(2) (1995):155–173.

Characterization of the Second Homology Group of a Stationary Type in a Stable Theory

John Goodrick

Department of Mathematics
Universidad de los Andes
Bogotá, Colombia
E-mail: jr.goodrick427@uniandes.edu.co

Byunghan Kim

Department of Mathematics
Yonsei University
Seoul, Republic of Korea
E-mail: bkim@yonsei.ac.kr

Alexei Kolesnikov

Department of Mathematics
Towson University, MD
USA
E-mail: akolesnikov@towson.edu

Let T be a stable theory. It was shown in Ref. 5 that one can define the notions of homology groups attached to a stationary type of T. It was also shown that if T fails to have an amalgamation property called 3-uniqueness, then for some stationary type p the homology group $H_2(p)$ has to be a nontrivial abelian profinite group. The goal of this paper is to show that for any abelian profinite group G there is a stable (in fact, categorical) theory and a stationary type p such that $H_2(p) \cong G$.

Keywords: The 2nd homology group, 3-uniqueness, groupoid, categorical theory

1. Preliminaries

The paper Ref. 5 introduces the definitions of homology groups H_n, $n \geq 0$, for stable first-order theories. These groups measure the failure of generalized amalgamation properties for $n \geq 2$. It was shown that, for a stationary type p in a stable theory, the group $H_2(p)$ must be abelian profinite. In the present paper, for a given abelian profinite group G, we provide a construction of a stable theory T_G and a stationary type p in that theory such that $H_2(p) \cong G$.

We refer the reader to Ref. 5 for the definitions of homology groups and the generalized amalgamation properties. The presentation in this paper is self-contained, modulo the following key result.

Fact 1.1 (Theorem 2.1 in Ref. 5). *If $T = T^{eq}$ is a stable theory and p is a stationary type, then the group $H_2(p)$ is isomorphic to the group $\mathrm{Aut}(\widetilde{ab}/\operatorname{acl}(a)\operatorname{acl}(b))$, where a, b are independent realizations of the type p and $\widetilde{ab} = \operatorname{acl}(ab) \cap \operatorname{dcl}(\operatorname{acl}(ac), \operatorname{acl}(bc))$ for some (equivalently, any) realization c of p such that c is independent from ab.*

Let $G = \varprojlim H_i$ be a given abelian profinite group. We construct a first-order theory T_G such that $\mathrm{Aut}(\widetilde{ab}/\operatorname{acl}(a)\operatorname{acl}(b)) \cong G$ for independent realizations of some stationary type. The needed theory T_G is a theory of a certain projective system of groupoids. We begin by presenting the definitions and facts about groupoids and projective systems of groupoids that will be needed to establish their model-theoretic properties.

1.1. *Groupoids*

Definition 1.1. A *groupoid* $\mathcal{G}$ is a category in which every morphism is invertible. A groupoid carries the following structure:

(1) the universe of $\mathcal{G}$ is partitioned into disjoint sets O and M of objects and morphisms;
(2) the domain and range maps, which we will denote by d and r, from M to O;
(3) the ternary relation $\circ$ that defines the composition operation;
(4) the identity map from O to M which selects the identity element in every group $\mathrm{Mor}(a, a)$.

A groupoid is *connected* if there is a morphism between any two of its objects.

It is well known (see, for example, Ref. 1) that if $\mathcal{G}$ is a connected groupoid, then for any $a, b \in \mathrm{Ob}(\mathcal{G})$, the groups $\mathrm{Mor}(a, a)$ and $\mathrm{Mor}(b, b)$ are isomorphic. The isomorphism is given by a conjugation by a morphism from a to b. The group $\mathrm{Mor}(a, a)$ in this case is called the *vertex group of* $\mathcal{G}$. It is also well known that the isomorphism type of a connected groupoid is determined by the vertex group and by the cardinality of the set of objects. We will need a more detailed information about isomorphisms (and in particular, automorphisms) of groupoids, so we state the following facts.

Let $\mathcal{G}$ be a connected groupoid and let $a \in \mathrm{Ob}(\mathcal{G})$. A *star at* a is a function $s : \mathrm{Ob}(\mathcal{G}) \setminus \{a\} \to \mathrm{Mor}(\mathcal{G})$ such that $s(b) \in \mathrm{Mor}_{\mathcal{G}}(a,b)$.

Fact 1.2. (1) Let $F : \mathcal{G} \to \mathcal{H}$ be an isomorphism of connected groupoids, let $a \in \mathrm{Ob}(\mathcal{G})$ be an arbitrary element and let s be a star at a in $\mathcal{G}$. Then F is uniquely determined by its restriction to the following three sets: $\mathrm{Ob}(\mathcal{G})$, $\mathrm{Mor}_{\mathcal{G}}(a,a)$, and the range of s.

(2) Conversely, given an arbitrary bijection σ between $\mathrm{Ob}(\mathcal{G})$ and $\mathrm{Ob}(\mathcal{H})$, a group isomorphism $\varphi : \mathrm{Mor}_{\mathcal{G}}(a,a) \to \mathrm{Mor}_{\mathcal{H}}(\sigma(a),\sigma(a))$, and arbitrary stars s in $\mathcal{G}$ at a and t in $\mathcal{H}$ at $\sigma(a)$, there is a unique isomorphism $F : \mathcal{G} \to \mathcal{H}$ that extends σ and φ and such that $F(s(b)) = t(F(b))$ for all $b \in \mathrm{Ob}(\mathcal{G}) \setminus \{a\}$.

(3) If $\mathcal{G}$ and $\mathcal{H}$ are connected groupoids such that $|\mathrm{Ob}(\mathcal{G})| = |\mathrm{Ob}(\mathcal{H})|$ and the vertex groups of $\mathcal{G}$ and $\mathcal{H}$ are isomorphic, then the groupoids $\mathcal{G}$ and $\mathcal{H}$ are isomorphic.

Proof. If $c, d \in \mathrm{Ob}(\mathcal{G}) \setminus \{a\}$, then $f \in \mathrm{Mor}_{\mathcal{G}}(c,d)$ can be written as $f = s(d) \circ h \circ s(c)^{-1}$ for a unique $h \in \mathrm{Mor}_{\mathcal{G}}(a,a)$. If $f \in \mathrm{Mor}_{\mathcal{G}}(a,b)$ ($f \in \mathrm{Mor}_{\mathcal{G}}(b,a)$) , $b \neq a$, then $f = s(b) \circ h$ ($f = h \circ s(b)^{-1}$, respectively) for a unique $h \in \mathrm{Mor}_{\mathcal{G}}(a,a)$. Thus, the for all $f \in \mathrm{Mor}(\mathcal{G})$, the value $F(f)$ is uniquely determined by $F \restriction O$, $F \restriction \mathrm{Mor}_{\mathcal{G}}(a,a)$ and by the restriction of F to the range of s.

The construction in (1) provides a way to define F extending σ, φ, and the map $s(b) \in \mathrm{Mor}_{\mathcal{G}}(a,b) \mapsto t(\sigma(b)) \in \mathrm{Mor}_{\mathcal{H}}(\sigma(a),\sigma(b))$. Associativity of groupoids ensures that the resulting map F preserves the composition. This establishes (2).

The third statement follows since we can choose at least one star in a connected groupoid (with at least two objects). $\square$

1.2. *Directed systems of groupoids*

Definition 1.2. Let $(I, <)$ be a directed partially ordered set and let O be a non-empty set. For each $i \in I$, let $\mathcal{G}_i$ be a connected groupoid with $\mathrm{Ob}(\mathcal{G}_i) = O$ and suppose that we are given a system $\{\chi_{j,i} \mid i \leq j \in I\}$ of functors $\chi_{j,i} : \mathcal{G}_j \to \mathcal{G}_i$ such that

(1) the functor $\chi_{j,i}$ is the identity map on objects and is full on morphisms;

(2) the system commutes: for all $i < j < k \in I$ we have $\chi_{k,i} = \chi_{j,i} \circ \chi_{k,j}$.

We call the system $\boldsymbol{\mathcal{G}} := \{\mathcal{G}_i, \chi_{j,i} \mid i < j \in I\}$ a *projective system of groupoids.* We denote the common set of objects by the symbol $\mathrm{Ob}(\boldsymbol{\mathcal{G}})$. The symbol $\mathrm{Mor}(\boldsymbol{\mathcal{G}})$ will denote the disjoint union $\bigcup_{i\in I}\{i\} \times \mathrm{Mor}(\mathcal{G}_i)$ and $\mathrm{Mor}_{\boldsymbol{\mathcal{G}}}(a,b)$ is the set $\bigcup_{i\in I}\{i\} \times \mathrm{Mor}_{\mathcal{G}_i}(a,b)$.

The assumption that for all $i, j \in I$, $\mathrm{Ob}(\mathcal{G}_i) = \mathrm{Ob}(\mathcal{G}_j)$ is made only to simplify the notation; it can be replaced by the requirement that $\chi_{j,i}$ is bijective on objects.

An alternative way to describe the projective system of groupoids is by saying that $\boldsymbol{\mathcal{G}}$ is a (contravariant) functor from $(I, <)$ to the subcategory of the category of all connected groupoids in which the only morphisms are functors that are identity maps on objects and are full on morphisms. (The partially ordered set $(I, <)$ is viewed as a category in the natural way.)

We will need a description of isomorphisms for the projective systems of groupoids.

Definition 1.3. Let $\boldsymbol{\mathcal{G}}$ and $\boldsymbol{\mathcal{H}}$ be projective systems of groupoids both indexed by a directed set I. We say that a family of functions $\{F_i \mid i \in I\}$ is an *isomorphism of the projective systems* if F_i is an isomorphism of groupoids $\mathcal{G}_i$ and $\mathcal{H}_i$ for each $i \in I$ and the isomorphisms F_i commute with the projection maps. That is, for all $i < j \in I$ we have $F_i \circ \chi^{\boldsymbol{\mathcal{G}}}_{j,i} = \chi^{\boldsymbol{\mathcal{H}}}_{j,i} \circ F_j$.

In category theory language, projective systems of groupoids are isomorphic if they are naturally isomorphic as functors. We "unwrap" this definition mostly to fix the notation for the component isomorphisms.

Definition 1.4. Let $\boldsymbol{\mathcal{G}} = \{\mathcal{G}_i, \chi_{j,i} \mid i < j \in I\}$ be a projective system of connected groupoids and fix $a \in \mathrm{Ob}(\boldsymbol{\mathcal{G}})$. A *star system at* a is a function $s : I \times (\mathrm{Ob}(\boldsymbol{\mathcal{G}}) \setminus \{a\}) \to \mathrm{Mor}(\boldsymbol{\mathcal{G}})$ such that $s(i,b) \in \mathrm{Mor}_{\mathcal{G}_i}(a,b)$ and $s(i,b) = \chi_{j,i}(s(j,b))$ for all $i < j \in I$.

The following is an easy generalization of Fact 1.2.

Proposition 1.1. *Let $\boldsymbol{\mathcal{G}}$ and $\boldsymbol{\mathcal{H}}$ be projective systems of groupoids both indexed by a directed set I.*

(1) Let $\boldsymbol{F} = \{F_i \mid i \in I\}$ be an isomorphism of the projective systems. Let $a \in \mathrm{Ob}(\boldsymbol{\mathcal{G}})$ be an arbitrary element and let s be a star system at a in $\boldsymbol{\mathcal{G}}$. Then $\boldsymbol{F}$ is uniquely determined by the restrictions of F_i, $i \in I$, to the sets $\mathrm{Ob}(\boldsymbol{\mathcal{G}})$, $\mathrm{Mor}_{\mathcal{G}_i}(a,a)$, and the range of s.

(2) Given an arbitrary bijection σ between $\mathrm{Ob}(\boldsymbol{\mathcal{G}})$ *and* $\mathrm{Ob}(\boldsymbol{\mathcal{H}})$*; a family* $\{\varphi_i \mid i \in I\}$ *of group isomorphisms* $\varphi_i : \mathrm{Mor}_{\mathcal{G}_i}(a,a) \to \mathrm{Mor}_{\mathcal{H}_i}(\sigma(a), \sigma(a))$ *that commute with projections; and arbitrary star systems s in $\boldsymbol{\mathcal{G}}$ at a and t in $\boldsymbol{\mathcal{H}}$ at $\sigma(a)$, there is a unique isomorphism $\{F_i \mid i \in I\}$ of projective systems such that for each $i \in I$ we have: $F_i : \mathcal{G}_i \to \mathcal{H}_i$ extends σ and φ_i; and $F_i(s(i,b)) = t(i,(F_i(b))$ for all $b \in \mathrm{Ob}(\boldsymbol{\mathcal{G}}) \setminus \{a\}$.*

(3) Suppose that $\boldsymbol{\mathcal{G}} = \{\mathcal{G}_i \mid i \in I\}$ and $\boldsymbol{\mathcal{H}} = \{\mathcal{H}_i \mid i \in I\}$ are projective systems of groupoids and that there is a system of isomorphisms $\{\varphi_i \mid i \in I\}$ between the vertex groups of $\mathcal{G}_i$ and $\mathcal{H}_i$ that commutes with the projection maps. If the object sets $\mathrm{Ob}(\boldsymbol{\mathcal{G}})$ *and* $\mathrm{Ob}(\boldsymbol{\mathcal{H}})$ *have the same cardinality, then there is a system of isomorphisms $\{F_i : \mathcal{G}_i \to \mathcal{H}_i \mid i \in I\}$ that commutes with the projection maps in the system.*

Proof. The first statement follows from Fact 1.2(2) "level-by-level." The existence of the functions F_i follows from Fact 1.2(3); it remains to verify that the system $\{F_i \mid i \in I\}$ commutes with the projections. For $i < j \in I$, take $(j,f) \in \mathrm{Mor}_{\boldsymbol{\mathcal{G}}}(c,d)$. We consider the case when $c, d \neq a$, the remaining cases are similar. Then $f = s(j,d) \circ h_j \circ s(j,c)^{-1}$ for a unique $h_j \in \mathrm{Mor}_{\mathcal{G}_j}(a,a)$. Let $h_i := \chi^{\boldsymbol{\mathcal{G}}}_{j,i}(h_j)$. Then we have

$$\begin{aligned} F_i(\chi^{\boldsymbol{\mathcal{G}}}_{j,i}(j,f) &= F_i(s(i,d) \circ h_i \circ s(i,c)^{-1}) = t(i, F_i(d)) \circ \varphi_i(h_i) \circ t(i, F_i(c))^{-1} \\ &= \chi^{\boldsymbol{\mathcal{H}}}_{j,i}(t(j, F_j(d)) \circ \chi^{\boldsymbol{\mathcal{H}}}_{j,i}(\varphi_j(h_j)) \circ \chi^{\boldsymbol{\mathcal{H}}}_{j,i}(t(j, F_j(c))^{-1}) = \chi^{\boldsymbol{\mathcal{H}}}_{j,i}(F(j,f)). \end{aligned}$$

The third statement follows from the existence of the star systems. □

2. Any profinite abelian group can occur as $H_2(p)$

In this section, we prove the main result of the paper.

Theorem 2.1. *Let $(I,<)$ be a directed partially ordered set and let $\{H_i, \varphi_{j,i} \mid i \leq j \in I\}$ be an inverse system of non-trivial finite abelian groups, where each of the group homomorphisms $\varphi_{j,i}$ is surjective. Let $G = \varprojlim H_i$. There is a stable theory T_G and a stationary type p of T_G such that the group $H_2(p)$ is isomorphic to G.*

The main idea is to axiomatize, in first-order logic, the class of projective systems of groupoids indexed by the partially ordered set I and such that the vertex group of the groupoid $\mathcal{G}_i$ is isomorphic to H_i. We need to be particularly careful to axiomatize in a way that fixes both the set I and the groups H_i. This is accomplished by coding I and $\{H_i \mid i \in I\}$ into the language of the structure.

2.1. *Description of T_G*

Language. The language L_G is a multisorted language with sorts O and $\{M_i \mid i \in I\}$ and containing the following:

(1) for each $i \in I$, the function symbols d_i and r_i, both from the sort M_i to the sort O;
(2) for each $i \in I$, the ternary relation $\circ_i$ on M_i;
(3) for each pair $i < j \in I$, function symbols $\chi_{j,i}$ from the sort M_j to the sort M_i;
(4) for each $i \in I$, a finite set $\{P_1^i, \ldots, P_{k_i}^i\}$ of unary predicates on the sort M_i, where $k_i = |H_i|$.

Standard structure. We describe a "standard" L_G-structure and then give a list of axioms T_G satisfied by this structure. We will then show that T_G is categorical in every infinite cardinal greater than $|I|$ (so we may assume that all the structures we are dealing with are standard).

Let O be an infinite set. Let $M_i = O^2 \times H_i$, and define the functions d_i, r_i, and the relation $\circ_i$ so that $(O, M_i, d_i, r_i, \circ_i)$ is a definable groupoid with the vertex group H_i. For each $i \in I$, fix an enumeration $\{h_1^i, \ldots, h_{k_i}^i\}$ of the group H_i. Define $P_\ell^i := \{(a, a, h_\ell^i) \mid a \in O\}$ for $\ell = 1, \ldots, k_i$. Define functions $\chi_{j,i} : M_j \to M_i$ as follows: $\chi_{j,i}(a, b, h) = (a, b, \varphi_{j,i}(h))$.

Axiomatization. Let T_G be the following list of $\forall\exists$-sentences:

(1) the sorts O, M_i, $i \in I$, are pairwise disjoint;
(2) O is an infinite set;
(3) for each $i \in I$, the structure $\mathcal{G}_i := (O, M_i, d_i, r_i, \circ_i)$ is a definable connected groupoid with the set O of objects and the set M_i of morphisms. For $a, b \in O$, we use the symbol $M_i(a, b)$ to denote the set $\{f \in M_i \mid d_i(f) = a,\ r_i(f) = b\}$;
(4) for each $i \in I$, for each $a \in O$ and $\ell = 1, \ldots, k_i$, there is a unique element $f_{a,\ell}^i \in M_i(a, a) \cap P_\ell^i$ and $P_\ell^i = \{f_{a,\ell}^i \mid a \in O\}$;
(5) for each $i \in I$, a sentence expressing that for all $a \in O$, the map $h_\ell^i \mapsto f_{a,\ell}^i$ is a group isomorphism between H_i and $M_i(a, a)$ (note however that the elements h_ℓ^i and the map are not a part of the structure);
(6) the system $\{\mathcal{G}_i, \chi_{j,i} \mid i < j \in I\}$ is a projective system of groupoids and $\chi_{j,i} \restriction M_j(a, a)$ agrees with $\varphi_{j,i}$.

It is straightforward to check that the following holds.

Claim 2.1. *Let $G = \varprojlim H_i$ be an abelian profinite group. Then for any infinite set O, the standard L_G-structure on O is a model of T_G.*

2.2. *Model-theoretic properties of T_G*

We establish that T_G is a complete and categorical theory. The latter fact allows us to work with the standard models of T_G we described above. We describe the algebraically closed sets in models of T_G and show that the theory has weak elimination of imaginaries.

Lemma 2.1. *The theory T_G is complete, categorical in every $\lambda > |I|$ and is model-complete.*

Proof. If M, N are models of T_G of cardinality $\lambda > |I|$, then $|O(M)| = |O(N)| = \lambda$.

The axioms of T_G guarantee that for any $M, N \models T_G$, for any $a \in \mathrm{Ob}(M)$ and $b \in \mathrm{Ob}(N)$, the groups $\mathrm{Mor}_i(a, a) \subset M$ and $\mathrm{Mor}_i(b, b) \subset N$ are isomorphic and that there is a family of isomorphisms of these groups that commutes with the maps $\chi^M_{j,i}$ and $\chi^N_{j,i}$. Thus, by Proposition 1.1(3), there is an isomorphism between the projective systems M and N.

Since T_G is categorical, it is complete. Since T_G is categorical, has no finite models, and is $\forall\exists$-axiomatizable, it is model-complete by Lindström's test. □

Definition 2.1. Let $\mathfrak{C}$ be a large model of T_G. If A is a small subset of $\mathfrak{C}$, then we say that the set of all $b \in O(\mathfrak{C})$ such that either $b \in A$ or b is the domain or the range of a morphism in A is *the support of* A and denoted $\mathrm{supp}(A)$.

Let $\mathfrak{C}$ be a large model of T_G. By Lemma 2.1, we may assume that $\mathfrak{C}$ is the standard model of T_G. Since T_G completely determines the vertex groups, by Proposition 1.1, every automorphism of $\mathfrak{C}$ is uniquely determined by a permutation of $O(\mathfrak{C})$ and by the image of a star system under the automorphism. We fix a specific star system.

Definition 2.2. Fix $a \in O(\mathfrak{C})$, where $\mathfrak{C}$ is a large standard model of T_G, and consider the star system s_0 at a that picks out the zeros: $s_0(i, b) = (a, b, 0_{H_i})$. We call this function the *zero star system at a*.

Lemma 2.2. *Let $\mathfrak{C}$ be a large model of T_G and let $O = O(\mathfrak{C})$.*

(1) If A is a small subset of $\mathfrak{C}$ and $A' = \mathrm{supp}(A)$, then for any $i \in I$ and any $f \in M_i$ if $\mathrm{supp}(f) \not\subset A'$, then there is an automorphism σ of $\mathfrak{C}$ that fixes pointwise $A \cup O$ and moves f.

(2) If A is a small subset of $\mathfrak{C}$ and $A' = \mathrm{supp}(A)$, then any permutation of O that fixes A' can be extended to an automorphism of $\mathfrak{C}$ that fixes A.

Proof. As we pointed out above, the theory T_G completely determines the structure of the vertex groups and the commuting system of maps between the vertex groups. Therefore, by Proposition 1.1, to specify an automorphism σ of $\mathfrak{C}$ we need to describe the permutation $\sigma \restriction O$ and the star system $\sigma \circ s_0$, where s_0 is the zero star system at some $a \in O$.

For the first statement, $\sigma \restriction O$ is the identity map. Choose an arbitrary element $a \in O$ and let s_0 be the zero star system at a. Let $f \in M_i(b, c)$, where at least one of b, c is not in A'. We may assume that $b \notin A'$; the argument in the case $c \notin A'$ is similar.

Let $h \in H_i$ be a non-zero element (it exists since we assume that all the groups H_i are non-trivial) and define the second star system s at a as follows. Let $s(j, d) = (a, d, 0)$ if $d \neq b$ or if $d = b$ but j is not $<_I$-comparable with i; let $s(j, b) = \chi_{i,j}(h_i)$ if $j \leq i$; and otherwise let $s(j, b) = (a, b, h_j)$, where $h_j \in H_j$ is an element such that $\chi_{j,i}(h_j) = h_i$ and such that $\chi_{k,j}(h_k) = h_j$ for all $i < j < k$. It is clear that the resulting automorphism will move f. It remains to show that for any $j \in I$ and any $f \in M_j \cap A$, we have $\sigma(f) = f$. This is immediate since neither $d_j(f)$ nor $r_j(f)$ are equal to b.

The statement (2) is clear when $A = \emptyset$. Otherwise, we can take $a \in A'$ and obtain an automorphism of $\mathfrak{C}$ from the following data: the given permutation of O that fixes A' and two zero star systems at a. □

The next step is to describe the definable closure and the algebraic closure of a subset of a large saturated model $\mathfrak{C}$ of T_G (for background on such models, called *monster models* in model theory, see, for example Section 10.1 of Ref. 6). Recall that the definable closure of a set A is the set of all elements that are fixed by $\mathrm{Aut}(\mathfrak{C}/A)$ (the group of automorphisms of $\mathfrak{C}$ that fix A pointwise); and the algebraic closure of A is the set of all elements that have a finite orbit under the action of $\mathrm{Aut}(\mathfrak{C}/A)$.

The general statement about the algebraic and definable closures is given in Proposition 2.1 below. Let us illustrate these model-theoretic notions in our context on the following simple example. Let $\mathcal{G}$ be a model of T_G and take $f \in \mathrm{Mor}_{\mathcal{G}_k}(a, b)$ for some $a \neq b \in \mathrm{Ob}(\mathcal{G})$. Then the definable closure of the set $\{f\}$ contains: a, b, the vertex groups $\mathrm{Mor}_{\mathcal{G}_i}(a, a)$ and $\mathrm{Mor}_{\mathcal{G}_i}(b, b)$ for all $i \in I$, as well as the hom-sets $\mathrm{Mor}_{\mathcal{G}_j}(a, b)$ for all $j \leq_I k$. The algebraic closure will also contain the sets $\mathrm{Mor}_{\mathcal{G}_i}(a, b)$ for all $i \in I$.

Proposition 2.1. *Let A be a subset of the monster model of T_G. Then* $\mathrm{dcl}(A) \cap O = \mathrm{supp}(A)$ *and* $\mathrm{acl}(A)$ *is the set* $\bigcup_{i \in I} \{M_i(a,b) \mid a, b \in \mathrm{supp}(A)\}$.

Proof. It is clear that every element of the support of A is definable from an element of A. For the reverse inclusion, take arbitrary elements $c \neq d \in O$ which are not in the support of A and define an automorphism σ of $\mathfrak{C}$ by the following data: the permutation of O that transposes c and d and fixes all other elements; and two zero star systems at a point $a \in O$, $a \neq c, d$. It is clear that σ fixes A pointwise. Thus, for any element $c \notin \mathrm{supp}(A)$, there is an automorphism σ such that $\sigma(c) \neq c$.

The argument for the algebraic closure of A is similar, using Lemma 2.2. □

Next we show that the theory T_G has *weak elimination of imaginaries* in the sense of Section 16.5 in Ref. 9: for every formula $\varphi(\overline{x}, \overline{a})$ defined over a model M, there is a smallest algebraically closed set $A \subseteq M$ such that $\varphi(\overline{x}, \overline{a})$ is equivalent to a formula with parameters in A.

This is a technical step; it is needed because the results of Ref. 5 hold for structures that have sorts (called *imaginary sorts*) for quotient spaces modulo all the definable equivalence relations (this is the role of the assumption $T = T^{eq}$). There is a general procedure of expanding any first order theory T so that its expansion T^{eq} has all the imaginary sorts (yet whose models have the same automorphism groups as the models of T). We are proving in the following two lemmas that the algebraic closure in T_G^{eq} is controlled by subsets of models of T_G.

Lemma 2.3. *The theory T_G has weak elimination of imaginaries.*

Proof. By Lemma 16.17 of Ref. 9, it suffices to prove the following two statements:

1. There is no strictly decreasing sequence $A_0 \supsetneq A_1 \supsetneq \ldots$, where every A_i is the algebraic closure of a finite set of parameters; and

2. If A and B are algebraic closures of finite sets of parameters in the monster model $\mathfrak{C}$, then $\mathrm{Aut}(\mathfrak{C}/A \cap B)$ is generated by $\mathrm{Aut}(\mathfrak{C}/A)$ and $\mathrm{Aut}(\mathfrak{C}/B)$.

Statement 1 follows immediately from the characterization of algebraically closed sets in Proposition 2.1 (that is, algebraic closures of finite sets are equal to algebraic closures of finite subsets of O).

To check statement 2, suppose that $\sigma \in \mathrm{Aut}(\mathfrak{C}/A \cap B)$, and assume that $A = \mathrm{acl}(A_0)$ and $B = \mathrm{acl}(B_0)$ where $A_0, B_0 \subseteq O(\mathfrak{C})$. By Lemma 2.2(2), any permutation of $O(\mathfrak{C})$ which fixes A_0 can be extended to an automorphism of $\mathrm{Aut}(\mathfrak{C}/A)$, and likewise for B_0 and B.

So as a first step, we can use the fact that $\mathrm{Sym}(O/A_0 \cap B_0)$ is generated by $\mathrm{Sym}(O/A_0)$ and $\mathrm{Sym}(O/B_0)$ to find an automorphism $\tau \in \mathrm{Aut}(\mathfrak{C})$ such that τ is in the subgroup generated by $\mathrm{Aut}(\mathfrak{C}/A)$ and $\mathrm{Aut}(\mathfrak{C}/B)$ and $\widetilde{\sigma} := \sigma \circ \tau^{-1}$ fixes O pointwise.

Now we present $\widetilde{\sigma}$ as the composition of two automorphisms $\sigma_A^0 \in \mathrm{Aut}(\mathfrak{C}/A)$ and $\sigma_B^0 \in \mathrm{Aut}(\mathfrak{C}/B)$. Take an arbitrary point $a \in A_0 \cap B_0$ (if the intersection is empty, choose $a \in A_0$). Let s_0 be the zero star system at a. Let $s = \widetilde{\sigma} \circ s_0$. Then s is a star system at a. Since $\widetilde{\sigma}$ fixes $A \cap B$ pointwise, for all $c \in A_0 \cap B_0$ and all $i \in I$ we have $s(i,c) = 0$. Define the star system s_A as follows: for all $c \in A_0$ and all $i \in I$, let $s_A(i,c) = 0$ and for $c \notin A_0$ let $s_A(i,c) = s(i,c)$ for all $i \in I$. Define a star system s_B by setting $s_B(i,c) = s(i,c)$ for all $c \in A_0$ and all $i \in I$ and otherwise let $s_B(i,c) = 0$. Note that for all $c \in B_0$ and all $i \in I$ we have $s_B(i,c) = 0$ (this is true by definition for $c \in B_0 \setminus (A_0 \cap B_0)$; and recall that $s(i,c) = 0$ for $c \in A_0 \cap B_0$).

Let $\sigma_A^0 \in \mathrm{Aut}(\mathfrak{C})$ be the automorphism determined by the identity permutation of O and the star system s_A and similarly let $\sigma_B^0 \in \mathrm{Aut}(\mathfrak{C})$ be the automorphism determined by s_B. Since the star systems $s_A(i,c) = 0$ for all $a \in A_0$ and $s_B(i,c) = 0$ for all $c \in B_0$, we have $\sigma_A^0 \in \mathrm{Aut}(\mathfrak{C}/A)$ and $\sigma_B^0 \in \mathrm{Aut}(\mathfrak{C}/B)$. Finally, since $s_A(i,c) + s_B(i,c) = s(i,c)$ for all $i \in I$ and $c \in O$, we have $\sigma_B^0 \circ \sigma_A^0 = \widetilde{\sigma}$. Finally, we get $\sigma_B^0 \circ \sigma_A^0 \circ \tau = \sigma$. □

Lemma 2.4. *Let $\mathfrak{C}$ be a monster model of T_G, let $O = O(\mathfrak{C})$. If X is a small subset of O, then*

$$\mathrm{acl}^{eq}(X) = \mathrm{dcl}^{eq}\left(\bigcup_{i\in I}\{M_i(a,b) \mid a,b \in X\}\right).$$

Proof. Suppose $g \in \mathrm{acl}^{eq}(X)$. Then $g = c/E$ for some X-definable finite equivalence relation $E(x,y)$. By Lemma 2.3, there is a smallest algebraically closed set A such that $E(x,c)$ is equivalent to a formula with parameters from A. By Theorem 16.15 in Ref. 9, the set A is the algebraic closure of a finite set of tuples. If the set $\mathrm{supp}(A)$ contained objects $a_1', \ldots, a_k' \notin X$, then (by the description of automorphisms) for any $b_1', \ldots, b_k' \notin X$, the formula $E(x,c)$ would be equivalent to a formula with parameters from $\mathrm{acl}(X \cup \{b_1', \ldots, b_k'\})$. In particular, we could choose $\{b_i' \mid i = 1, \ldots, k\}$ to be disjoint from $\{a_i' \mid i = 1, \ldots, k\}$, contradicting the minimality of A. □

2.3. *Computing the automorphism group*

Let a_0, a_1, a_2 be distinct elements in $O(\mathfrak{C})$. Let

$$\Gamma_2 := \mathrm{Aut}(\widetilde{a_0 a_1}/\,\mathrm{acl}^{eq}(a_0), \mathrm{acl}^{eq}(a_1)),$$

where $\widetilde{a_0 a_1} = \mathrm{acl}^{eq}(a_0 a_1) \cap \mathrm{dcl}^{eq}(\mathrm{acl}^{eq}(a_0 a_2), \mathrm{acl}^{eq}(a_1 a_2))$.

Proposition 2.2. *The group Γ_2 described above is isomorphic to the group $G = \varprojlim H_i$.*

Proof. By Lemma 2.4 and the fact that any morphism in $M_i(a_0, a_1)$ is a composition of morphisms in $M_i(a_0, a_2)$ and $M_i(a_2, a_1)$, it follows that the set $\widetilde{a_0 a_1}$ is interdefinable with $\bigcup_{i \in I}(M_i(a_0, a_1) \cup (M_i(a_0, a_0) \cup M_i(a_1, a_1))$. By construction, any automorphism of $\mathfrak{C}$ fixes $M_i(a, a)$ for all $a \in O$ and all $i \in I$. So to compute the group Γ_2, it is enough to describe all the automorphisms of $\mathrm{acl}(a_0, a_1)$ that fix a_0 and a_1.

We define an isomorphism $\Phi : G \to \Gamma_2$. Take $g \in G$. By Lemma 2.2, to define an automorphism of $\mathrm{acl}(a_0, a_1)$ over $a_0 a_1$, it is enough to specify a star system at a_0 over $\{a_0, a_1\}$. Let $s(i, a_1) = g(i)$. This induces a unique automorphism of $\mathrm{acl}(a_0, a_1)$ that sends the zero star system to s. If $s_1(i, a_1) = g_1(i)$ and $s_2(i, a_1) = g_2(i)$, the automorphism determined by s_2 maps $0 \in M_i(a, b)$ to $g_2(i)$ and therefore, it maps $g_1(i)$ to $g_2(i) + g_1(i)$. Thus, the function Φ is a injective homomorphism. Surjectivity is also clear: given an automorphism $\psi \in \Gamma_2$, the function $i \in I \mapsto \psi(s_0(i, a_1)) \in H_i$ is an element of G because an automorphic image of a star system is a star system. □

2.4. *Proof of the main theorem*

To complete the proof of the main theorem, we need to check that the type of an element in $O(\mathfrak{C})$ is stationary and that distinct elements of $O(\mathfrak{C})$ form an independent set.

Proposition 2.3. *Let $\mathfrak{C}$ be a large model of T_G, let A be a small subset of $\mathfrak{C}$ and let $c \in O(\mathfrak{C})$. If $c \notin \mathrm{supp}(A)$, then $\mathrm{tp}(c/A)$ does not fork over $\emptyset$.*

Proof. By finite character of non-forking, it is enough to prove the statement for a finite set A. If $c, d \notin \mathrm{supp}(A)$, then $\mathrm{tp}(c/A) = \mathrm{tp}(d/A)$ by Lemma 2.2(2). So the type of c over the finite set A is isolated by the formula expressing $c \notin \mathrm{supp}(A)$. It is clear that this formula does not divide over $\emptyset$, so $\mathrm{tp}(c/A)$ does not fork over $\emptyset$. □

From Proposition 2.3 and Lemma 2.2(2), we immediately obtain the following.

Corollary 2.1. *Let* $c \in O(\mathfrak{C})$. *Then the type* $\mathrm{tp}(c)$ *is stationary.*

Now we can complete the proof of the main result.

Proof of Theorem 2.1. Let $\mathfrak{C}$ be a large model of T_G, take $a \in O(\mathfrak{C})$. By Corollary 2.1, the type $p = \mathrm{tp}(a)$ is a stationary type. By Proposition 2.3, any distinct realizations of p form an independent set over $\emptyset$. By Theorem 1.1, it is enough to show that G is isomorphic to the group $\Gamma_2(p)$, and this was done in Proposition 2.2. □

References

1. Ronald Brown. From groups to groupoids: a brief survey. *Bull. London Math. Soc.*, **19**, (1987), 113-134.
2. Tristram de Piro, Byunghan Kim, and Jessica Millar. Constructing the type-definable group from the group configuration. *Journal of Mathematical Logic*, **6** (2006), 121–139.
3. John Goodrick and Alexei Kolesnikov. Groupoids, covers, and 3-uniqueness in stable theories. *Journal of Symbolic Logic*, **75** (2010), 905–929.
4. John Goodrick, Byunghan Kim, and Alexei Kolesnikov. Amalgamation functors and boundary properties in simple theories. *Israel Journal of Mathematics*, **193** (2013), 169–207.
5. John Goodrick, Byunghan Kim, and Alexei Kolesnikov. Homology groups of types in model theory and the computation of $H_2(p)$. *Journal of Symbolic Logic*, **78** (2013), 1086–1114.
6. Wilfrid Hodges. Model theory. Cambridge University Press, Cambridge. 1993.
7. Ehud Hrushovski. Groupoids, imaginaries and internal covers. *Turkish Journal of Mathematics* **36** (2012), 173–198.
8. Byunghan Kim and Hyeung-Joon Kim. Notions around tree property 1. *Annals of Pure and Applied Logic*, **162** (2011), 698–709.
9. Bruno Poizat. A course in model theory: an introduction to contemporary mathematical logic. Springer, 2000.

Some Questions Concerning Ab Initio Generic Structures

Koichiro Ikeda *

Faculty of Business Administration, Hosei University,
Chiyoda-ku, Tokyo 102-8160, Japan
E-mail: ikeda@hosei.ac.jp

We introduce some questions and results related to stability class and model completeness of *ab initio* generic structures. Then we give a characterization for the model completeness of generic structures.

Keywords: Generic structure; Stability class; Model completeness

In the late 1980s, Hrushovski modified the Ehrenfeucht-Fraïssé homogeneous universal relational structures to obtain counter-examples to two famous conjectures by Lachlan and Zilber[10,11]. His method is called the Hrushovski construction, and a structure constructed by his construction is called a generic structure. The Hrushovski construction has been applied in various ways, and many model theorists have constructed interesting examples[2,7–9,13,18]. Our original concern was to consider the following questions.

Question 0.1. Which stability class does each generic structure belongs to?

Question 0.2. Which generic structure is model complete?

In this short note, we introduce some results and questions related to these two questions.

1. Generic structures

There are several ways to define a generic structure. Here we adopt one of most concrete definitions. Our notation and definition mainly follow from Baldwin-Shi[5] and Wagner[19].

For simplicity, let L be a language which consists of one n-ary relation R with irreflexivity and symmetricity for $n \geq 2$, and let $A, B, C, ...$ be L-structures or (hyper-)graphs with (hyper-)edge R.

*The author is supported by Grants-in-Aid for Scientific Research (No.23540164).

For a real number $\alpha \in (0, 1]$, we define a predimension $\delta(A)$ of a finite A as follows:

$$\delta(A) = |A| - \alpha|R^A|.$$

We denote $\delta(B/A) = \delta(B \cup A) - \delta(A)$.

For finite $A \subset B$, A is said to be closed (or strong) in B (in symbol, $A \leq B$), if $\delta(X/A) \geq 0$ for any $X \subset B - A$. When A, B are not necessarily finite, $A \leq B$ is defined by $A \cap X \leq X$ for any finite $X \subset B$.

For $A \subset B$, there is a smallest set $C \leq B$ containing A. Such a C is denoted by $\mathrm{cl}_B(A)$.

Let $\mathbf{K}_\alpha = \{A \text{ finite} : \delta(A') \geq 0 \text{ for all } A' \subset A\}$.

Definition 1.1. Let $\mathbf{K} \subset \mathbf{K}_\alpha$. Then a countable L-structure M is said to be a $(\mathbf{K}, \leq)$-generic structure, if it satisfies the following:

(1) $A \in \mathbf{K}$ for any finite $A \subset M$;
(2) If $A \leq B \in \mathbf{K}$ and $A \leq M$, then there is a $B'(\cong_A B)$ with $B' \leq M$;
(3) M has finite closures, i.e., $\mathrm{cl}_M(A)$ is finite for any finite $A \subset M$.

By the definition, the generic M has finite closures, however any model of $\mathrm{Th}(M)$ does not always have finite closures. We say that the theory of M has finite closures, if any model of $\mathrm{Th}(M)$ has finite closures.

$(\mathbf{K}, \leq)$ is said to have the amalgamation property (AP), if whenever $A \leq B \in \mathbf{K}$ and $A \leq C \in \mathbf{K}$, then there are $B'(\cong_A B)$ and $C'(\cong_A C)$ with $B', C' \leq B' \cup C' \in \mathbf{K}$. If $(\mathbf{K}, \leq)$ is closed under substructures and has AP, then there exists a $(\mathbf{K}, \leq)$-generic structure.

By the back-and-forth method, if M, N are $(\mathbf{K}, \leq)$-generic then $M \cong N$. Moreover, it can be seen that the generic M is ultra-homogeneous over finite closed sets, i.e., if A, B are finite with $A \cong B$ and $A, B \leq M$, then $\mathrm{tp}(A) = \mathrm{tp}(B)$.

We say that the theory T of the generic M is ultra-homogeneous over finite closed sets, if any model of T is ultra-homogeneous over finite closed sets. Then it is easily checked that M is saturated if and only if $\mathrm{Th}(M)$ has finite closures and is ultra-homogeneous over finite closed sets.

2. Stability class

Recall that, in our setting, each generic structure is basically constructed by using one relation R and a real number $\alpha \in (0, 1]$. The following are well-known examples of generic structures:

Example 2.1.

(1) Hrushovski's strongly minimal structure[11]. This is a counter-example to Zilber's conjecture which says that any non locally modular $\aleph_1$-categorical theory is bi-interpretable with an algebraically closed field. In this example, arity$(R) = 3$ and $\alpha = 1$.
(2) Hrushovski's ω-categorical pseudoplane[10]. This is a counter-example to Lachlan's conjecture which says that any stable ω-categorical theory is ω-stable. In this example, arity$(R) = 2$ and α is a well-chosen irrational number.
(3) Baldwin's projective plane[2]. This projective plane is non-desarguesian and almost strongly minimal, and then it is a counter-example to another conjecture of Zilber. In this example, arity$(R) = 2$ and $\alpha = 1/2$.
(4) Shelah-Spencer's random graph[6,17]. Shelah and Spencer[17] proved that, for any irrational $\alpha \in (0, 1)$, the class of finite graphs with edge probability $n^{-\alpha}$ satisfies a 0-1 law, and hence the almost sure theory is complete. Baldwin and Shelah[6] proved that the theory is equal to the theory of the $(\mathbf{K}_\alpha, \leq)$-generic structure with arity$(R) = 2$.

Examples 1-3 are saturated, but example 4 is not saturated because the theory does not have finite closures. Note that the theory of example 4 is ultra-homogeneous over closed sets, i.e., if $A \cong B$ and $A, B \leq \mathcal{M}$ then $\mathrm{tp}(A) = \mathrm{tp}(B)$.

Fact 2.1. Let M be a generic structure whose theory is ultra-homogeneous over closed sets. Then

(1) $\mathrm{Th}(M)$ is stable;
(2) If α is rational, then $\mathrm{Th}(M)$ is ω-stable.

By this fact, examples 1-4 is stable, and in particular, both of examples 1 and 3 are ω-stable. Moreover, we can check that 2 and 4 are not superstable. On the contrary, all known examples were either strictly stable or ω-stable. In other words, there was no generic structure whose theory is strictly superstable (i.e., superstable but not ω-stable). So the following question arose naturally.

Question 2.1 (Baldwin[1,3]). *Is there a generic structure whose theory is strictly superstable?*

For this question, I first obtained the following theorem.

Theorem 2.1 (Ikeda[13]). *Any saturated generic structure is strictly stable or ω-stable.*

After that, I tried to remove "saturated" from the assumption of Theorem 2.1, but I could not. Instead I obtained the following result, which gave an answer to Question 2.1.

Theorem 2.2 (Ikeda[13]). *There is a generic structure whose theory is strictly superstable.*

The structure constructed above is not saturated because the theory does not have finite closures. In this example,

$$\text{arity}(R) = 2 \text{ and } \alpha = 1,$$

and hence the forking relation is trivial, i.e., any pairwise forking independent set is forking independent. On the other hand, all well-known examples of generic structures have the non-trivial forking relation. Therefore we obtained the following result.

Theorem 2.3 (Ikeda-Kikyo[16]). *There is a generic structure whose theory is strictly superstable and has the non-trivial forking relation.*

Roughly speaking, we need

$$\text{arity}(R) > 2 \text{ or } \alpha < 1$$

to construct a generic structure whose forking relation is non-trivial. Indeed, in Theorem 2.3 we constructed two kinds of generic structures as follows:

(1) $\text{arity}(R) = 2$ and α is rational with $\alpha < 1$.
(2) $\text{arity}(R) > 2$ and $\alpha = 1$.

Each construction is similar to that of Theorem 2.2.

A theory T is said to be small, if $|S(T)| = \omega$. Baldwin asked me the next question, which is still open.

Question 2.2 (Baldwin). *Is there a generic structure whose theory is strictly superstable and small?*

3. Model completeness

A theory T is said to be model complete, if whenever $M, N \models T$ and $M \subset N$, then $M \prec N$. A structure M is said to be model complete, if $\mathrm{Th}(M)$ is model complete. For model completeness of generic structures, the following results are known.

Theorem 3.1 (Holland[12]). *Hrushovski's strongly minimal structure is model complete.*

Theorem 3.2 (Baldwin-Holland[4]). *Baldwin's projective plane is model complete in a language with additional constant symbols.*

Theorem 3.3 (Baldwin-Shelah[17]). *Shelah-Spencer's random graph is not model complete.*

The following question may be still open.

Question 3.1. Is Baldwin's projective plane model complete?

Moreover, we don't know whether the following question is open or not.

Question 3.2. Is Hrushovski's ω-categorical pseudoplane model complete?

We want to give an answer to each of Question 3.1 and 3.2, but we have difficulty in solving these questions. (However, recently Kikyo and I have obtained some result[15] related to Question 3.2.)

In what follows, we concentrate to give a characterization of model completeness of saturated generic structures (Corollary 3.2).

Definition 3.1. Let $A \leq B \in \mathbf{K}$ and $A \subset C \in \mathbf{K}$. Then we say that B can be amalgamated to C over A, if there is a $B'(\cong_A B)$ with $C \leq B' \cup C \in \mathbf{K}$.

The following proposition is inspired by the proof of Theorem 1.37 in Baldwin-Shelah[5].

Proposition 3.1. *Let M be a $(\mathbf{K}, \leq)$-generic structure. Suppose that $\mathrm{Th}(M)$ is model complete. Then, if $A \subset C \in \mathbf{K}$ and $A \not\leq C$, then there is a $B \in \mathbf{K}$ with $A \leq B$ which cannot be amalgamated to C over A.*

Proof. Assume otherwise. Then there are $A, C \in \mathbf{K}$ with $A \subset C$ and $A \not\leq C$ such that any $B \in \mathbf{K}$ with $A \leq B$ can be amalgamated to C over A. Since M is generic, we can take an A_0 satisfying

$$A_0 \cong A \text{ and } A_0 \leq M.$$

Similarly we can take A_1, C_1 satisfying

$$A_1C_1 \cong AC \text{ and } C_1 \leq M.$$

Since $\mathrm{Th}(M)$ is model complete, we have $\mathrm{tp}_\exists(A_i) \vdash \mathrm{tp}(A_i)$, where $\mathrm{tp}_\exists(A_i) = \{\psi(\bar{x}) : \mathcal{M} \models \psi(A_i), \psi \text{ is an } \exists\text{-formula}\}$. Then it suffices to show that

- $\mathrm{tp}(A_0) \neq \mathrm{tp}(A_1)$;
- $\mathrm{tp}_\exists(A_0) \subset \mathrm{tp}_\exists(A_1)$.

First, since $A_0 \leq M$ and $A_1 \not\leq M$, clearly we have $\mathrm{tp}(A_0) \neq \mathrm{tp}(A_1)$. Next we show that

$$\mathrm{tp}_\exists(A_0) \subset \mathrm{tp}_\exists(A_1).$$

Take any $\exists Y \varphi(Y, X) \in \mathrm{tp}_\exists(A_0)$, where φ is a quantifier-free formula. We can assume that $X \subset Y$. Take a realization B_0 of $\varphi(Y, A_0)$ in M. Clearly $A_0 \leq B_0$. Take a $B_1 \in \mathbf{K}$ with $B_1A_1 \cong B_0A_0$. By our assumption, B_1 can be amalgamated to C_1 over A_1. Since M is generic, we can take a $B_1' \subset M$ with $B_1' \cong_{C_1} B_1$. So we have $\models \varphi(B_1', A_1)$. It follows that $\exists Y \varphi(Y, X) \in \mathrm{tp}_\exists(A_1)$. □

We call that $(\mathbf{K}, \leq)$ is trivial, if $A \subset B \in \mathbf{K}$ implies $A \leq B$. We call that $(\mathbf{K}, \leq)$ has the strong amalgamation property (SAP), if whenever $A \leq B \in \mathbf{K}$ and $A \subset C \in \mathbf{K}$ then there is a $B' \cong_A B$ with $C \leq B' \cup C \in \mathbf{K}$. Then we have the following corollary.

Corollary 3.1. *Let M be a $(\mathbf{K}, \leq)$-generic structure. If $(\mathbf{K}, \leq)$ is non-trivial and has SAP, then $\mathrm{Th}(M)$ is not model complete.*

Proof. Since $(\mathbf{K}, \leq)$ is non-trivial, there are $A, C \in \mathbf{K}$ with $A \subset C$ and $A \not\leq C$. Moreover, since $(\mathbf{K}, \leq)$ has SAP, any $B \in \mathbf{K}$ with $A \leq B$ can be amalgamated to C over A. By Proposition 3.1, $\mathrm{Th}(M)$ is not model complete. □

It is easily checked that the amalgamation class of Shelah-Spencer's random graph is non-trivial and has SAP. So Theorem 3.3 is also derived from Corollary 3.1.

Proposition 3.2. *Let M be a saturated $(\mathbf{K}, \leq)$-generic structure. Suppose that if $A \subset C \in \mathbf{K}$ and $A \not\leq C$, then there is a $B \in \mathbf{K}$ with $A \leq B$ which cannot be amalgamated to C over A. Then $\mathrm{Th}(M)$ is model complete.*

Proof. To prove the model completeness, it is enough to show that, for any finite $F \subset M$,

$$\mathrm{tp}_{\exists}(F) \vdash \mathrm{tp}(F).$$

Take any finite $F \subset M$. Since M is generic, $A = \mathrm{cl}(F)$ is finite. Let $(C_i)_{i \in \omega}$ be an enumeration of $\{C \in \mathbf{K} : A \subset C \text{ and } A \not\leq C\}$. By our assumption, for each $i \in \omega$, there is $B_i \in \mathbf{K}$ with $A \leq B_i$ which cannot be amalgamated to C_i over A. Then

$$\Sigma(Z) = \{\exists X \exists Y_0 ... \exists Y_n \textstyle\bigwedge_{i \leq n} (XY_iZ \cong AB_iF) : n \in \omega\}.$$

is consistent. (Indeed, since $A = \mathrm{cl}(F)$, each B_i can be embedded into M over A. So F is a realization of Σ.) Since Σ is a set of $\exists$-formulas, it is enough to show that

$$\Sigma \vdash \mathrm{tp}(F).$$

Take any realization F' of Σ in M. Then

$$\Gamma(X) = \{\exists Y_i (XY_iF' \cong AB_iF) : i \in \omega\}$$

is consistent. Since M is saturated, we can take a realization $A'(\subset M)$ of Γ. To show that $\mathrm{tp}(F') = \mathrm{tp}(F)$, it is enough to prove that

$$A' \leq M.$$

So suppose that $A' \not\leq M$. Let $C' = \mathrm{cl}(A')$, and take C with $CA \cong C'A'$. Clearly $A \not\leq C$, and so there is an $i \in \omega$ with $C = C_i$. Since A' is a realization of Γ, there is a $B'_i \subset M$ with $A'B'_iF' \cong AB_iF$. Then $C' \leq B'_i \cup C' \in \mathbf{K}$. Hence B'_i can be amalgamated to C'_i over A'. A contradiction. □

Question 3.3. Can the condition "saturated" be removed from the assumption of Proposition 3.2?

By Propositions 3.1 and 3.2, we have the following.

Corollary 3.2. *Let M be a saturated $(\mathbf{K}, \leq)$-generic structure. Then the following are equivalent.*

(1) $\mathrm{Th}(M)$ is model complete;
(2) If $A \subset C \in \mathbf{K}$ and $A \not\leq C$, then there is a $B \in \mathbf{K}$ with $A \leq B$ which cannot be amalgamated to C over A.

Both of Theorems 3.1 and 3.2 were proved by using Lindström's theorem, which says that if a theory is $\forall\exists$-axiomatizable and categorical for some (infinite) cardinal then it is model complete.

Question 3.4. Can the model completeness of Hrushovski's strongly minimal structure be proved by using Corollary 3.2?

Question 3.5. Can the model completeness of Baldwin's projective plane be proved by using Corollary 3.2?

References

1. J. T. Baldwin, Problems on pathological structures, In Helmut Wolter Martin Weese, editor, *Proceedings of 10th Easter Conference in Model Theory* 1–9 (1993).
2. J. T. Baldwin, An almost strongly minimal non-Desarguesian projective plane, Trans. Am. Math. Soc. 342, 695–711 (1994).
3. J. T. Baldwin, A field guide to Hrushovski's constructions, http://www.math.uic.edu/jbaldwin/pub/hrutrav.pdf (2009).
4. J. T. Baldwin and K. Holland, Constructing ω-stable structures: model completeness, Annals of Pure and Applied Logic 125, 159–172 (2004).
5. J. Baldwin and N. Shi, Stable generic structures, Annals of Pure and Applied Logic 79, 1–35 (1996).
6. J. T. Baldwin and S. Shelah, Randomness and semigenericity, Transactions of the American Mathematical Society 349, 1359–1376 (1997).
7. A. Baudisch, A new uncountably categorical group, Transactions of the American Mathematical Society 348, 889–940 (1995).
8. A. Nesin and M. J. De Bonis, There are $2^{\aleph_0}$ many almost strongly minimal generalized n-gons that do not interpret an infinite group, Journal of Symbolic Logic 63, 485–508 (1998).
9. B. Herwig, Weight ω in stable theories with few types, Journal of Symbolic Logic 60, 353–373 (1995).
10. E. Hrushovski, A stable $\aleph_0$-categorical pseudoplane, preprint (1988).
11. E. Hrushovski, A new strongly minimal set, Annals of Pure and Applied Logic 62, 147–166 (1993).
12. K. Holland, Model completeness of the new strongly minimal sets, J. Symbolic Logic 64, 946–962 (1999).
13. K. Ikeda, Minimal but not strongly minimal structures with arbitrary finite dimension, Journal of Symbolic Logic 66, 117–126 (2001).
14. K. Ikeda, Ab initio generic structures which are superstable but not ω-stable, Archive for Mathematical Logic 51, 203–211 (2012).

15. K. Ikeda and H. Kikyo, Model complete generic structures, preprint.
16. K. Ikeda and H. Kikyo, On superstable generic structures, Archive for Mathematical Logic 51, 591–600 (2012).
17. S. Shelah and J. Spencer, Zero-one laws for sparse random graphs, Journal of the American Mathematical Society 1,97–115 (1988).
18. K. Tent, Very homogeneous generalized n-gons of finite Morley rank, Journal of the London Mathematical Society 62, 1–15 (2000).
19. F. O. Wagner, Relational structures and dimensions, In *Automorphisms of first-order structures*, Clarendon Press, Oxford, 153–181 (1994).

Model Complete Generic Structures

Koichiro Ikeda

Faculty of Business Administration, Hosei University
2-17-1 Fujimi, Chiyoda, Tokyo 102-8160, Japan
E-mail: ikeda@hosei.ac.jp

Hirotaka Kikyo

Graduate School of System Informatics, Kobe University
1-1 Rokkodai, Nada, Kobe 657-8501, Japan
E-mail: kikyo@kobe-u.ac.jp

Generic structures constructed from a certain kind of amalgamation classes are shown to have a model complete countably categorical theory.

1. Introduction

Generic structures constructed by the Hrushovski's amalgamation construction are known to have theories which are nearly model complete. If an amalgamation class has the full amalgamation property then its generic structure has a theory which is not model complete[5]. On the other hand, Hrushovski's strongly minimal structure constructed by the amalgamation construction which refuted a Zilber's conjecture has a model complete theory[8].

In this paper, we show that some countably categorical structures constructed by the amalgamation construction similar to the construction of Hrushovski's pseudoplane have model complete theories. It seems to be possible that countably categorical structures constructed by the amalgamation construction have model complete theories.

Let $\mathcal{L}$ be the language $\{R\}$ with R a ternary relation symbol. The relation R will be symmetric and irreflexive in the structures we consider. So, we can consider an $\mathcal{L}$-structure as a 3-hypergraph. $e_R(A)$ will be the number of the hyperedges in A.

For a finite $\mathcal{L}$-structure A, define a predimension function δ by $\delta(A) = |A| - e_R(A)$ where $|A|$ denotes the cardinality of A, and $e_R(A)$ denotes the number of subsets (hyperedges) realising relation R. Let $\mathbf{K}$ be a class of finite $\mathcal{L}$-structures whose non-empty substructures have positive predimension. If $\mathbf{K}$ has the amalgamation property then an infinite structure can be constructed by amalgamating all elements of $\mathbf{K}$.

We essentially use notation and terminology from Wagner[13].

For a set X, $[X]^n$ denotes the set of all n-element subsets of X. For an $\mathcal{L}$-structure A, $\mathrm{dom}(A)$ denotes the domain of A. We often write A for the domain of A if there is no ambiguity. For $t = \{t_1, t_2, t_3\} \in [\mathrm{dom}(A)]^3$, we write $A \models R(t)$ if $A \models R(t_1, t_2, t_3)$. $R(A)$ denotes the set $\{t \in [\mathrm{dom}(A)]^3 \mid A \models R(t)\}$. Points in $R(A)$ will be called *R-relations*.

For a set X, $|X|$ denotes the cardinality of X. If $X \subseteq \mathrm{dom}(A)$, $A|X$ denotes the substructure of A with the domain X. If there is no ambiguity, X denotes $A|X$. $B \subseteq A$ means that B is a substructure of A. If B and C are substructures of some structure, then $B \cap C$ denotes the substructure $B|(\mathrm{dom}(B) \cap \mathrm{dom}(C)) = C|(\mathrm{dom}(B) \cap \mathrm{dom}(C))$. Structures $B \cup C$ and $B - C$ are defined similarly.

Suppose $A \subseteq B$ with A finite. $A \leq B$ if whenever $A \subseteq X \subseteq B$ with X finite then $\delta(A) \leq \delta(X)$.

We write $A < B$ if whenever $A \subsetneq X \subseteq B$ with X finite then $\delta(A) < \delta(X)$. A is *closed* in B if $A < B$. Note that $A < A$.

Suppose $A < B$ and $A < C$. An $\mathcal{L}$-isomorphism $f : B \to C$ is called a *closed* embedding of B into C over A if $f(B) < C$ and $f(x) = x$ for any $x \in A$.

With this notation, put

$$\mathbf{K}_1 = \{A : \text{finite} \mid A > \emptyset\}.$$

Definition 1.1. Let $\mathbf{K} \subseteq \mathbf{K}_1$ be an infinite class. $\mathbf{K}$ has the *amalgamation property* if for any $A, B, C \in \mathbf{K}$, whenever $A < B$ and $A < C$ then there is $D \in \mathbf{K}$ such that $A < D$ and there is a closed embedding of B into D over A and a closed embedding of C into D over A.

$\mathbf{K}$ has the *hereditary property* if for any finite structures A, B, $A \subseteq B \in \mathbf{K}$ implies $A \in \mathbf{K}$.

$\mathbf{K}$ is called an *amalgamation class* if $\mathbf{K}$ has the hereditary property and the amalgamation property.

Definition 1.2. Suppose $\mathbf{K} \subseteq \mathbf{K}_1$. A countable $\mathcal{L}$-structure M is a *generic structure* of $\mathbf{K}$ if the following conditions are satisfied:

(1) If A is a finite substructure of M then there exists a finite substructure B of M such that $A \subseteq B < M$.
(2) If A is a finite substructure of M then $A \in \mathbf{K}$.

(3) For any A, $B \in \mathbf{K}$, if $A < M$ and $A < B$ then there is a closed embedding of B into M over A.

There is a smallest B satisfying (1), written $\mathrm{cl}(A)$. We have $A \subseteq \mathrm{cl}(A) < M$ and if $A \subseteq B < M$ then $\mathrm{cl}(A) \subseteq B$. The set $\mathrm{cl}(A)$ is called a *closure* of A in M. Apparently, $\mathrm{cl}(A)$ is unique for given finite set A.

In general, if A and D are $\mathcal{L}$-structures and $A \subseteq D$, we write $\mathrm{cl}_D(A)$ for the smallest substructure B such that $A \subseteq B < D$.

Proposition 1.1. *Let* $\mathbf{K}$ *be an amalgamation class and* M *the generic structure for* $(\mathbf{K}, <)$. *Assume that* M *is saturated. If* A *is a finite subset of* M *then* $\mathrm{tp}(A)$ *is determined by* $\mathrm{qftp}(\mathrm{cl}(A))$ *of the closure* $\mathrm{cl}(A)$ *of* A *in* M.

Suppose X, Y are sets and $j : X \to Y$ a map. For $Z \subseteq [X]^m$, put $j(Z) = \{\{j(x_1), \ldots, j(x_m)\} \mid \{x_1, \ldots, x_m\} \in Z\}$.

Let B, C be $\mathcal{L}$-structures and assume $X \subseteq \mathrm{dom}(B) \cap \mathrm{dom}(C)$.

We write $D = B \rtimes_X C$ if the following hold:

(1) There is an embedding $f : B \to D$.
(2) There is an embedding $g : C \to D$.
(3) If $x \in X$ then $f(x) = g(x)$.
(4) $D = f(B) \cup g(C)$.
(5) $f(B) \cap g(C) = f(X) = g(X)$.
(6) $R(D) = f(R(B)) \cup g(R(C) - R(C|X))$.

B and $D|f(\mathrm{dom}(B))$ are isomorphic as $\mathcal{L}$-structures by f, but C and $D|g(\mathrm{dom}(C))$ are not necessarily isomorphic as $\mathcal{L}$-structures by g. B is embedded to D as an $\mathcal{L}$-substructure but $C|X$ is not necessarily so.

We have $\delta(B \rtimes_X C) = \delta(B) + \delta(C) - \delta(C|X)$.

Note also that if $A < B$ holds then $A' < A' \rtimes_{\mathrm{dom}(A)} B$ holds for any $\mathcal{L}$-structure A' with $\mathrm{dom}(A') = \mathrm{dom}(A)$.

If $A = B|X = C|X$, we write $B \rtimes_X C$ as $B \otimes_X C$, or $B \otimes_A C$. $B \otimes_A C$ is called a *free amalgam of* B *and* C *over* A.

Definition 1.3. Suppose $\mathbf{K} \subseteq \mathbf{K}_1$. $\mathbf{K}$ has the *free amalgamation property* if for any $A, B, C \in \mathbf{K}$, whenever $A < B$ and $A < C$ then $B \otimes_A C \in \mathbf{K}$.

Note that if $A < B$ and $A < C$ then $B, C < B \otimes_A C$. Therefore, if a class $\mathbf{K}$ has the free amalgamation property then it has the amalgamation property.

Definition 1.4. Suppose $h : \mathbf{R}^+ \to \mathbf{R}^+$ is a monotone increasing concave (convex upward) unbounded function. Assume that $h(0) \leq 0$, and $h(1) \leq 1$.

$h(x) = \log_3(x+1)$ and $h(x) = \log_2(x+2) - 1$ are examples. Define $\mathbf{K}_h$ as follows:

$$\mathbf{K}_h = \{A \in \mathbf{K}_1 \mid B \subseteq A \Rightarrow \delta(B) \geq h(|B|)\}.$$

The following is the main theorem.

Theorem 1.1. *Suppose $h : \mathbf{R}^+ \to \mathbf{R}^+$ is a monotone increasing concave unbounded function. Assume that $h(0) \leq 0$, $h(1) \leq 1$, $h(3) \leq 2$ and $h(x) + 1 \geq h(2x+2)$ for any positive integer x.*

Then $(\mathbf{K}_h, <)$ has the free amalgamation property.

If M is a generic structure for $(\mathbf{K}_h, <)$ then $Th(M) = \{\varphi \in \mathcal{L} \mid M \models \varphi\}$ is model complete.

In the rest of the paper, we assume that the assumption of Theorem 1.1 holds:

Assumption 1.1.

(1) $h : \mathbf{R}^+ \to \mathbf{R}^+$ is a monotone increasing concave unbounded function.
(2) $h(0) \leq 0$, $h(1) \leq 1$, and $h(3) \leq 2$.
(3) $h(x) + 1 \geq h(2x+2)$ for any positive integer x.

Proposition 1.2. *$(\mathbf{K}_h, <)$ has the free amalgamation property.*

Proof. Let A, B, C be members of $\mathbf{K}_h$ such that $A < B$ and $A < C$. Let $D = B \otimes_A C$. We can assume that $B \subseteq D$, $C \subseteq D$ and $B \cap C = A$. Suppose $U \subseteq D$. If $U \subseteq B$ or $U \subseteq C$ then $U \in \mathbf{K}_h$ by $B, C \in \mathbf{K}_h$.

Assume $B \subsetneq U$ and $C \subsetneq U$. We have $U = U \cap B \otimes_A U \cap C$, $\delta(U \cap B) > \delta(A)$, and $\delta(U \cap C) > \delta(A)$.

By symmetry, we can also assume that $|U \cap C| \leq |U \cap B|$. Hence, $|U| \leq 2|U \cap B|$.

We have $\delta(U) = \delta(U \cap B) + \delta(U \cap C) - \delta(A) > \delta(U \cap B)$. Hence,

$$\begin{aligned}\delta(U) &\geq \delta(U \cap B) + 1 \\ &\geq h(|U \cap B|) + 1 \\ &\geq h(2(|U \cap B| + 1)) \\ &\geq h(|U|).\end{aligned}$$

Therefore, $D \in \mathbf{K}_h$. □

2. Extensions by T-Paths and T-Cycles

From now on, we just write "structure" for an "$\mathcal{L}$-structure."

Definition 2.1. Let $D(a, b, b')$ be an structure with domain $\{a, b, b'\}$ and one R-relation. We call $D(a, b, b')$ a *triangle.*

Let $\{b_0, b_1, \ldots, b_k\}$ be a set of $k+1$ distinct points and $\{a_0, a_1, \ldots, a_{k-1}\}$ a set of k distinct points. Assume further that $a_i \neq b_j$ for any i, j. An structure

$$D(a_0, b_0, b_1) \otimes_{b_1} D(a_1, b_1, b_2) \otimes_{b_2} \cdots \otimes_{b_{k-1}} D(a_{k-1}, b_{k-1}, b_k)$$

is called a *t-path.* We call k the *length* of the t-path, and $\{a_0, a_1, \ldots, a_{k-1}\}$ the set of *feet.* By identifying b_0 and b_k, we get a cycle called a *t-cycle.* In this case, k is called the length of the t-cycle and $\{a_0, a_1, \ldots, a_{k-1}\}$ the set of *feet* also. Note that a t-cycle has length at least 2.

For a triangle, choosing two points from the domain, we call it the set of feet.

Definition 2.2. Let A be an structure. B is an *extension of A by a t-path* if $B = A \rtimes_X D$ for some t-path D with the set of feet X.

B is an *extension of A by a t-cycle* if $B = A \rtimes_X C$ for some t-cycle C with the set of feet X.

B is an *extension of A by a triangle* if $B = A \rtimes_X C$ for some triangle C with a two point subset X of $\mathrm{dom}(C)$.

B is an *extension of A by an independent point* if $B = A \otimes P$ for some one point structure P.

Note that if B is an extension of A by a t-cycle or by a triangle then B is a 0-extension of A, i.e., $\delta(B/A) := \delta(B) - \delta(A) = 0$.

Definition 2.3. B is an *p-extension* of A if there is a sequence $\{U_i \mid 0 \leq i \leq l\}$ of structures such that $U_0 = A$, U_i is an extension of U_{i-1} by a t-path or by a independent point for $i \geq 1$, and $U_l = B$.

The following two lemmas are immediate.

Lemma 2.1.

(1) If B is an extension of A by an independent point then $|B| = |A|+1$ and $\delta(B) = \delta(A) + 1$.

(2) If B is an extension of A by a t-path then $|B| \leq 2|A| + 1$ and $\delta(B) = \delta(A) + 1$.

(3) *If B is an extension of A by a t-cycle then* $A \leq B$, $|B| \leq 2|A|$ *and* $\delta(B) = \delta(A)$.

(4) *If B is an extension of A by a triangle then* $A \leq B$, $|B| = |A| + 1$ *and* $\delta(B) = \delta(A)$.

Lemma 2.2. *Suppose B is a p-extension of A and* $A \subsetneq B$.

(1) $A < B$.

(2) *If* $A \in \mathbf{K}_h$ *then* $B \in \mathbf{K}_h$.

(3) *If* $A \subseteq A'$ *then* $A' \otimes_A B$ *is a p-extension of* A'.

(4) *If* B_1 *and* B_2 *are p-extensions of A then* $B_1 \otimes_A B_2$ *is a p-extension of A.*

Lemma 2.3. *Let C be a t-cycle or a triangle with the set of feet X, A a structure such that* $X \subseteq \mathrm{dom}(A)$. *Suppose* $U \subseteq A_X \rtimes C$. *If* $\mathrm{dom}(C) \not\subseteq \mathrm{dom}(U)$, *then U is a p-extension of* $U \cap A$.

Proof. If C is a t-cycle or a triangle, it breaks up into t-paths and independent points in U.

Note that if C is a triangle, then $U = (U \cap A)$ or $U = (U \cap A) \otimes P$ for some one-point structure P. □

Lemma 2.4. *Suppose* $A \in \mathbf{K}_h$. *Let P be a one-point structure with* $\mathrm{dom}(P) = \{p\}$. *Let C be a t-cycle with the set of feet* $X \cup \{p\}$ *for some* $X \subseteq \mathrm{dom}(A)$ *or a triangle (3 points with one 3-hyperedge) with* $\mathrm{dom}(C) = \{p, x, y\}$, $X = \{x\} \subseteq A$, *and* $y \notin \mathrm{dom}(A) \cup \{p\}$. *Put* $D = (A \otimes P) \rtimes_{X \cup \{p\}} C$. *Then* $D \in \mathbf{K}_h$ *and* $A < D$.

Proof. Suppose $U \subseteq D$.

Consider the case $\mathrm{dom}(C) \subseteq \mathrm{dom}(U)$. Then $U = ((U \cap A) \otimes P) \rtimes_{X \cup \{p\}} C$. Since U is a extension of $(U \cap A) \otimes P$ by a t-cycle or by a triangle, $|U| \leq 2(|U \cap A| + 1)$ by Lemma 2.1 (3). Hence, $\delta(U) = \delta(U \cap A) + 1 \geq h(|U \cap A|) + 1 \geq h(2(|U \cap A| + 1)) \geq h(|U|)$. In the case $A \subseteq U$, we have $U = D$. $\delta(U) = \delta(D) = \delta(A \otimes P) = \delta(A) + 1 > \delta(A)$.

Now, consider the case $\mathrm{dom}(C) \not\subseteq \mathrm{dom}(U)$.

By Lemma 2.3, U is a p-extension of $U \cap (A \otimes P)$. Since $A \otimes P \in \mathbf{K}_h$, $U \cap (A \otimes P) \in \mathbf{K}_h$. Therefore, $U \in \mathbf{K}_h$ by Lemma 2.2 (2). Since $U \cap (A \otimes P) < U$, we have $A < U$ if $A \subseteq U$. □

Lemma 2.5. *Suppose* $A \in \mathbf{K}_h$. *Let P be a one-point structure with* $\mathrm{dom}(P) = \{p\}$, *and m a natural number. Let* C_0 *be a t-cycle with the set of feet* $\mathrm{dom}(A) \cup \{p\}$ *and* C_1 *a t-cycle or a triangle with the set of feet*

$X \cup \{p\}$ for some $X \subseteq \mathrm{dom}(A)$. Put $D_i = (A \otimes P) \rtimes_{\mathrm{dom}(A) \cup \{p\}} C_0$ for $i < m$, $D_m = (A \otimes P) \rtimes_{X \cup \{p\}} C_1$, and $D = D_1 \otimes_{A \otimes P} D_2 \otimes_{A \otimes P} \cdots \otimes_{A \otimes P} D_m$.

If $\delta(D) \geq h(|D|)$ then $D \in \mathbf{K}_h$ and $A < D$.

Proof. We prove the lemma by induction on m. In case $m = 1$, the lemma holds by Lemma 2.4.

Suppose $m > 1$ and $\delta(D) \geq h(|D|)$. Note that $\delta(D) = \delta(A) + 1 > \delta(A)$. It is enough to show that if $U \subsetneq D$ then $U \in \mathbf{K}_h$ and $A \subseteq U$ implies $A < U$.

Suppose $U \subsetneq D$.

We can assume that $\mathrm{dom}(C_1)$ is identical to its isomorphic image in D_m.

Case $\mathrm{dom}(C_1) \subseteq \mathrm{dom}(U)$. We can argue this case at once, but we divide the case into subcases to make it easier to understand.

Subcase $A \otimes P \not\subseteq U$. In this case, $U \cap D_i$ is a p-extension of $U \cap (A \otimes P)$ for $i \leq m - 1$ by Lemma 2.3. Also, $U \cap D_m = (U \cap A) \rtimes_{X \cup \{p\}} C_1 \in \mathbf{K}_h$ by Lemma 2.4. Therefore, U is a p-extension of $U \cap D_m$ and $U \in \mathbf{K}_h$ by Lemma 2.2. In this case, $A \not\subseteq U$.

Subcase $A \otimes P \subseteq U$. In this case, $U \cap D_m = D_m$. Let $I = \{i \mid D_i \subseteq U\} = \{i_1, \ldots, i_k\}$ where $i_k = m$. Since $U \subsetneq D$, $|I| < m$. Put

$$D'' = D_{i_1} \otimes_{A \otimes P} \cdots \otimes_{A \otimes P} D_{i_k}.$$

Then $D'' \in \mathbf{K}_h$ and $A < D''$ by the induction hypothesis. Since $U \cap D_j$ is a p-extension of $A \otimes P$ for each $j \notin I$, U is a p-extension of D''. Hence $U \in \mathbf{K}_h$ and $D'' < U$. Thus, $A < U$.

Case $\mathrm{dom}(C_1) \not\subseteq \mathrm{dom}(U)$.

By the induction hypothesis, $D' = D_1 \otimes_{A \otimes P} D_2 \otimes_{A \otimes P} \cdots \otimes_{A \otimes P} D_{m-1}$ belongs to $\mathbf{K}_h$ and $A < D'$. Hence $U \cap D' \in \mathbf{K}_h$. Also, $U \cap D_m$ is a p-extension of $U \cap (A \otimes P)$ by Lemma 2.3. Therefore, U is a p-extension of $U \cap D'$ and thus $U \in \mathbf{K}_h$ and $U \cap D' < U$ by Lemma 2.2. So, we also have $A < U$ if $A \subseteq U$. □

3. Model Completeness

Definition 3.1. An structure B is *critical* for h if $\delta(B) \geq h(|B|)$ but $\delta(B) < h(|B| + 1)$. B is also called h-critical.

The following is the main proposition.

Proposition 3.1. *Suppose $A \in \mathbf{K}_h$. Then there is a h-critical structure $B \in \mathbf{K}_h$ such that $A < B$.*

Proof. If $|A| = 1$ then $A < A$, $A \in \mathbf{K}_h$, and A is critical for h. We can assume that $|A| \geq 2$. We are going to construct a critical structure B such that $A < B$ with $\delta(B) = \delta(A) + 1$.

Let P be a one-point structure, and p the unique point of P. Let N be the maximum value of a integer x such that $\delta(A \otimes P) \geq h(x)$. Suppose $N = m \cdot (|A| + 1) + r$ with $0 \leq r < |A| + 1$.

Let C be a t-cycle of length $|A| + 1$. Let C' be a t-cycle of length r if $r \geq 2$. If $r = 1$ then let C' be a triangle.

Put $D_i = (A \otimes P) \rtimes_{\mathrm{dom}(A) \cup \{p\}} C$ for $i = 1, \ldots, m$ and $D_{m+1} = (A \otimes P) \rtimes_{X \cup \{p\}} C'$ where $X \cup \{p\}$ is the set of feet of C' with $X \subseteq A$.

Then put $B = D_1 \otimes_{A \otimes P} D_2 \otimes_{A \otimes P} \cdots \otimes_{A \otimes P} D_m \otimes_{A \otimes P} D_{m+1}$.

By Lemma 2.5, $B \in \mathbf{K}_h$ and $A < B$. □

Now, we can prove the main theorem.

Proof of Theorem 1.1. Let M be a generic structure for $(\mathbf{K}_h, <)$.

Let A be a finite subset of $\mathrm{dom}(M)$. It is enough to show that $\mathrm{tp}(A)$ is isolated by an existential formula. Let $\mathrm{cl}(A)$ be the closure of A in M. $\mathrm{cl}(A)$ is also finite. By Proposition 3.1, there is an h-critical structure $B \in \mathbf{K}_h$ such that $\mathrm{cl}(A) < B$. Since M is a generic structure for $(\mathbf{K}_h, <)$, there is a closed $\mathcal{L}$-embedding of B into M over $\mathrm{cl}(A)$. We can assume that $\mathrm{cl}(A) < B < M$. There is an existential formula in $\mathrm{tp}(A)$ saying that "$\exists X$ *s.t.* $X \cong_A B$." Since $\mathrm{qftp}(B)$ determines $\mathrm{qftp}(\mathrm{cl}(A))$, thus it determines $\mathrm{tp}(A)$ as well.

Therefore, $Th(M)$ is model complete. □

4. Examples

We need some more definitions for arguments in this section.

Definition 4.1. Let M be a generic structure for $(\mathbf{K}_h, <)$. For a finite subset A of M, put

$$d(A) = \delta(\mathrm{cl}(A)).$$

For finite subset A, B of M, put

$$d(A/B) = d(A \cup B) - d(B).$$

Definition 4.2. Let M be a generic structure for $(\mathbf{K}_h, <)$. Let A, B, C be finite subsets of M. A and B are *d-independent over* C if $d(A/B \cup C) = d(A/C)$.

Proposition 4.1. *Put $h(x) = \log_3(x+1)$. Then $\mathbf{K}_h$ has the free amalgamation property and the generic structure for $(\mathbf{K}_h, <)$ has a model complete theory which is supersimple of SU-rank 1.*

$h(x) = \log_3(x+1)$ satisfies Assumption 1.1. Model completeness follows from Theorem 1.1. The proof for the rest of the property is in Wagner[14] and in Kim[12]. The proof there are for $h(x) = \log_3(x) + 1$, but they are essentially the same for this case.

Proposition 4.2. *Put $h(x) = \log_2(x+2) - 1$. Then $\mathbf{K}_h$ has the free amalgamation property and the generic structure for $(\mathbf{K}_h, <)$ has a model complete theory which is unsimple.*

Proof. $h(x) = \log_2(x+2) - 1$ satisfies Assumption 1.1. Model completeness follows from Theorem 1.1.

To show that the generic structure for $(\mathbf{K}_h, <)$ has an unsimple theory, we show that $\mathbf{K}_h$ is not closed under an independence theorem diagram (ITD)[7]. We refer to Evans and Wong[7] for the definition and facts about an ITD.

Consider the structure D with the domain

$$\{a, b, c, p_1, p_2, p_3, p_4, q_1, q_2, q_3, q_4, r_1, r_2, r_3, r_4\}$$

and the R-relations

$$\begin{gathered}\{a,b,p_1\}, \{a,b,p_2\}, \{a,b,p_3\}, \{a,b,p_4\},\\ \{b,c,q_1\}, \{b,c,q_2\}, \{b,c,q_3\}, \{b,c,q_4\},\\ \{c,a,r_1\}, \{c,a,r_2\}, \{c,a,r_3\}, \{c,a,r_4\}.\end{gathered}$$

Put

$$\begin{aligned}&D_0 = \emptyset, \quad D_1 = D|\{a\}, \quad D_2 = D|\{b\}, \quad D_3 = D|\{c\},\\ &D_{12} = \mathrm{cl}_D(a,b) = D|\{a,b,p_1,p_2,p_3,p_4\},\\ &D_{23} = \mathrm{cl}_D(b,c) = D|\{b,c,q_1,q_2,q_3,q_4\}, \text{ and}\\ &D_{13} = \mathrm{cl}_D(c,a) = D|\{c,a,r_1,r_2,r_3,r_4\}.\end{aligned}$$

By inspection, D_i for $i = 0,1,2,3,4$ and D_{ij} for $1 \le i < j \le 3$ belong to $\mathbf{K}_h$. e.g. $\delta(D_{12}) = 2$, $|D_{12}| = 6$, and $h(6) = \log_2(6+2) - 1 = 2$. On the other hand, $\delta(D) = 3$, $|D| = 15$, but $h(15) = \log_2(17) - 1 > 4 - 1 = 3$. Hence, (D, D_i, D_{ij}) is an ITD but $D \notin \mathbf{K}_h$. Therefore, the generic structure for $(\mathbf{K}_h, <)$ has an unsimple theory. □

References

1. J.T. Baldwin, Problems on pathological structures, In Helmut Wolter, Martin Weese, editor, *Proceedings of 10th Easter Conference in Model Theory* **1993**, 1–9 (1993).
2. J.T. Baldwin, An almost strongly minimal non-Desarguesian projective plane, Trans. Am. Math. Soc. **342**, 695–711 (1994).
3. J.T. Baldwin, A field guide to Hrushovski's constructions, http://www.math.uic.edu/jbaldwin/pub/hrutrav.pdf (2009).
4. J.T. Baldwin and K. Holland, Constructing ω-stable structures: model completeness, Ann. Pure Appl. Log. **125**, 159–172 (2004).
5. J.T. Baldwin and S. Shelah, Randomness and semigenericity, Trans. Am. Math. Soc. **349**, 1359–1376 (1997).
6. J.T. Baldwin and N. Shi, Stable generic structures, Ann. Pure Appl. Log. **79**, 1–35 (1996).
7. D.M. Evans and M.W.H. Wong, Some remarks on generic structures, J. Symb. Log. **74**, 1143–1154 (2009).
8. K. Holland, Model completeness of the new strongly minimal sets, J. Symb. Log. **64**, 946–962 (1999).
9. E. Hrushovski, A stable $\aleph_0$-categorical pseudoplane, preprint (1988).
10. E. Hrushovski, A new strongly minimal set, Ann. Pure Appl. Log. **62**, 147–166 (1993).
11. K. Ikeda, Simplicity of generic structures (in Japanese), Kokyuroku of RIMS **1390** (Kyoto University), 9-17 (2004).
12. B. Kim, *Simplicity Theory*, Oxford (2014).
13. F.O. Wagner, Relational structures and dimensions, In *Automorphisms of first-order structures*, Clarendon Press, Oxford, 153–181 (1994).
14. F.O. Wagner, *Simple Theories*, Kluwer (2000).

On Categorical Relationship among Various Fuzzy Topological Systems, Fuzzy Topological Spaces and Related Algebraic Structures

Purbita Jana
Department of Pure Mathematics, University of Calcutta, Kolkata, West Bengal 711106, India
E-mail: purbita_presi@yahoo.co.in
www.caluniv.ac.in

Mihir K. Chakraborty
Indian Statistical Institute & Jadavpur University
Kolkata, West Bengal, India
E-mail: mihirc4@gmail.com

This paper generalizes the categorical relationship among the topological systems, topological spaces and frames as shown by Vickers. Here similar relationship among fuzzy topological systems (on fuzzy sets), fuzzy topological spaces (on fuzzy sets) and frames has been obtained. Connections of this categorical study with future development of logics are indicated.

Keywords: Topological system; Fuzzy topological system; Fuzzy topological space; Category theory; Logic of finite observations.

1. Introduction

Fuzzy sets (Zadeh, 1965) [15] represent concepts in which belongingness of objects has several possible degrees rather than having only two: viz. 1 (belonging to) and 0 (not belonging to). Almost from the advent of fuzzy set theory the concept of fuzzy topological spaces were introduced and developed [3,4,6,8,10,11]. At the beginning the concept of fuzzy topology was developed on crisp sets. Afterwards in 1992 M.K. Chakraborty and T.M.G. Ashannullah proposed the concept of fuzzy topological spaces on fuzzy sets [3]. In that paper the advantages of this kind of extension had been discussed.

Steve Vickers in 1989 introduced the concept of topological system [14], which is defined as follows:

Definition 1.1. A **topological system** is a triple $(X, \models, A)$ where X is a nonempty set, A is a frame and $\models$, is a binary relation (i.e. $\models \subseteq X \times A$), satisfying the following conditions.

(1) if S is a finite subset of A, then $x \models \bigwedge S \Leftrightarrow x \models a$ for all $a \in S$,
(2) if S is any subset of A, then $x \models \bigvee S \Leftrightarrow x \models a$ for some $a \in S$.

Definition 1.2. A **frame** is a partially ordered set such that

(1) every subset has a join,
(2) every finite subset has a meet, and
(3) binary meets distribute over arbitrary joins: i.e. $x \wedge \bigvee Y = \bigvee\{x \wedge y : y \in Y\}$.

The relation $\models$ is called the satisfaction relation.

Vickers [14] established an interrelation among topological systems, topological spaces and frames. More precisely the categorical relationships among categories of topological systems, topological spaces and frames are established.

This study is linked with the logic of finite observations or geometric logic [14] in that the satisfaction relation matches with this logic.

The notion of satisfaction may, however, be graded [2]. It is reasonable to assume that in some situation x satisfies a to some extent or degree i.e. the satisfaction relation $\models$ is a fuzzy relation [15]. Consequently we arrived at the notion of fuzzy topological system [9]. The extent to which x satisfies a shall be denoted by $gr(x \models a)$ which is an element of some suitable value set. In our earlier paper [9] the notion of fuzzy topological system has been defined in the following way.

Definition 1.3 (Fuzzy Topological System). A **fuzzy topological system** is a triple $(X, \models, A)$, where X is a set, A is a frame and $\models$ is a $[0, 1]$- fuzzy relation from X to A such that

(1) if S is a finite subset of A, then $gr(x \models \bigwedge S) = inf\{gr(x \models s) : s \in S\}$,
(2) if S is any subset of A, then $gr(x \models \bigvee S) = sup\{gr(x \models s) : s \in S\}$.

This notion in a different name and in more generalized context has been defined by some other researchers almost simultaneously and independently[7,12,13].

In Ref. 9, a category of fuzzy topological system is defined to establish a relationship among categories of fuzzy topological systems, fuzzy topological spaces[6] and frames.

In this paper our aim is to generalize the above results obtained in Ref. 9. Instead of fuzzy topological spaces on ordinary sets, fuzzy topological spaces on fuzzy sets[3] is considered. That these objects form category was established in Ref. 4. We shall first define fuzzy topological systems with respect to fuzzy sets and show that they form a category with respect to proper morphisms. Then we will establish the categorical relationship among the three categories viz. category of fuzzy topological systems with respect to fuzzy sets ($\mathscr{F}$-**TopSys**), category of fuzzy topological spaces on fuzzy sets ($\mathscr{F}$-**Top**) and category of frames (**Frm**).

This paper is organized as below:
in section 2 necessary preliminaries are presented, section 3 contains the main definitions and categorical results. In section 4 as concluding remarks we have included some future directions of our research which will link the categorical study with many valued and fuzzy logics of finite observations.

2. Preliminaries

The basic notions of category theory are mostly taken from Ref. 1. Let $G : \mathbb{A} \longrightarrow \mathbb{B}$ be a functor, and let B be a $\mathbb{B}$-object.

Definition 2.1. For any category $\mathbb{A} = (O, hom_{\mathbb{A}}, id, \circ)$ the **dual** (or **opposite**) category of $\mathbb{A}$ is the category $\mathbb{A}^{op} = (O, hom_{\mathbb{A}^{op}}, id, \circ^{op})$ where $hom_{\mathbb{A}^{op}}(A, B) = hom_{\mathbb{A}}(B, A)$ and $f \circ^{op} g = g \circ f$. (Thus $\mathbb{A}$ and $\mathbb{A}^{op}$ have the same objects and,except for their direction, the same morphisms).

Definition 2.2 (G-structured arrow and G-costructured arrow).

(1) A G-**structured arrow** with domain B is a pair (f, A) consisting of an $\mathbb{A}$-object A and a $\mathbb{B}$-morphism $f : B \longrightarrow GA$.
(2) A G-**costructured arrow** with codomain B is a pair (A, f) consisting of an $\mathbb{A}$-object A and a $\mathbb{B}$-morphism $f : GA \longrightarrow B$.

Definition 2.3 (G-universal arrow and G-couniversal arrow).

(1) A G-structured arrow (g, A) with domain B is called G-**universal** for B provided that for each G-structured arrow (g', A') with domain B there exists a unique $\mathbb{A}$-morphism $\hat{f} : A \longrightarrow A'$ with $g' = G(\hat{f}) \circ g$ i.e., s.t. the triangle

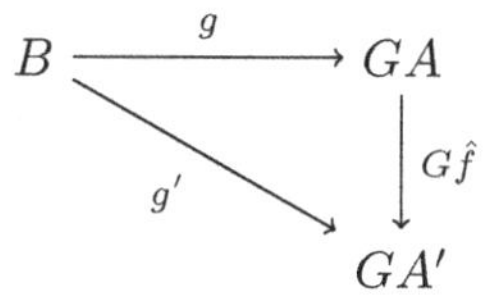

commutes.

(2) A G-costructured arrow (A, g) with codomain B is called G-**couniversal** for B provided that for each G-costructured arrow (A', g') with codomain B there exists a unique $\mathbb{A}$-morphism $\hat{f}$: $A' \longrightarrow A$ with $g' = g \circ G(\hat{f})$ i.e., s.t. the triangle

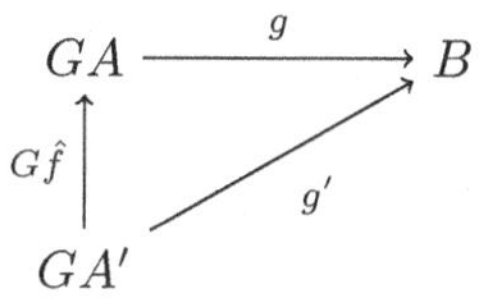

commutes.

Definition 2.4 (Left Adjoint and Right Adjoint).

(1) A functor $G : \mathbb{A} \longrightarrow \mathbb{B}$ is said to be **left adjoint** provided that for every $\mathbb{B}$-object B there exists a G-couniversal arrow with codomain B.
i.e. there exist a natural transformation $\eta : A \longrightarrow FGA$ where $F : \mathbb{B} \longrightarrow \mathbb{A}$ is a functor, s.t. for given $f : A \longrightarrow FB$ there exist a unique $\mathbb{B}$-morphism $\hat{f} : GA \longrightarrow B$ s.t. the triangle

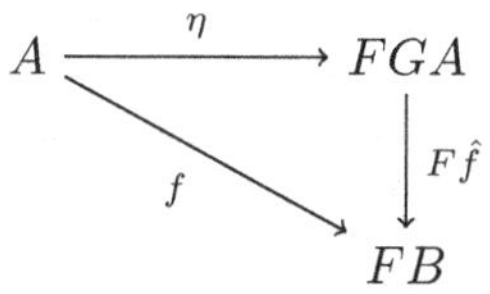

commutes. This η is called the unit of the adjunction. Hence we have the diagram of unit as follows:

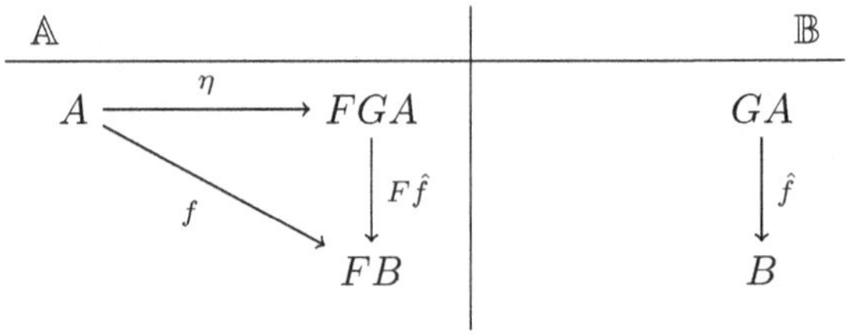

The diagram above indicates the fact that $\eta : A \longrightarrow FGA$ is the F-universal arrow provided that for given $f : A \longrightarrow FB$ there exist a unique $\mathbb{B}$-morphism $\hat{f} : GA \longrightarrow B$ s.t. the triangle commutes.

(2) A functor $G : \mathbb{A} \longrightarrow \mathbb{B}$ is said to be **right adjoint** provided that for every $\mathbb{B}$-object B there exists a G-universal arrow with domain B.
From the definition above it follows that there exist a natural transformation $\xi : FGA \longrightarrow A$, where $F : \mathbb{B} \longrightarrow \mathbb{A}$ is a functor s.t. for given $f' : FB \longrightarrow A$ there exist a unique $\mathbb{B}$-morphism $\hat{f} : B \longrightarrow GA$ s.t the triangle

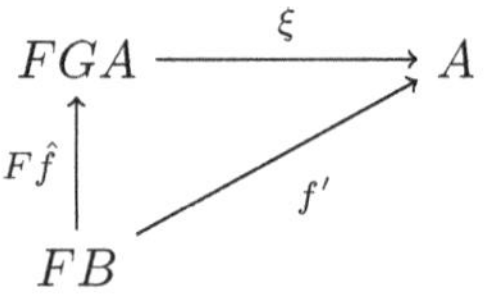

commutes. This ξ is called the co-unit of the adjunction. Hence we have the diagram of co-unit as follows:

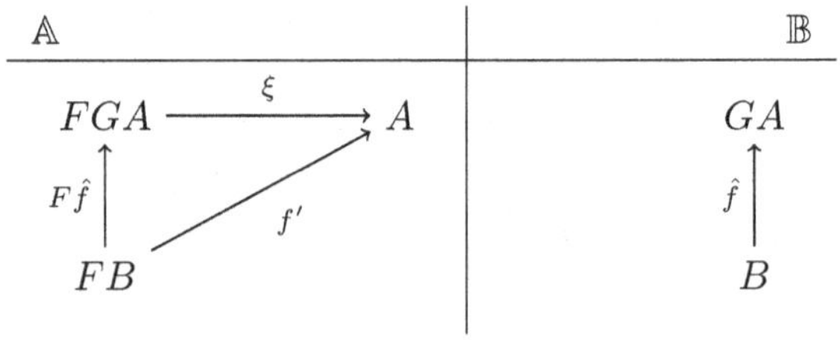

The diagram above indicates the fact that $\xi : FGA \longrightarrow A$ is the F-couniversal arrow provided that for given $f' : FB \longrightarrow A$ there exist a unique $\mathbb{B}$-morphism $\hat{f} : GA \longrightarrow B$ s.t. the triangle commutes.

3. Relationships among $\mathscr{F}$-TopSys, $\mathscr{F}$-Top and Frm

Definition 3.1. A **$\mathscr{F}$-topological system** is a quadruple $(X, \tilde{A}, \models, P)$, where $(X, \tilde{A})$ is a non-empty fuzzy set, P is a frame and $\models$ is a $[0, 1]$- fuzzy relation from X to P such that

(1) $gr(x \models p) \leq \tilde{A}(x)$,
(2) if S is a finite subset of P, then $gr(x \models \bigwedge S) = inf\{gr(x \models s) : s \in S\}$,
(3) if S is any subset of P, then $gr(x \models \bigvee S) = sup\{gr(x \models s) : s \in S\}$.

Note 1: Because of condition 1, $\models$ is a fuzzy relation on the fuzzy set $(X, \tilde{A})$[5].

Note 2: The notion of topological system introduced in Ref. 14 was defined by crisp set and crisp relation whereas the notion of fuzzy topological system defined in Ref. 9 consists of crisp set and fuzzy relation. In our new setting the notion of $\mathscr{F}$-topological system is defined by fuzzy set and fuzzy relation.

The notion of continuous map between these $\mathscr{F}$-topological systems is defined as follows:

Definition 3.2. Let $D = (X, \tilde{A}, \models, P)$ and $E = (Y, \tilde{B}, \models', Q)$ be $\mathscr{F}$ topological systems. A **continuous map** $f : D \longrightarrow E$ is a pair (f_1, f_2) where,

(1) $f_1 : (X, \tilde{A}) \longrightarrow (Y, \tilde{B})$ is a proper function (Definition 3.3) from $(X, \tilde{A})$ to $(Y, \tilde{B})$,
(2) $f_2 : Q \longrightarrow P$ is a frame homomorphism and
(3) $gr(x \models f_2(q)) = gr(f_1(x) \models' q)$, for all $x \in X$ and $q \in Q$.

Definition 3.3. [4] f_1 is a **proper function** from $(X, \tilde{A})$ to $(Y, \tilde{B})$. So f_1 is a relation from $(X, \tilde{A})$ to $(Y, \tilde{B})$ s.t. $\forall x \in \mid \tilde{A} \mid$, $\exists$ *unique* $y \in \mid \tilde{B} \mid$ for which $\tilde{A}(x) = f_1(x, y)$ and $f_1(x, y') = 0$ if $y' \neq y \in \mid \tilde{B} \mid$, where $\mid \tilde{A} \mid = \{x \in X : \tilde{A}(x) > 0\}$ and $\mid \tilde{B} \mid = \{y \in Y : \tilde{B}(y) > 0\}$. For a fixed $x \in \mid \tilde{A} \mid$, we will denote that *unique* $y \in \mid \tilde{B} \mid$ by $f_1(x)$.

Let us define identity map and composition of two maps as follows:

Definition 3.4. Let $D = (X, \tilde{A}, \models, P)$ be a $\mathscr{F}$ topological system. The **identity map** $I_D : D \longrightarrow D$ is a pair (I_1, I_2) defined by $I_1 : (X, \tilde{A}) \longrightarrow (X, \tilde{A})$ s.t. $I_1(x_1, x_2) = \tilde{A}(x)$ iff $x_1 = x_2$, otherwise $I_1(x_1, x_2) = 0$ and $I_2 : P \longrightarrow P$ *is identity morphism of* P. Let $D = (X, \tilde{A}, \models', P)$, $E = (Y, \tilde{B}, \models'', Q)$, $F = (Z, \tilde{C}, \models''', R)$. Let $(f_1, f_2) : D \longrightarrow E$ and $(g_1, g_2) : E \longrightarrow F$ be continuous maps. The **composition** $(g_1, g_2) \circ (f_1, f_2) : D \longrightarrow F$ is defined by $g_1 \circ f_1 : (X, \tilde{A}) \longrightarrow (Z, \tilde{C})$, $f_2 \circ g_2 : R \longrightarrow P$ i.e. $(g_1, g_2) \circ (f_1, f_2) = (g_1 \circ f_1, f_2 \circ g_2)$.

Remark: In fact we can show that the identity map and the composition of two continuous maps are indeed continuous maps. Hence we propose the category $\mathscr{F}$-**TopSys** whose objects are $\mathscr{F}$-topological systems and the morphisms are the above mentioned continuous maps. Thus we get a category of fuzzy topological systems whose underlying sets are fuzzy sets.

Definition 3.5 ($\mathscr{F}$-topological space). [4] Let $(X, \tilde{A})$ be a fuzzy set and τ a collection of fuzzy subsets of $(X, \tilde{A})$ such that

(1) $(X, \tilde{\varphi})$ and $(X, \tilde{A})$ are in τ, where $\tilde{\varphi} : X \longrightarrow [0,1]$ is such that $\tilde{\varphi}(x) = 0$, for all $x \in X$;
(2) $(X, \tilde{A_1})$, $(X, \tilde{A_2})$ are in τ implies $(X, \tilde{A_1} \cap \tilde{A_2})$ is in τ, where $\tilde{A_1} \cap \tilde{A_2}(x) = \tilde{A_1}(x) \wedge \tilde{A_2}(x)$, for all $x \in X$;
(3) $(X, \tilde{A_i}) \in \tau$ implies $(X, \cup_{i \in I} \tilde{A_i}) \in \tau$, where $\cup_{i \in I} \tilde{A_i} : X \longrightarrow [0,1]$ is such that $(\cup_{i \in I} \tilde{A_i})(x) = \vee_{i \in I} \tilde{A_i}(x)$, for all $x \in X$. Then $(X, \tilde{A}, \tau)$ is a $\mathscr{F}$**-topological space**.

Definition 3.6 ($\mathscr{F}$-Top). [4] The category $\mathscr{F}$-**Top** is defined thus.

- The objects are $\mathscr{F}$-topological spaces on fuzzy sets $(X, \tilde{A}, \tau)$, $(Y, \tilde{B}, \tau')$ etc.
- The morphisms are proper functions satisfying the following property: If $f : (X, \tilde{A}, \tau) \longrightarrow (Y, \tilde{B}, \tau')$ and $(Y, \tilde{B_1})$ is a subset of $(Y, \tilde{B})$ such that $(Y, \tilde{B_1}) \in \tau'$ then $(X, f^{-1}(\tilde{B_1})) \in \tau$.
- The identity on $(X, \tilde{A}, \tau)$ is the identity proper function on the fuzzy set $(X, \tilde{A})$. This is a $\mathscr{F}$-**Top** morphism can be proved.
- If $f : (X, \tilde{A}, \tau) \longrightarrow (Y, \tilde{B}, \tau')$ and $g : (Y, \tilde{B}, \tau') \longrightarrow (Z, \tilde{C}, \tau'')$ are morphisms in $\mathscr{F}$-**Top**, their composition $g \circ f$ is the composition of proper functions between fuzzy sets. It can be verified that $g \circ f$ is a morphism in $\mathscr{F}$-**Top**.

Note that the category $\mathscr{F}$-**Top** is denoted by **Fuzz-Top** in Ref. 3 and Fuz–Top $[0,1]$ in Ref. 4.

Definition 3.7. Frames together with frame homomorphisms form the category **Frm**.

The interrelation among the categories: $\mathscr{F}$-**TopSys**, $\mathscr{F}$-**Top**, **Frm** via some suitable functors are now established. To define the suitable functors at first we propose a notion of extent as follows:

Definition 3.8. Let $(X, \tilde{A}, \models, P)$ be a $\mathscr{F}$- topological system and $p \in P$. For each p, its **extent** in $(X, \tilde{A}, \models, P)$ is given by $ext(p) = (X, ext^*(p))$

where $ext^*(p)$ is a mapping from X to $[0,1]$ given by $ext^*(p)(x) = gr(x \models p)$ for all $x \in X$.
i.e. $ext^*(p) : X \longrightarrow [0,1]$ such that $ext^*(p)(x) = gr(x \models p)$ for all $x \in X$.
Also $ext(P) = \{(X, ext^*(p))\}_{p \in P} = (X, ext^* P)$ where $ext^* P = \{ext^* p\}_{p \in P}$.

Now the functor Ext is defined as follows:

Definition 3.9. **Ext** is a (forgetful) functor from $\mathscr{F}$-**TopSys** to $\mathscr{F}$-**Top** defined thus.
Ext acts on the object $(X, \tilde{A}, \models', P)$ as $Ext(X, \tilde{A}, \models', P) = (X, \tilde{A}, ext(P))$ and on the morphism (f_1, f_2) as $Ext(f_1, f_2) = f_1$.

Next we define another functor J as follows:

Definition 3.10. **J** is a functor from $\mathscr{F}$-**Top** to $\mathscr{F}$-**TopSys** defined thus.
J acts on the object $(X, \tilde{A}, \tau)$ as $J(X, \tilde{A}, \tau) = (X, \tilde{A}, \in, \tau)$ where $gr(x \in \tilde{T}) = \tilde{T}(x)$ for $\tilde{T} \in \tau$ and on the morphism f as $J(f) = (f, f^{-1})$.

To make a connection between $\mathscr{F}$-**TopSys** and **Frm**^{op} (opposite category of **Frm**) we introduce two functors viz. Loc, S in the following ways:

Definition 3.11. **Loc** is a functor from $\mathscr{F}$-**TopSys** to **Frm**^{op} defined thus.
Loc acts on the object $(X, \tilde{A}, \models, P)$ as $Loc(X, \tilde{A}, \models, P) = P$ and on the morphism (f_1, f_2) as $Loc(f_1, f_2) = f_2$.

Definition 3.12. **S** is a functor from **Frm**^{op} to $\mathscr{F}$-**TopSys** defined thus.
S acts on the object P as $S(P) = (Hom(P, [0,1]), \tilde{P}, \models_*, P)$, where
$Hom(P, [0,1]) = \{frame\ hom\ v : P \longrightarrow [0,1]\}$, $gr(v \models_* p) = v(p)$ and $\tilde{P}(v) = \bigvee_{p \in P} v(p)$, and on the morphism f as $S(f) = (\_ \circ f, f)$.

Finally it can be shown that the following theorems hold.

Theorem 3.1. *Ext is the right adjoint to the functor J.*

Proof. We will prove the theorem by presenting the co-unit of the adjunction. Recall that $J(X, \tilde{A}, \tau) = (X, \tilde{A}, \in \tau)$ and $Ext(X, \tilde{A}, \models, P) = (X, \tilde{A}, ext(P))$.
$So, J(Ext(X, \tilde{A}, \models, P)) = (X, \tilde{A}, \in, ext(P))$.
Let us draw the diagram of co-unit-

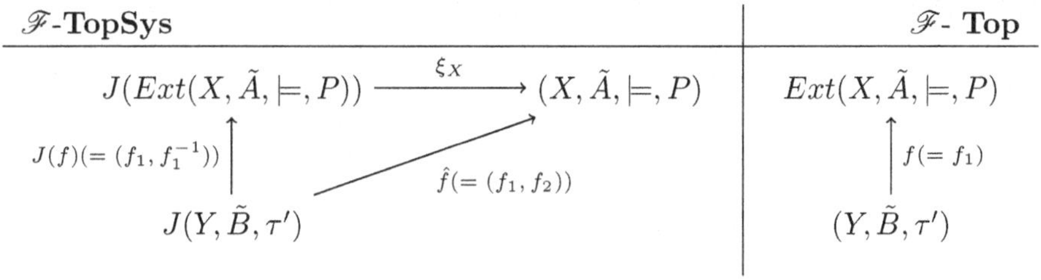

𝒯-TopSys
𝒯- Top
$J(Ext(X,\tilde{A},\models,P))$
ξ_X
$(X,\tilde{A},\models,P)$
$Ext(X,\tilde{A},\models,P)$
$J(f)(=(f_1,f_1^{-1}))$
$\hat{f}(=(f_1,f_2))$
$f(=f_1)$
$J(Y,\tilde{B},\tau')$
$(Y,\tilde{B},\tau')$

Let us define co-unit by $\xi_X = (id_X, ext')$, i.e.

$$(X, \tilde{A}, \subset, ext(P)) \xrightarrow[(id_X, ext')]{\xi_X} (X, \tilde{A}, \models, P)$$

where ext' is a mapping from P to $ext(P)$ such that, $ext'(p) = (X, ext^*(p))$ for all $p \in P$. It can be shown that $(id_X, ext') : J(Ext(X, \tilde{A}, \models, P)) \longrightarrow (X, \tilde{A}, \models, P)$ is indeed a continuous map of $\mathscr{F}$-Topological System as follows.
According to the definition $ext^*(p)(x) = gr(x \models p)$.
Hence $ext'(p)(x) = gr(x \models p)$.
Consequently $gr(x \in ext'(p)) = gr(id_X(x) \models p)$.

Now define f as follows.
Given $(f_1, f_2) : J(Y, \tilde{B}, \tau') \longrightarrow (X, \tilde{A}, \models, P)$, then $f = f_1$.
Now we will prove that the diagram on the left commutes.
Here $J(f) = (f_1, f_1^{-1})$ and $(f_1, f_2) = \xi_X \circ J(f) = (id_X, ext') \circ (f_1, f_1^{-1}) = (id_X \circ f_1, f_1^{-1} \circ ext')$ Clearly $id_X \circ f_1 = f_1$.
Also we have $f_1^{-1} ext'(p) = f_1^{-1}(p) = f_2(p)$. So, $f_2 = f_1^{-1} \circ ext'$.
Hence $\xi_X(= (id_X, ext')) : J(Ext(X, \tilde{A}, \models, P)) \longrightarrow (X, \tilde{A}, \models, P)$ is the co-unit, consequently Ext is the right adjoint to the forgetful functor J. □

Theorem 3.2. *Loc is the left adjoint to the functor S.*

Proof. We will prove the theorem by presenting the unit of the adjunction.

Recall that $S(Q) = (Hom(Q, [0,1]), \tilde{Q}, \models_*, Q)$ where $gr(v \models_* q) = v(q)$.

Hence $S(Loc(X, \tilde{A}, \models, P)) = (Hom(P, [0,1]), \tilde{P}, \models_*, P)$.

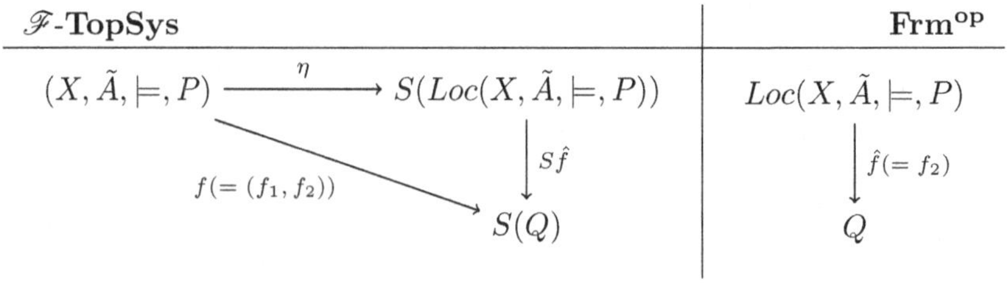

𝒯-TopSys
Frm^op
$(X,\tilde{A},\models,P)$
η
$S(Loc(X,\tilde{A},\models,P))$
$Loc(X,\tilde{A},\models,P)$
$S\hat{f}$
$f(=(f_1,f_2))$
$\hat{f}(=f_2)$
$S(Q)$
Q

Then unit is defined by $\eta = (p*, id_P)$, i.e.

$$(X, \tilde{A}, \models, P) \xrightarrow[(p^*, id_P)]{\eta} S(Loc(X, \tilde{A}, \models, P))$$

where $p^* : (X, \tilde{A}) \longrightarrow (Hom(P, [0,1]), \tilde{P})$ s.t. for any $x \in| \tilde{A} |$, $p^*(x)$ is a mapping from P to $[0,1]$ and $p^*(x)(p) = gr(x \models p)$. We can show that $(p^*, id_P) : (X, \tilde{A}, \models, P) \longrightarrow S(Loc(X, \tilde{A}, \models, P))$ is a continuous map of $\mathscr{F}$-Topological System in the following way.

Here it will be enough to show that $gr(x \models id_P(p)) = gr(p^*(x) \models p)$. We have $gr(x \models p) = p * (x)(p) = gr(p^*(x) \models p)$.

Let us define $\hat{f}$ as follows-
$(f_1, f_2) : (X, \tilde{A}, \models, P) \longrightarrow (Hom(P, [0,1]), \tilde{P}, \models_*, P)$ then $\hat{f} = f_2$[as f_2 is the frame homomorphism].
Recall that $S(\hat{f}) = (- \circ f_2, f_2)$. Now we have to show that the triangle on the left commute.
We have to show that $(f_1, f_2) = (-\circ f_2, f_2)\circ(p^*, id_P) = ((-\circ f_2)p^*, id_P \circ f_2)$
Clearly $f_2 = id_P \circ f_2$.
It is only left to show that $f_1 = (- \circ f_2)p^* = p_x \circ f_2$.
We have for all $q \in Q$

$$\begin{aligned} p^*(x) \circ f_2(q) &= p^*(x)(f_2(q)) \\ &= gr(x \models f_2(q)) \\ &= gr(f_1(x) \models_* q) \\ &= f_1(x)(q). \end{aligned}$$

So, $(\_ \circ f_2)p^* = f_1$. □

Theorem 3.3. *$Ext \circ S$ is the right adjoint to the functor $Loc \circ J$.*

Proof. Proof follows from Theorem 3.1 and Theorem 3.2. □

4. Concluding Remarks

- In this paper the interconnection among the category of fuzzy topological spaces on fuzzy sets[3], category of fuzzy topological system whose underlying sets are fuzzy sets and the category of frames is established.
- It is clear that the category of topological systems (**TopSys**)[14] is a subcategory of the category of fuzzy topological systems ($[0,1]$-**TopSys**)[9], and $[0,1]$-**TopSys** is a subcategory of the category $\mathscr{F}$-**TopSys**.

- It may be observed that the categorical relationship among topological systems, topological spaces and frames [14] is a restricted case of the categorical relationship among fuzzy topological systems, fuzzy topological spaces and frames by choosing proper subcategories and restricted functors. Also the categorical relationship among fuzzy topological systems, fuzzy topological spaces and frames is a restricted case of the categorical relationship among $\mathscr{F}$-topological systems, $\mathscr{F}$-topological spaces and frames by choosing proper subcategories and restricted functors.
- In future we hope to introduce many valued geometric logic. We wish to show that the $[0,1]$-fuzzy relation i.e. the fuzzy satisfaction relation of fuzzy topological system will match the proposed logic. Also we intend to propose a concept of fuzzy geometric logic. We will show that the $[0,1]$-fuzzy relation of $\mathscr{F}$-topological system matches fuzzy geometric logic.

References

1. J. Adámek, G. E. Strecker, H. Herrlich, *Abstract and Concrete Categories* (John Wiley & Sons, ISBN 0-471-60922-6, 1990).
2. M.K. Chakraborty, Graded consequence: further studies, *Journal of Applied Non-classical Logics* **5(2)**, 127 (1995).
3. M.K. Chakraborty and T.M.G. Ahsannullah, Fuzzy topology on fuzzy sets and tolerance topology, *Fuzzy Sets and Systems* **45**, 103 (1992).
4. M.K. Chakraborty and M. Banerjee, A new category for fuzzy topological spaces, *Fuzzy Sets and Systems* **51**, 227 (1992).
5. M.K. Chakraborty and M. Das, Studies in fuzzy relations over fuzzy subsets, *Fuzzy Sets and Systems* **9**, 79 (1983).
6. C.L. Chang, Fuzzy topological spaces, *Journal of Mathematical Analysis and Applications* **24**, 182 (1968).
7. J.T. Denniston, A. Melton, and S.E. Rodabough, Interweaving algebra and topology: Lattice-valued topological systems, *Fuzzy Sets and Systems* **192**, 58 (2012).
8. U. Höhle, Fuzzy topologies and topological space objects in a topos, *Fuzzy Sets and Systems* **19**, 299 (1986).
9. P. Jana and M.K. Chakraborty, Categorical relationships of fuzzy topological systems with fuzzy topological spaces and underlying algebras, *Annals of Fuzzy Mathematica and Informatics* (to appear).
10. R. Löwen, Fuzzy topological spaces and fuzzy compactness, *J. Math.*

Annal. Appl. **56**, 621 (1976).

11. S.E. Rodabaugh, A categorical accommodation of various notions of fuzzy topology, *Fuzzy Sets and Systems* **9**, 241 (1983).
12. Apostolos Syropoulos and Valeria de Pavia, Fuzzy topological systems, *8th Panhellenic Logic Symposium*, Ioannina, Greece (July 2011).
13. S. Solovyov, Categorical foundations of variety-based topology and topological systems, *Fuzzy Sets Syst.* **192**, 176 (2012).
14. Steven J. Vickers, *Topology Via Logic*, volume 5, Cambridge Tracts in Theoretical Computer Science University Press (1989).
15. L.A. Zadeh, Fuzzy sets, *Information and Control* **8**, 338 (1965).

Realizability and Existence Property of a Constructive Set Theory with Types

Farida Kachapova
School of Computer and Mathematical Sciences, Auckland University of Technology, Auckland 1142, New Zealand
E-mail farida.kachapova@aut.ac.nz

This paper describes an axiomatic theory BT with operations and infinitely many types for sets intended for formalising constructive mathematics. In addition to previously known metamathematical properties of BT, we introduce a realizability for BT and use it to prove its existence property. Due to a predicative comprehension axiom, sets in BT can be interpreted recursively as external terms. This method is used in constructing the realizability and a countable set-theoretical model for BT. The latter can be applied to establish further metamathematical properties of the theory BT.

Keywords: realizability; existence property; constructive mathematics; intuitionistic; type theory; predicative

1. Introduction

Formal systems for constructive mathematics were introduced by several authors (see, for example, Refs. 1–4). Axiomatic theories containing operations and sets were introduced by Feferman[5,6] and Beeson[3,7,8]. Each of these theories contains a rather weak comprehension axiom that does not allow to construct sets of high types. Kashapova[9] introduced a theory BT with operations and infinitely many types of sets by generalising the Beeson's theory $BEM + (CA)$; $BEM + (CA)$ makes the bottom fragment of BT with types $\leqslant 1$. Many theorems of classical mathematics can be expressed in the theory BT, due to its rich language.

In Ref. 9 we constructed an interpretation of BT with classical logic in BT that helps to produce constructive versions of some theorems of classical mathematics. In Ref. 9 we also studied metamathematical properties of BT, in particular, we showed that BT is proof-theoretically stronger than the second-order arithmetic with a predicative comprehension axiom, that BT is interpretable in the second-order arithmetic with Δ_1^1 comprehension axiom and that BT is consistent with the following weak form of the formal

Church thesis:

$$(\forall f \in \mathbb{N}^{\mathbb{N}})\, \exists e \forall n (fn \simeq \{e\}(n)).$$

In this paper we construct a realizability for the theory BT and use it to prove the existence property of BT, which is another evidence of the constructive nature of BT. This realizability uses the X^* method introduced by Kreisel and Troelstra[10] and used in Refs. 5,7. Unlike the previous cases, our realizability interprets sets of high types using recursive approach.

In the rest of the Introduction we explain some notations and terminology.

A *logical-mathematical language of first order* (or *logical-mathematical language,* or just *language* in short) is defined as $\Omega = \langle Srt, Cnst, Fn, Pr \rangle$, where

Srt is a non-empty set of sorts of objects, and for each sort $\pi \in Srt$ there is a countable collection of variables of this sort;

$Cnst$ is the set of all constants of the language;

Fn is the set of all functional symbols of the language;

Pr is the set of all predicate symbols of the language.

In the language Ω formulas are constructed from atomic formulas and logical constant $\bot$ using logical connectives $\wedge$ (conjunction), $\vee$ (disjunction) and $\supset$ (implication), and quantifiers $\forall$ and $\exists$. The negation $\neg\varphi$ is an abbreviation for $\varphi \supset \bot$. The logical connective $\equiv$ (equivalence) is defined as $(\varphi \supset \psi) \wedge (\psi \supset \varphi)$. The complexity of a formula φ is the number of occurrences of logical symbols (the main three connectives and quantifiers) in φ.

Next we define an *axiomatic theory* (or just *theory* in short) as $Th = \langle \Omega, l, A \rangle$, where each of the three objects is described as follows.

Ω is a logical-mathematical language.

l is the logic of the theory. We will use only the intuitionistic logic HPC (Heyting's predicate calculus) and the classical logic CPC (classical predicate calculus).

A is some set of closed formulas of the language Ω called the non-logical axioms of Th. When axioms are stated as non-closed formulas, it means that they must be closed by universal quantifiers over all parameters.

The notation $Th \vdash \varphi$ (formula φ is derivable in the theory Th) means that φ is derivable in the logic l from a finite subset of the axiom set A. The theory Th is consistent if it is not true that $Th \vdash \bot$.

In our axiomatic theory variables have indices for types. We will often omit an index of a variable when its value is clear; usually it is the same index as in a previous occurrence of the variable.

We denote $\varphi[x_1, \ldots, x_n/t_1, \ldots, t_n]$ the result of proper substitution of terms $t_1, \ldots, t_n$ instead of variables $x_1, \ldots, x_n$ in the formula φ. We denote $\bar{\varphi}$ the closure of formula φ, that is the formula φ with universal quantifiers over all its parameters. Symbol $\doteq$ denotes the graphical identity of two expressions.

2. Axiomatic Theory *BT*

The language of the theory BT has the following variables:

$m, n, \ldots$ over natural numbers (variables of type ω) and
$X^k, Y^k, Z^k, \ldots$ of type k ($k \geqslant 0$).

We will identify the variables $X^0, Y^0, Z^0, \ldots$ of type 0 with variables $x, y, z, \ldots$, respectively, which we call *operation variables.* We consider the type ω smaller than any other type. Variables of type 0 are interpreted as operations and variables of types $\geqslant 1$ are interpreted as sets.

BT has a numerical constant 0 and the following operation constants:

combinatorial constants $\underline{k}, \underline{s}, \underline{d}, \underline{p}, \underline{p}_1, \underline{p}_2$

and constants $\underline{c}_n$ ($n \geqslant 0$), which are used for constructing sets.

There are no functional symbols in BT.

Predicate symbols:

$Ap(f, x, y)$, $x =_{0\omega} m$, $x =_{0k} Y^k (k \geqslant 0)$, $X^k \in_k Y^{k+1} (k \geqslant 0)$.

$Ap(f, x, y)$ means that y is the result of application of operation f to x.

Atomic formulas are obtained from predicate symbols by substituting constants and variables of corresponding types. Formulas are constructed from atomic formulas and $\bot$ using logical connectives and quantifiers.

The language of BT is defined. A formula of BT is called *n-elementary* if it contains only types $\leqslant n$, no quantifiers over variables of type n and no predicate symbol $=_{0n}$.

External terms are defined recursively as follows.

(1) Every constant and variable is an external term.
(2) If t and τ are external terms, then $t\tau$ is an external term.

$t\tau$ is interpreted as the result of application of operation t to τ. External terms are generally not part of the language BT. The notation $t_1 t_2 t_3 \ldots t_n$ means $(\ldots((t_1 t_2)t_3 \ldots)t_n$.

As usual in combinatorial logic, we consider each operation to have one argument. A function f with n arguments can be written as an operation that is applied n times, i.e. instead of $f(x_1, x_2, \ldots, x_n)$ we write $(\ldots((f(x_1)x_2)\ldots)x_n)$. For example, for addition of two numbers we can use operation f such that $f(m)(n) = m+n$; it is represented with λ-terms: $f(m) = \lambda n.(m+n), f = \lambda m.\lambda n.(m+n)$.

For an external term t we define a relation $t \simeq x$ by induction on the construction of t as follows.

(1) If t is a constant or a variable of type s, then $t \simeq x \leftrightharpoons x =_{0s} t$.
(2) If $t \doteq t_1 t_2$, then $t \simeq x \leftrightharpoons \exists y, z(t_1 \simeq y \wedge t_2 \simeq z \wedge Ap(y, z, x))$.

Some more notations for external terms:

$$t \downarrow \leftrightharpoons \exists x(t \simeq x);$$

$$t \simeq \tau \leftrightharpoons \exists x(t \simeq x \wedge \tau \simeq x);$$

$$t \cong \tau \leftrightharpoons \forall x(t \simeq x \equiv \tau \simeq x).$$

We fix Gödel numbering of all expressions of the language BT and denote $\llcorner q \lrcorner$ the Gödel number of an expression q in this numbering.

The theory BT has the following axioms.

1. Intuitionistic predicate logic.

2. Equality axioms

1) $x =_{00} x$.
2) $u =_{0k} X^k \wedge v =_{0k} X \wedge u =_{0n} Y^n \supset v =_{0n} Y (k \geqslant 0, n \geqslant 0)$.
3) $u =_{0\omega} m \wedge v =_{0\omega} m \supset u =_{00} v$.
4) $u =_{0\omega} m \wedge u =_{00} v \supset v =_{0\omega} m$.
5) $Ap(f, x, y) \wedge f =_{00} g \wedge x =_{00} u \wedge y =_{00} v \supset Ap(g, u, v)$.
6) $X^k \in_k Y^{k+1} \wedge X^k =_k U^k \wedge Y^{k+1} =_{k+1} Z^{k+1} \supset U \in_k Z (k \geqslant 0)$.

3. Combinatorial axioms

1) $Ap(f, x, y) \wedge Ap(f, x, z) \supset y = z$.
2) $\underline{k}xy \simeq x$.
3) $\underline{s}xy \downarrow$. 4) $\underline{s}xyz \cong xz(yz)$.
5) $\underline{p}xy \downarrow$. 6) $\neg(\underline{p}xy \simeq 0)$.
7) $\underline{p}_i x \downarrow (i = 1, 2)$. 8) $\underline{p}_i(\underline{p}x_1x_2) \simeq x_i (i = 1, 2)$.
9) $\exists m(\underline{p}n0 \simeq m)$. 10) $\exists Z^k(\underline{p}xY^k \simeq Z)$.
11) $n = m \supset \underline{d}xynm \simeq x$. 12) $n \neq m \supset \underline{d}xynm \simeq y$.

13) $\exists x(x =_{0\omega} n)$. 14) $\exists x(x =_{0k} Y^k)$.

In BT the successor of a natural number n is given by $\underline{p}n0$.

Any natural number m is represented by a term $\underbrace{\underline{p}(\underline{p}(\ldots(\underline{p}}_{m}0)\ldots))$, which we denote $\bar{m}$.

4. Induction over natural numbers

$$\frac{\varphi[n/0], \quad \varphi \supset \exists m\,(\underline{p}n0 \simeq m \wedge \varphi[n/m])}{\varphi}$$

Finite sequences are introduced in BT using the pair operation:

$$\langle x_1 \rangle \leftrightharpoons x_1;$$
$$\langle x_1, x_2, \ldots, x_{n+1} \rangle \leftrightharpoons \underline{p}(\langle x_1, x_2, \ldots, x_n \rangle)x_{n+1}.$$

5. Comprehension axiom

$$\exists U^{k+1}\left[\, \underline{c}_n(\bar{X}) \simeq U \wedge \forall Z^k (Z \in_k U \equiv \varphi)\right],$$

where $n = \llcorner Z^k.\bar{X}.\varphi \lrcorner$, $\bar{X}$ is a list of variables of arbitrary types, and φ is a $(k+1)$-elementary formula with all its parameters in the list $Z^k, \bar{X}$.

This completes the definition of the theory BT.

In an axiomatic set theory comprehension axiom states what kind of sets can be constructed in the theory. In BT the comprehension axiom has a constructive character: it is predicative (because the formula φ does not contain quantifiers over variables of types $\geqslant k+1$) and the existence of a set of objects with property φ means that this set is produced by an operation on parameters of φ, which is consistent with the constructive meaning of existence. On one hand, the theory BT has a predicative, quite restrictive comprehension; on the other hand, it has a rich language with infinitely many types of sets, so many parts of constructive mathematics can be developed in BT.

Denote BT_s the fragment of BT containing only types $\leqslant s$ $(s \geqslant 0)$.

The only terms in BT are constants and variables. So for brevity we will call external terms just terms. Also for brevity we will denote the term $\tau(\langle t_1, t_2, \ldots, t_n \rangle)$ as $\tau(t_1, t_2, \ldots, t_n)$.

We denote $\tau\,[X_1, \ldots, X_n / t_1, \ldots, t_n]$ the result of proper substitution of terms $t_1, \ldots, t_n$ instead of variables $X_1, \ldots, X_n$ in the term τ.

Lemma 2.1. *$BT \vdash t \simeq Y \supset (\varphi \equiv \varphi[Y/t])$.*

Proof. Proof is by induction on the complexity of φ. □

As usual in combinatorial logic, for any term t and variable x we can construct a term $\lambda x.t$ that does not contain x and has the property:

$$BT \vdash \lambda x.t \downarrow \wedge (\lambda x.t)x \cong t.$$

$\lambda x_1 \ldots x_n.t$ is an abbreviation for $\lambda x_1.(\ldots(\lambda x_n.t)\ldots)$.

Using λ-terms we can define in BT_0 recursion operator and μ-operator, and hence all primitive recursive functions. It means that BT_0 contains the intuitionistic arithmetic HA.

Using primitive recursion, we introduce terms $\underline{p}_k^m$ for projection operation on m-tuples such that:

$$BT \vdash \underline{p}_k^m(x) \downarrow \text{ and } BT \vdash \underline{p}_k^m(\langle x_1, \ldots, x_m \rangle) \simeq x_k (1 \leqslant k \leqslant m).$$

3. Realizability

We construct a realizability in BT by modifying the realizability in Ref. 11 for our language with infinitely many types of sets. To every variable X_{2i}^k we assign the variable X_{2i+1}^k, which we will denote X_{2i}^{k*}. We do not define X_{2i+1}^{k*}. We will consider only formulas that contain no variables with odd lower indices.

If $\bar{X}$ is a list of variables $X_1, \ldots, X_m$, then $\bar{X}^*$ denotes the list $X_1^*, \ldots, X_m^*$.

For any variable a we define its interpretation a^* as follows:

$$\underline{0}^* = 0,$$

$$\underline{k}^* = \lambda x x^* y y^*.x^*,$$

$$\underline{s}^* = \lambda x x^* y y^* z z^*.x^* z z^*(yz)(y^* z z^*),$$

$$\underline{d}^* = \lambda x x^* y y^* u u^* v v^*.d x^* y^* u^* v^*,$$

$$\underline{p}^* = \lambda x x^* y y^*.\underline{p} x^* y^*,$$

$$\underline{p}_i^* = \lambda z z^*.\underline{p}_i z^*, \ (i = 1, 2).$$

Next we define realizability for a formula φ and the interpretation of constant $\underline{c}_n$ by simultaneous induction on their Gödel numbers.

1. Realizability $fr\varphi$.

(1) $fr Ap(g, x, y) \leftrightharpoons gx \simeq y \wedge g^* x x^* \simeq y^*.$

(2) $fr(x =_{0\omega} m) \leftrightharpoons x =_{0\omega} m \wedge x^* =_{0\omega} m.$

(3) $fr(x =_{0k} Y^k) \leftrightharpoons x =_{0k} Y \wedge x^* =_{0k} Y^*.$

(4) $fr(Z^k \in_k Y^{k+1}) \leftrightharpoons \langle f, Z, Z^* \rangle \in_k Y^* \wedge Z \in_k Y$.

(5) $fr\bot \leftrightharpoons \bot$.

(6) $fr(\psi \wedge \chi) \leftrightharpoons \underline{p}_1 fr\psi \wedge \underline{p}_2 fr\chi$.

(7) $fr(\psi \vee \chi) \leftrightharpoons \exists m, u\,[f \simeq \langle m, u \rangle \wedge (m = 0 \supset ur\psi) \wedge (m \neq 0 \supset ur\chi)]$.

(8) $fr(\psi \supset \chi) \leftrightharpoons \forall g(gr\psi \supset fg\downarrow \wedge fg\, r\chi) \wedge (\psi \supset \chi)$.

(9) $fr(\forall n\psi) \leftrightharpoons \forall n(fn\downarrow \wedge fn\, r\psi)$.

(10) $fr(\forall X^k\psi) \leftrightharpoons \forall X^k X^{k*}(f(X, X^*)\downarrow \wedge f(X, X^*)r\psi)$.

(11) $fr(\exists n\psi) \leftrightharpoons \exists n, u(f \simeq \langle n, u \rangle \wedge ur\psi)$.

(12) $fr(\exists X^k\psi) \leftrightharpoons \exists X^k, X^{k*}, u(f \simeq \langle u, X, X^* \rangle \wedge ur\psi)$.

2. Interpretation of $\underline{c}_n$.

If n is not of the form $n = \llcorner Z^k.\bar{X}.\varphi \lrcorner$, then $\underline{c}_n^* = \underline{c}_n$.

Suppose $n = \llcorner Z^k.\bar{X}.\varphi \lrcorner$ and $\bar{X}$ is a list of m variables. We take variables Y^k and v, which are not in the list $Z, \bar{X}$ and denote

$$\chi \leftrightharpoons \exists v, Z^k, Z^{k*}(Y^k \simeq \langle v, Z, Z^* \rangle \wedge vr\varphi).$$

Since the Gödel number of $\underline{c}_n$ is greater than the Gödel number of φ, the formula χ is already defined.

For $m = 0$ we define $\underline{c}_n^* = \underline{c}_{\llcorner Y^k.\chi \lrcorner}$.

For $m \geqslant 1$ we define $\underline{c}_n^* = \lambda uu^*.\underline{c}_{\llcorner Y^k.\bar{X},\bar{X}^*.\chi \lrcorner}(u, \underline{p}_1^m(u^*), \ldots, \underline{p}_m^m(u^*))$.

This completes the definition of the realizability.

Next two lemmas state basic properties of the realizability.

Lemma 3.1. $BT \vdash fr\varphi \supset \varphi$.

Proof. Proof is by induction on the complexity of φ. □

The interpretation t^* of a term t is defined by induction on its construction as follows.

(1) If t is a numerical variable m, then $t^* = m$.
(2) If t is another variable or constant, then t^* is already defined.
(3) If $t \doteq \tau\eta$, then $t^* = \tau^*\eta\eta^*$.

Lemma 3.2. *For terms t and τ the following holds.*

1. $BT \vdash gr(t \simeq \tau) \supset t \simeq \tau \wedge t^* \simeq \tau^*$.

2. There exists a term $g_{t,\tau}$ *such that*

$$BT \vdash t \simeq \tau \wedge t^* \simeq \tau^* \supset g_{t,\tau} \downarrow \wedge g_{t,\tau} r(t \simeq \tau).$$

Proof. Each proof is by induction on the construction of the terms. □

Theorem 3.1. *(Soundness of the realizability).*

If $BT \vdash \varphi$*, then for some term* t*,* $BT \vdash t \downarrow \wedge\, tr\varphi$*.*

Proof. Proof is by induction on the derivation of φ. We will consider only the case of comprehension axiom, other cases are simpler.

We assume that φ is the formula

$$\exists U^{k+1} \left[\underline{c}_n(\bar{X}) \simeq U \wedge \forall Z^k (Z \in_k U \equiv \varphi) \right].$$

Here the formula φ is $(k+1)$-elementary, all its parameters are in the list $Z^k, \bar{X}$ and $n = \llcorner Z^k.\bar{X}.\varphi \lrcorner$. Denote m the length of the list $\bar{X}$.

The formula $tr\varphi$ is equivalent to the following:

$$\exists U^{k+1}, U^{k+1*}, g, h \left\{ t \simeq \langle g, h, U, U^* \rangle \wedge gr \left[\underline{c}_n(\bar{X}) \simeq U \right] \right.$$
$$\left. \wedge \forall Z^k, Z^{k*} [h(Z, Z^*) \downarrow \wedge h(Z, Z^*) r(Z \in_k U \equiv \varphi)] \right\}. \quad (1)$$

Denote

$$\chi \leftrightharpoons \exists v, Z^k, Z^{k*} (Y^k \simeq \langle v, Z, Z^* \rangle \wedge vr\psi).$$

All parameters of χ are in the list $Y^k, \bar{X}, \bar{X}^*$, and both formulas $vr\psi$ and χ are $(k+1)$-elementary. Denote $q = \llcorner Y^k.\bar{X}, \bar{X}^*.\chi \lrcorner$. By the comprehension axiom, there exist U^{k+1} and U^{k+1*} such that:

$$\underline{c}_n(\bar{X}) \simeq U \wedge \forall Z^k (Z \in_k U \equiv \psi), \quad (2)$$

$$\underline{c}_q(\bar{X}, \bar{X}^*) \simeq U^* \wedge \forall Z^k, Z^{k*}, v(\langle v, Z, Z^* \rangle \in_k U^* \equiv vr\psi). \quad (3)$$

Denote $\tau = \underline{c}_n(\bar{X})$ and $h = \lambda y.\langle \lambda v.v, \lambda v.v \rangle$. By (3) and the definition of $\underline{c}_n^*$:

$$\tau^* = \underline{c}_n(\bar{X})^* \cong \underline{c}_n^*(\bar{X})(\bar{X}^*) \cong \underline{c}_q(\bar{X}, \bar{X}^*) \simeq U^*.$$

So by (2),

$$\tau \simeq U \wedge \tau^* \simeq U^*$$

and by Lemma 3.2.2, there exists a term g with

$$gr(\tau \simeq U).$$

Then the term $t = \langle g, h, \tau, \tau^* \rangle$ satisfies (1) and therefore $tr\varphi$. □

4. Existence Property of BT

Theorem 4.1. *Suppose Y is a variable of any type.*
If $BT \vdash \exists Y \varphi$, then there exists a term t such that

$$BT \vdash \exists Y(t \simeq Y) \wedge \varphi[Y/t].$$

Proof. We will consider only the case when Y is a variable X^k; for a numerical variable the proof is similar.

Suppose $BT \vdash \exists Y \varphi$. By Theorem 3.1, there exists a term τ such that

$$BT \vdash \exists Y, Y^*, u[\tau \simeq \langle u, Y, Y^* \rangle \wedge ur\varphi].$$

By Lemma 3.1, $BT \vdash \exists Y, Y^*, u[\tau \simeq \langle u, Y, Y^* \rangle \wedge \varphi]$ and for $t = \underline{p}_2^3(\tau)$ we get $BT \vdash \exists Y(t \simeq Y \wedge \varphi[Y/t])$ by Lemma 2.1. □

5. Set-Theoretical Model of BT

When realizing our predicative comprehension axiom, we assigned a term to every set in BT. Here we will use this approach to generate a countable set-theoretical model of BT, where the domain for each type of variables is a set of terms.

First we introduce normal terms in BT, similar to Ref. 11.

Definition 5.1. Consider terms t and τ.

1. t *contracts* to τ if one of the following is satisfied.

(1) $t \doteq \underline{k}\tau t_1$,

(2) $t \doteq \underline{s}t_1t_2t_3$ and $\tau \doteq t_1t_3(t_2t_3)$,

(3) $t \doteq \underline{d}\tau t_1\bar{n}\bar{n}$ or $t \doteq \underline{d}t_1\tau\bar{n}\bar{m}$ for $n \neq m$,

(4) $t \doteq \underline{p}_1(\underline{p}\tau t_1)$ or $t \doteq \underline{p}_2(\underline{p}t_1\tau)$.

2. t *immediately reduces* to τ if τ is obtained from t by contracting one occurrence of a sub-term in t.

3. $t \succeq \tau$ (t reduces to τ) if there is a sequence of terms $t \doteq t_0, t_1, \ldots, t_n \doteq \tau$ such that each t_i immediately reduces to t_{i+1}.

4. t is a *normal term* if no sub-term of t can be contracted.

Lemma 5.1. *1. If $t \succeq \tau$, then*

$$BT \vdash t\downarrow \supset (t \simeq \tau).$$

2. Any term t reduces to a normal term.

Proof. 1. Follows from the definitions.

2. Proof is by induction on the number of occurrences of constants in t. If t has no constants, then t is a normal term, since only terms starting with constants can be contracted.

Assume that any term with n occurrences of constants can be reduced to a normal term, and consider a term t with $n+1$ occurrences of constants. If none of sub-terms of t can contract, then t is a normal term. Suppose t immediately reduces to t' when sub-term τ contracts to r. Then there are fewer occurrences of constants in r than in τ. So t' has $\leqslant n$ occurrences of constants and by the inductive assumption it reduces to some normal term $s : t' \succeq s$. Then $t \succeq s$. □

Theorem 5.1. *Any term t reduces to a unique normal term.*

Proof. We need to prove the uniqueness. The proof is a modification of the proof in Ref. 10, pp. 108-109.

First we introduce terminology and notations 1) - 3).

1) We say that pair of terms (t, τ) *is obtained by rule* (*) from pairs (t_1, τ_1) and (t_2, τ_2) if $t \doteq t_1 t_2$ and $\tau \doteq \tau_1 \tau_2$.

2) We say that term t *partially reduces* to term τ ($t \succeq^* \tau$) if there exists a sequence of pairs of terms $(t_0, \tau_0), \ldots, (t_n, \tau_n)$ such that $t_n \doteq t, \tau_n \doteq \tau$, and for any $k \leqslant n$:

> $t_k \doteq \tau_k$ or
> t_k contracts to τ_k, or
> for some $i, j < k$, the pair (t_k, τ_k) is obtained by rule (*) from (t_i, τ_i) and (t_j, τ_j).

The length n of the sequence is called the *reduction length* of t to τ.

3) We denote $t \succeq^*_n \tau$ if there exists a sequence of terms $t_0, \ldots, t_n$ such that $t_0 \doteq t, t_n \doteq \tau$ and for any $i < n, t_i \succeq^* t_{i+1}$.

The proof of uniqueness consists of the following steps.

$$\text{If } t \succeq^* t' \text{ and } t \succeq^* t'', \text{then for some term } \tau, t' \succeq^* \tau \text{ and } t'' \succeq^* \tau. \quad (4)$$

$$t \succeq \tau \Leftrightarrow \exists n (t \succeq^*_n \tau). \quad (5)$$

$$\text{If } t \succeq^*_n t' \text{ and } t \succeq^*_m t'', \text{then for some term } \tau, t' \succeq^*_m \tau \text{ and } t'' \succeq^*_n \tau. \quad (6)$$

$$\text{If } t' \text{ and } t'' \text{ are normal terms, } t \succeq t' \text{ and } t \succeq t'', \text{ then } t' \doteq t''. \quad (7)$$

Proof of (4)

Suppose $t \succeq^* t'$ and $t \succeq^* t''$.

If $t \doteq t'$, we can take $\tau \leftrightharpoons t''$.

If $t \doteq t''$ or $t' \doteq t''$, we can take $\tau \leftrightharpoons t'$.

Next we assume $\neg(t \doteq t' \vee t \doteq t'' \vee t' \doteq t'')$. The proof is by induction on the reduction length n of t to t'.

Basis of induction: $n = 0$. Then t contracts to t'. Any term contracts in a unique way. So t cannot contract to t'', otherwise we would have $t' \doteq t''$. Since $\neg(t \doteq t'')$, in the reduction sequence of t to t'' the pair (t, t'') is obtained by the rule (*). There are several possible cases depending on the form of t.

Case 1: $t \doteq \underline{k}t't_1$.

Then (t, t'') is obtained by rule (*) from pairs $(\underline{k}t', \tau_0)$ and (t_1, τ_1). Since terms of the form $\underline{k}t'$ do not contract, then $\tau_0 \doteq \underline{k}t'$ or for some τ_2, $(\underline{k}t', \tau_0)$ is obtained by rule (*) from $(\underline{k}, \underline{k})$ and (t', τ_2). In both cases τ_0 has the form $\underline{k}\tau_3$ and $t' \succeq^* \tau_3$. Then $t'' \doteq \underline{k}\tau_3\tau_1$ and we can take $\tau \leftrightharpoons \tau_3$.

Case 2: $t \doteq \underline{s}t_1t_2t_3$ and $t' \doteq t_1t_3(t_2t_3)$.

Then similarly to Case 1, there exist τ_1, τ_2 and τ_3 such that $t_1 \succeq^* \tau_1$, $t_2 \succeq^* \tau_2$, $t_3 \succeq^* \tau_3$ and $t'' \doteq \underline{s}\tau_1\tau_2\tau_3$. Let us take $\tau \leftrightharpoons \tau_1\tau_3(\tau_2\tau_3)$. Then t'' contracts to τ; $t_1t_3 \succeq^* \tau_1\tau_3$, $t_2t_3 \succeq^* \tau_2\tau_3$ and $t' \doteq t_1t_3(t_2t_3) \succeq^* \tau$.

Case 3: $t \doteq \underline{d}t't_1\bar{n}\bar{n}$.

Since term 0 and terms of the form $\underline{p}k0$ do not contract, then $\bar{n} \succeq^* r$ implies $r \doteq \bar{n}$. Therefore there exist τ_1, τ_2 such that $t' \succeq^* \tau_1$, $t_1 \succeq^* \tau_2$ and $t'' \doteq \underline{d}\tau_1\tau_2\bar{n}\bar{n}$. We can take $\tau \leftrightharpoons \tau_1$.

Case 4: $t \doteq \underline{p}_1(\underline{p}t't_1)$.

Then there exist τ_1, τ_2 such that $t' \succeq^* \tau_1, t_1 \succeq^* \tau_2$ and $t'' \doteq \underline{p}_1(\underline{p}\tau_1\tau_2)$. We can take $\tau \leftrightharpoons \tau_1$.

Cases $t \doteq \underline{d}t't_1\bar{n}\bar{m}$ for $n \neq m$ and $t \doteq \underline{p}_2(\underline{p}t_1t')$ are similar to Cases 3 and 4, respectively.

Inductive step. Assume that (4) holds for any reduction length $< n$. We will prove it for n. The case when t contracts to t' is proven the same way as the basis of induction.

It remains to consider the case when in the sequence reducing t to t' the pair (t, t') is obtained by rule (*) from pairs (t_1, t_2) and (t_3, t_4). Then $t_1 \succeq^* t_2$, $t_3 \succeq^* t_4$, $t \doteq t_1t_3$ and $t' \doteq t_2t_4$. The case when t contracts to t'' is proven the same way as the basis of induction.

So we can assume that in the sequence reducing t to t'' the pair (t, t'') is obtained by rule (*) from some pairs (t'_1, t'_2) and (t'_3, t'_4). Then $t'_1 \succeq^* t'_2$, $t'_3 \succeq^* t'_4$, $t \doteq t'_1 t'_3$ and $t'' \doteq t'_2 t'_4$. So $t_1 \doteq t'_1$, $t_3 \doteq t'_3$ and by the inductive assumption there exist τ_1 and τ_2 such that $t_2 \succeq^* \tau_1$, $t'_2 \succeq^* \tau_1$, $t_4 \succeq^* \tau_2$ and $t'_4 \succeq^* \tau_2$. Let us take $\tau \leftrightharpoons \tau_1 \tau_2$. Then by rule (*) we obtain $t' \succeq^* \tau$ and $t'' \succeq^* \tau$.

Proof of (5)

$\Rightarrow$ It is sufficient to prove:

$$\text{if } t \text{ immediately reduces to } \tau, \text{ then } \exists n (t \succeq^*_n \tau).$$

Suppose $t \doteq s[x/r]$, $\tau \doteq s[x/r']$ and r contracts to r'. Proof is by induction on the complexity of the term s.

If x is not a parameter of s, then $t \doteq \tau$ and $t \succeq^*_0 \tau$.

If $s \doteq x$, then $t \doteq r$, $\tau \doteq r'$ and $r \succeq^*_1 r'$.

Suppose $s \doteq s_1 s_2$. Denote $t_0 \leftrightharpoons s_1[x/r]$, $t' \leftrightharpoons s_1[x/r']$, $\tau_0 \leftrightharpoons s_2[x/r]$ and $\tau' \leftrightharpoons s_2[x/r']$. By the inductive assumption for some m, k we have $t_0 \succeq^*_m t'$ and $\tau_0 \succeq^*_k \tau'$. So there exist $t_1, \ldots, t_m, \tau_1, \ldots, \tau_k$ such that $t_m \doteq t'$, $\tau_k \doteq \tau'$ and for any $i < m, j < k$, $t_i \succeq^* t_{i+1}$, $\tau_j \succeq^* \tau_{j+1}$. Then $t \succeq^*_{m+k} \tau$ with the sequence: $t_0\tau_0, t_1\tau_0, \ldots, t_m\tau_0, t_m\tau_1, \ldots, t_m\tau_k$.

$\Leftarrow$ It is sufficient to prove:

$$t \succeq^* \tau \quad \Rightarrow \quad t \succeq \tau.$$

We prove it by induction on the reduction length of t to τ.

If $t \doteq \tau$ or t contracts to τ, then $t \succeq \tau$ is obvious.

Suppose (t, τ) is obtained by rule (*) from (t', τ') and (t'', τ''). By the inductive assumption $t' \succeq \tau'$ and $t'' \succeq \tau''$. Then there exist $t_0, \ldots, t_q$, $\tau_0, \ldots, \tau_r$ such that $t' \doteq t_0$, $\tau' \doteq t_q$, $t'' \doteq \tau_0$, $\tau'' \doteq \tau_r$; for any $i < q, j < r$, t_i immediately reduces to t_{i+1} and τ_j immediately reduces to τ_{j+1}. Then t reduces to τ with the sequence: $t_0\tau_0, t_1\tau_0, \ldots, t_q\tau_0, t_q\tau_1, \ldots, t_q\tau_r$.

Proof of (6)

Proof is by induction on $n + m$.

If $n = 0$, then we take $\tau \leftrightharpoons t''$.

If $m = 0$, then we take $\tau \leftrightharpoons t'$.

If $n = m = 1$, then $t \succeq^* t', t \succeq^* t''$ and the rest follows from (4).

Assume the statement holds for the sums less than $n+m$. We can also assume that $n > 1$ or $m > 1$, since the case $n \leqslant 1$ & $m \leqslant 1$ has already been considered. Without loss of generality we can assume $n > 1$.

Then for some t_1 we have $t \succeq^*_{n-1} t_1$ and $t_1 \succeq^* t'$. By the inductive assumption there exists τ_1 such that $t_1 \succeq^*_m \tau_1$ and $t'' \succeq^*_{n-1} \tau_1$. By the inductive assumption applied to $t_1 \succeq^*_1 t'$ and $t_1 \succeq^*_m \tau_1$ there exists τ such that $t' \succeq^*_m \tau$ and $\tau_1 \succeq^*_1 \tau$. Then $t' \succeq^*_m \tau$ and $t'' \succeq^*_n \tau$.

Proof of (7)

Suppose the premises. Then by (5), for some n and m we have $t \succeq^*_n t'$ and $t \succeq^*_m t''$. By (6), there exists τ such that $t' \succeq^*_m \tau$ and $t'' \succeq^*_n \tau$. By (5), $t' \succeq \tau$ and $t'' \succeq \tau$. Since t' and t'' are normal terms, we have $t' \doteq \tau \doteq t''$. □

Next we use normal terms to construct a set-theoretical model for BT.

Definition 5.2. 1. Set $\mathcal{H} = \{\bar{m} \mid m \in \mathbb{N}\}$ is the domain for numerical variables.

2. The set M^0 of all closed normal terms is the domain for operation variables. The domain M^n for variables of type n will be defined in next definition.

3. Every constant is a closed normal term; it is interpreted as itself.

4. *Evaluation* is a finite function of the form $\gamma = \begin{pmatrix} Y_1 \dots Y_m \\ t_1 \dots t_m \end{pmatrix}$, where $Y_1, \dots, Y_m$ are distinct variables of any types and $t_1, \dots, t_m$ are elements of M^0.

We denote $dom(\gamma) = \{Y_1, \dots, Y_m\}$ and $\gamma(Y_i) = t_i$, $i = 1, 2, \dots, m$. For any constant a, $\gamma(a) = a$.

We define $\gamma(Y/t) = \begin{pmatrix} Y_1 \dots Y_{i-1}\ Y_i\ Y_{i+1} \dots Y_m \\ t_1 \dots t_{i-1}\ t\ t_{i+1} \dots t_m \end{pmatrix}$ if $Y \doteq Y_i$ for some $i = 1, \dots, m$, and $\gamma(Y/t) = \begin{pmatrix} Y_1 \dots Y_m\ Y \\ t_1 \dots t_m\ t \end{pmatrix}$ otherwise.

Further we will use the notation $\diamond$ for an arbitrary logical connective and Q for a quantifier $\forall$ or $\exists$.

Definition 5.3. By induction on n we define a triple of objects $M^n, E^n, \models^n$.

Step 1: $n = 0$.

1. M^0 is already defined.

2. E^0 is the set of all pairs (γ, φ), where:

(a) φ is a 0-elementary formula;
(b) γ is an evaluation,
(c) all parameters of φ are in $dom(\gamma)$,
(d) $m \in dom(\gamma) \Rightarrow \gamma(m) \in \mathcal{H}$,
(e) $x \in dom(\gamma) \Rightarrow \gamma(x) \in M^0$.

3. $\models^0$ is a relation on E^0.

For any $(\gamma, \varphi) \in E^0$ we define $\gamma \models^0 \varphi$ by induction on the complexity of φ as follows.

$$\gamma \models^0 Ap(f, x, y) \leftrightharpoons \gamma(f)\gamma(x) \succeq \gamma(y),$$
$$\gamma \models^0 (x =_{0\omega} m) \leftrightharpoons \gamma(x) \doteq \gamma(m),$$
$$\gamma \models^0 \bot \leftrightharpoons \bot,$$
$$\gamma \models^0 (\psi \diamond \chi) \leftrightharpoons (\gamma \models^0 \psi) \diamond (\gamma \models^0 \chi),$$
$$\gamma \models^0 (Qm\psi) \leftrightharpoons (Qt \in \mathcal{H}) \left[\gamma(m/t) \models^0 \psi\right].$$

Step 2: $n \geqslant 1$.

Assume $M^j, E^j, \models^j$ are defined for any $j < n$.

First by induction on k we define a triple of objects $M^n_k, E^n_k, \models^n_k$.

1. $k = 0$.

$$M^n_0 = \emptyset.$$
$$E^n_0 = E^{n-1}_0.$$
$$\gamma \models^n_0 \varphi \leftrightharpoons \gamma \models^{n-1} \varphi.$$

2. $k \geqslant 1$. Assume the triple $M^n_{k-1}, E^n_{k-1}, \models^n_{k-1}$ is defined.

(1) A term $t \in M^n_k$ if it satifies one of the following three conditions:

(a) $t \in M^n_{k-1}$;

(b) $t \doteq \underline{p}t_1t_2$, where $t_1 \in M^0$ and $t_2 \in M^n_{k-1}$;

(c) $t \doteq c_{\llcorner Z^{n-1}.\bar{X}.\psi \lrcorner}(\theta(X_1), \ldots, \theta(X_m))$, where $\bar{X}$ is the list $X_1, \ldots, X_m$;
$\bar{X} \subseteq dom(\theta)$; $(\theta, \forall Z^{n-1}\psi) \in E^n_{k-1}$ and all parameters of formula ψ
are in the list $Z^{n-1}, \bar{X}$.

(2) E^n_k is the list of all pairs (γ, φ) such that

(a) φ is an n-elementary formula;
(b) γ is an evaluation,

(c) all parameters of φ are in $dom(\gamma)$,

(d) $m \in dom(\gamma) \Rightarrow \gamma(m) \in \mathcal{H}$,

(e) $X^j \in dom(\gamma), j < n \Rightarrow \gamma(X^j) \in M^j$,

(f) $X^n \in dom(\gamma) \Rightarrow \gamma(X^n) \in M_k^n$.

(3) For any $(\gamma, \varphi) \in E_k^n$ we define $\gamma \models_k^n \varphi$ by induction on the complexity of φ as follows.

(a) If φ is $\perp$ or an atomic formula with no variables of type n and no predicate symbol $=_{0,n-1}$, then $(\gamma \models_k^n \varphi) \leftrightharpoons (\gamma \models_{k-1}^n \varphi)$.

(b) $\gamma \models_k^n (x =_{0,n-1} Y^{n-1}) \leftrightharpoons \gamma(x) \doteq \gamma(Y)$.

(c) Suppose $\varphi \doteq Y^{n-1} \in_{n-1} U^n$. Denote $t = \gamma(U^n)$.

i. If $t \in M_{k-1}^n$, then $(\gamma \models_k^n \varphi) \leftrightharpoons (\gamma \models_{k-1}^n \varphi)$.

ii. If t satisfies the condition (b) in the definition of M_k^n, then

$$(\gamma \models_k^n \varphi) \leftrightharpoons \perp.$$

iii. If t satisfies the condition (c) in the definition of M_k^n, then

$$(\gamma \models_k^n \varphi) \leftrightharpoons [\theta(Z/\gamma(Y)) \models_{k-1}^n \psi].$$

(d) $\gamma \models_k^n (\psi \diamond \chi) \leftrightharpoons (\gamma \models_k^n \psi) \diamond (\gamma \models_k^n \chi)$.

(e) $\gamma \models_k^n (Qm\psi) \leftrightharpoons (Qt \in \mathcal{H}) [\gamma(m/t) \models_k^n \psi]$.

(f) For $j < n$, $(\gamma \models_k^n (QX^j\psi)) \leftrightharpoons (Qt \in M^j) \left[\gamma(X^j/t) \models_k^n \psi\right]$.

Finally,

$$M^n = \bigcup_{k=0}^{\infty} M_k^n; \qquad E^n = \bigcup_{k=0}^{\infty} E_k^n,$$

$$\gamma \models^n \varphi \leftrightharpoons \exists k \left[(\gamma, \varphi) \in E_k^n \wedge \gamma \models_k^n \varphi\right].$$

This completes Definition 5.3.

Next lemmas describe basic properties of the aforementioned concepts.

Lemma 5.2. *1. For $n \geqslant 1, k \geqslant 0$: $M_k^n \subseteq M_{k+1}^n$.*

2. $M^n \subseteq M^0 (n \geqslant 0)$.

3. For any $n \geqslant 0, k \geqslant 1$ and $(\gamma, \varphi) \in E_{k-1}^n$:

$$(\gamma \models_k^n \varphi) \Leftrightarrow (\gamma \models_{k-1}^n \varphi).$$

4. For any $n \geqslant 0, k \geqslant 0$ *and* $(\gamma, \varphi) \in E^n_k$*:*

$$(\gamma \models^n \varphi) \Leftrightarrow (\gamma \models^n_k \varphi).$$

5. If $n \geqslant 1$*,* $(\gamma, \varphi) \in E^n$ *and* φ *is an* $(n-1)$*-elementary formula, then:*

$$(\gamma \models^n \varphi) \Leftrightarrow (\gamma \models^{n-1} \varphi).$$

Proof. 1. Follows from the definitions.

2. Using the definitions, we prove by induction on k: $M^n_k \subseteq M^0$.

3. Proof is by induction on k using the definitions.

4. Follows from part 3.

5. Since φ is an $(n-1)$-elementary formula, we have $(\gamma, \varphi) \in E^{n-1} \subseteq E^n_0$. By part 4,

$$(\gamma \models^n \varphi) \Leftrightarrow (\gamma \models^n_0 \varphi) \Leftrightarrow (\gamma \models^{n-1} \varphi).$$

□

Lemma 5.3. *For* $n \geqslant 0, (\gamma, \varphi) \in E^n$ *the following hold.*

1. If $\varphi \doteq Ap(f, x, y)$*, then*

$$(\gamma \models^n \varphi) \Leftrightarrow \gamma(f)\gamma(x) \succeq \gamma(y).$$

2. If $\varphi \doteq (x =_{0\omega} m)$*, then*

$$(\gamma \models^n \varphi) \Leftrightarrow \gamma(x) \doteq \gamma(m).$$

3. If $\varphi \doteq (x =_{0k} Y^k)$*, then*

$$(\gamma \models^n \varphi) \Leftrightarrow \gamma(x) \doteq \gamma(Y^k).$$

4. $(\gamma \models^n \bot) \Leftrightarrow \bot$.

5. If $\varphi \doteq \psi \diamond \chi$*, then*

$$(\gamma \models^n \varphi) \Leftrightarrow (\gamma \models^n \psi) \diamond (\gamma \models^n \chi).$$

6. If $\varphi \doteq Qm\psi$*, then*

$$(\gamma \models^n \varphi) \Leftrightarrow (Qt \in \mathcal{H})(\gamma(m/t) \models^n \psi).$$

7. If $\varphi \doteq QX^k\psi$*, then*

$$(\gamma \models^n \varphi) \Leftrightarrow (Qt \in M^k)(\gamma(X^k/t) \models^n \psi).$$

Proof. Follows from the previous lemma and the definitions. □

Lemma 5.4. *Suppose* $\varphi \doteq (t \simeq x)$, $\gamma = \begin{pmatrix} Y_1 \dots Y_m \\ \tau_1 \dots \tau_m \end{pmatrix}$ *and* $(\gamma, \varphi) \in E^p$. *Then*

$$(\gamma \models^p \varphi) \Leftrightarrow t[Y_1, \dots, Y_m/\tau_1, \dots, \tau_m] \succeq \gamma(x).$$

Proof. Proof is by induction on the construction of t using Lemma 5.3. □

Definition 5.4.

$$1.\ E = \bigcup_{n=0}^{\infty} E^n.$$

2. For any evaluation $(\gamma, \varphi) \in E$ we define:

$$(\gamma \models \varphi) \ \leftrightharpoons\ \exists p\,[(\gamma, \varphi) \in E^p \wedge \gamma \models^p \varphi]\,.$$

This definition is valid because by to Lemma 5.2.5,

$$\text{for any } (\gamma, \varphi) \in E^p \cap E^n: \qquad (\gamma \models^p \varphi) \Leftrightarrow (\gamma \models^n \varphi).$$

Our set-theoretical model for the theory BT consists of the domains $\mathcal{H}, M^n (n \geqslant 0)$ and the relation $\models$.

Lemma 5.5. *For any* $(\gamma, \varphi) \in E$ *the following hold.*

1. If $\varphi \doteq Ap(f, x, y)$, *then*

$$(\gamma \models \varphi) \Leftrightarrow \gamma(f)\gamma(x) \succeq \gamma(y).$$

2. If $\varphi \doteq (x =_{0\omega} m)$, *then*

$$(\gamma \models \varphi) \Leftrightarrow \gamma(x) \doteq \gamma(m).$$

3. If $\varphi \doteq (x =_{0k} Y^k)$, *then*

$$(\gamma \models \varphi) \Leftrightarrow \gamma(x) \doteq \gamma(Y^k).$$

4. $(\gamma \models \bot) \Leftrightarrow \bot$.

5. If $\varphi \doteq \psi \diamond \chi$, *then*

$$(\gamma \models \varphi) \Leftrightarrow (\gamma \models \psi) \diamond (\gamma \models \chi).$$

6. If $\varphi \doteq Qm\psi$, *then*

$$(\gamma \models \varphi) \Leftrightarrow (Qt \in \mathcal{H})(\gamma(m/t) \models \psi).$$

7. If $\varphi \doteq QX^k\psi$, *then*

$$(\gamma \models \varphi) \Leftrightarrow (Qt \in M^k)(\gamma(X^k/t) \models \psi).$$

8. Suppose $\varphi \doteq Y^{n-1} \in_{n-1} U^n$. *Denote* $t = \gamma(U^n)$.

i) Suppose t *has the form* $c_{\llcorner Z^{n-1}.\bar{X}.\psi \lrcorner}(\theta(X_1), \dots, \theta(X_m))$, *where* $\bar{X}$ *is the list* $X_1, \dots, X_m$; $\bar{X} \subseteq dom(\theta)$; $(\theta, \forall Z^{n-1}\psi) \in E^n_{k-1}$ *and all parameters of formula* ψ *are in the list* $Z^{n-1}, \bar{X}$. *Then*

$$(\gamma \models \varphi) \Leftrightarrow [\theta(Z/\gamma(Y)) \models \psi)].$$

ii) If t *does not have the form as in part i), then*

$$(\gamma \models \varphi) \Leftrightarrow \bot.$$

Proof. Follows from the definitions and Lemma 5.3. □

Denote BT^{cl} the theory BT with classical logic and BT^{cl}_p the fragment BT_p with classical logic.

Theorem 5.2. *(Soundness of the model). Suppose* γ *is the empty evaluation,* $p > 0$.

1. $BT^{cl}_{p-1} \vdash \varphi \;\Rightarrow\; (\gamma \models^p \overline{\varphi})$.
2. $BT^{cl} \vdash \varphi \;\Rightarrow\; (\gamma \models \overline{\varphi})$.

Proof. 1. Proof is by induction on the derivation of φ. Clearly, axioms and rules of classical logic hold in the set-theoretical model. Of non-logical axioms we will consider only the comprehension axiom, other cases are simpler.

So we assume that φ is the formula:

$$\exists U^{k+1}\left[\, \underline{c}_n(\bar{X}) \simeq U \wedge \forall Z^k(Z \in_k U \equiv \psi)\right],$$

where $n = \llcorner Z^k.\bar{X}.\psi \lrcorner$, $\bar{X}$ is a list of variables $X_1, \dots, X_m$, and ψ is a $(k+1)$-elementary formula with all its parameters in the list $Z^k, \bar{X}$. Clearly, $k+1 \leqslant p-1$.

Suppose θ is an evaluation with $dom(\theta) = \{\bar{X}\}$ and $(\theta, \forall Z^k\psi) \in E^{p-1}$. Then $(\theta, \forall Z^k\psi) \in E^{k+1}$. Denote $t_i = \theta(X_i), i = 1, 2, \dots, m$, and denote $\bar{t}$ the list of terms $t_1, \dots, t_m$.

According to Lemmas 5.3 and 5.4, it is sufficient to find $\eta \in M^{k+1}$ such that:

$$\underline{c}_n(\bar{t}) \succeq \eta \wedge (\forall \tau \in M^k) \left[\begin{pmatrix} Z & U \\ \tau & \eta \end{pmatrix} \models^p (Z \in U) \Leftrightarrow (\theta(Z/\tau) \models^p \psi) \right]. \quad (8)$$

According to Lemma 5.2.1, there exists a number s such that for any variable X_i of type $k+1$ the corresponding term $t_i = \theta(X_i)$ belongs to M_s^{k+1}. We take $\eta = \underline{c}_{\llcorner Z.\bar{X}.\psi \lrcorner}(\bar{t})$ and prove that it satisfies (8).

Clearly $\eta \in M_{s+1}^{k+1} \subseteq M^{k+1}$. By the definition of $\models_{s+1}^{k+1}$, for any $\tau \in M^k$ we have:

$$\begin{pmatrix} Z & U \\ \tau & \eta \end{pmatrix} \models_{s+1}^{k+1} (Z \in U) \Leftrightarrow \left[\theta(Z/\tau) \models_s^{k+1} \psi\right].$$

By Lemma 5.2.4, 5, for any $\tau \in M^k$:

$$\begin{pmatrix} Z & U \\ \tau & \eta \end{pmatrix} \models^p (Z \in U) \Leftrightarrow \begin{pmatrix} Z & U \\ \tau & \eta \end{pmatrix} \models_{s+1}^{k+1} (Z \in U)$$

$$\Leftrightarrow \left[\theta(Z/\tau) \models_s^{k+1} \psi\right] \Leftrightarrow \left[\theta(Z/\tau) \models^{k+1} \psi\right] \Leftrightarrow \left[\theta(Z/\tau) \models^p \psi\right],$$

which implies (8).

2. Follows from part 1, since any derivation in BT is finite, therefore it is a derivation in BT_{p-1} for some $p > 0$. □

The described set-theoretical model can have useful applications in proving further metamathematical properties of BT.

6. Discussion

We described an axiomatic theory BT for constructive mathematics, which contains operations and infinitely many types of sets. The theory BT is consistent with classical logic and many theorems of classical mathematics can be expressed in its language. In this paper we constructed a realizability for the theory BT and used it to prove the existence property of BT. Using the same recursive approach to sets as in the realizability, we constructed a countable set-theoretical model of BT where domains are made of its terms.

Among known metamathematical properties of BT are its consistency with a weak form of Church thesis and interpretability in a weak second-order arithmetic. Next we are planning to use the described realizability and set-theoretical model to investigate other metamathematical properties of the theory BT, in particular, the disjunction property and the consistency with a stronger form of Church thesis.

Acknowledgements

The author thanks the referee for valuable comments and suggestions that helped to improve this paper.

References

1. E. Bishop, *Foundations of Constructive Analysis* (McGraw-Hill, 1967).
2. E. Bishop and D. Bridges, *Constructive Analysis* (Springer, 1985).
3. M. Beeson, *Foundations of Constructive Mathematics. Metamathematical Studies* (Springer, 1985).
4. G. Sommaruga, *History and Philosophy of Constructive Type Theory*, Synthese Library (Book 290), Vol. 344 (Springer, 2010).
5. S. Feferman, A language and axioms for explicit mathematics, *Lecture Notes in Math.* **450**, 87 (1975).
6. S. Feferman, Constructive theories of functions and classes, in *Logic Colloquium 78 (Mons)*, (Amsterdam-New York, North-Holland, 1979).
7. M. Beeson, A type-free Gödel interpretation, *The Journal of Symbolic Logic* **43**, 213 (1978).
8. M. Beeson, Some relations between classical and constructive mathematics, *The Journal of Symbolic Logic* **43**, 228 (1978).
9. F. Kashapova, Isolation of classes of constructively derivable theorems in a many-sorted intuitionistic set theory equivalent to a second-order arithmetic, *Soviet Math. Dokl.* **29**, 583 (1984).
10. A. S. Troelstra, *Metamathematical Investigation of Intuitionistic Arithmetic and Analysis* (Springer Lecture Notes, 1973).
11. M. Beeson, Principles of continuous choice and continuity of functions in formal systems for constructive mathematics, *Ann. Math. Logic* **12**, 249 (1977).

Goal-directed Unbounded Coalitional Game and Its Complexity*

Hu Liu

The Institute of Logic and Cognition, Sun Yat-sen University,
Guangzhou, 510275, China
**E-mail: liuhu2@mail.sysu.edu.cn*
www.sysu.edu.cn

This paper deals with goal-directed unbounded coalitional games (GUCGs), in which each player is assigned a propositional formula as her goal. Players in a game may form coalitions for their own interests. The end condition of a game is reached if those players whose goals are satisfied can cooperate to prevent other players from reaching their goals. We study various solution concepts of GUCGs and their computational complexity.

Keywords: Extended Coalitional Game; Goal-directed Game

1. Introduction

Agents who coordinate their actions can often be more rewarded. The topic of cooperative agents has occupied many researchers in AI, logic and economics. Much attention has been paid to the field of coalitional game theory. This paper studies a family of coalitional games, called *goal-directed unbounded coalitional games* (henceforth, GUCGs). A GUCG is a game played in an infinite sequence of rounds, in which each player is assigned a propositional formula as her goal. At each round, the next state of the game is determined by all players' joint moves and the current state.

In such games, players need to reason about future possibilities, so that they can make reasonable decisions on what they will do in the future, such as which coalition should they join, and when they should stop their actions and end a game. We study solution concepts for such games and the fundamental question of coalition formation. The work in this paper is based on the idea that, a player concerns not only how to reach her goal, but also how to keep her goal satisfied in a potentially infinite game.

*This research is supported by the National Fund of Social Science (No. 13BZX066) and Guangdong 12th Five-year Planning Fund of Philosophy and Social Science.

Players whose goals are not satisfied at the current state are likely to make further moves to change that situation. On the other hand, players whose goals are satisfied at the current state would like to stay at the state. They will react against those whose actions may jeopardize their goals, possibly by cooperation with others.

Game theory, as originated in economics, usually takes utility as the measurement of players' interests[1]. There are also considerable researches on goal-directed qualitative games, in which players' interests are not to maximize their utilities, but to fulfill certain pre-defined goals. Goals could be propositional formulas[2,3], certain target states[4], or abstract entities[5,6]. Agents with goals have been a focus in various areas of AI, see for example Refs. 7,8.

There does not exist unique right game model. Because the contexts where games are played are diverse, both goal models and utility models have their merits. Utility models are in general more expressive, while goal models are usually simpler and easier to deal with. Goal models are appropriate in contexts where goals are very clear. (See Refs. 5,9 for more arguments on comparing goals and utilities.)

A GUCG is based on the game-like structure that was independently developed by Alur et al. in Refs. 10,11 and Pauly in Refs. 12,13. It was called concurrent game structure (CGS) in the former and called game frame (GF) in the latter. They are different in motivations and notations, but their semantics are formally equivalent as shown in Ref. 14. CGS has logical formalisms called ATL and ATL*, which are extensions of the well-known model checking logic CTL. They are intended to provide model checking tools for open computing systems. GF is more game theoretically driven that uses notions in game theory. It was argued in Refs. 12,13 that GF and its logic CL have a solid relationship with social choice, both conceptually and technically. It is safe for us to use the definition of either CGS or GF. We will take the notational convention of CGS. It is simpler for our representation because we will use some technical results of ATL and ATL* model checking.

CGSs are game-like structures, whereas they are not real games in the meaning of game theory. A CGS does not contain any measurement of players' interests. A game in this paper, i.e. a GUCG, is simply extended from a CGS by assigning a propositional goal to each player. GUCGs are extensive games with simultaneous moves played in an (potentially) infinite sequence of rounds. (Turn-based games can be defined as a special case of GUCGs.)

The study of repeated games has been a well-established area of game theory. A repeated game is played by the same set of players on the same strategic game over a long (typically, infinite) time horizon[15]. The focus of repeated games is to find equilibriums where players' overall utilities are evaluated over possible runs of the game, which are infinite sequence of states of the game. A pre-defined discount factor is used to discount future utilities.

Little has been done on repeated coalitional games[16]. In Refs. 4,17, a qualitative repeated game similar to a GUCG was studied, where players are with so-called ω-regular goals, which are certain properties of runs of a game. The work in this paper is different from Refs. 4,17. Games in Refs. 4,17 are two persons zero-sum games, while ours are multi-player games, and are not zero-sum. Moreover, goals in Refs. 4,17 are to reach certain given states of a game, while the purpose of players in a GUCG is to make sure that the game ends at a state that satisfies their goals. The ending condition is the focus of GUCG.

Usually, a real world game is complicated in many aspects. GUCG is an ideal model that assumes the following simplifications: (1) Players have perfect information about the game they play. They know everyone's possible moves at every state; they know the outcome of every collective moves at every state. Players do not know the actual actions chosen by other players, as usually assumed in simultaneous games. (2) Players have unlimited reasoning powers. This means, players can reason about the game they play for unlimited steps. We do not consider resource-bounded players in this paper. (3) Players do not reason about their goals. This means, players cannot modify their goals during a game. It would be interesting to see how it works by relaxing any of these assumptions. We leave these to future work.

In the study of one-step qualitative games[5,6 2,3], solution concepts were defined in a similar manner as in the traditional utilitarian games. The situation for unbounded games is more complicated.

For example, consider a game with two players, a and b, with respective goals g_a and g_b. The game is as shown in Figure 1:

Five states, q_0 to q_4, are identified. Transitions are represented by arrows, such as the arrow from q_0 to q_2 means that at state q_0, there is a move of a and a move of b such that q_2 is the next state if they make the moves. We do not label arrows with particular moves because they are irrelevant for now. First, consider it as an one-step game played at q_0. In such a game, q_2 will be a solution (a Nash equilibrium) because for all the

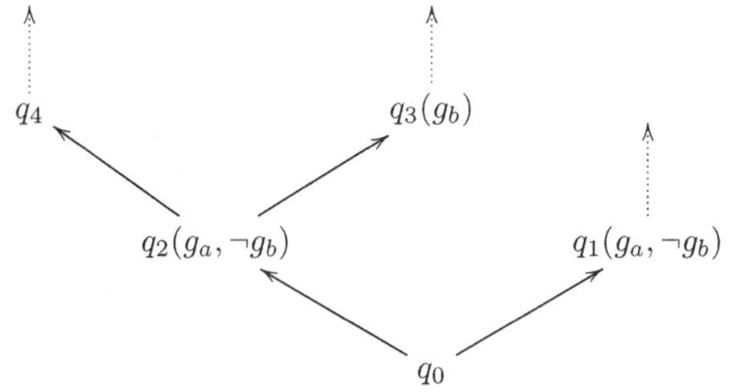

Fig. 1. An example of extensive game.

outcomes of the one-step game, i.e. q_1 and q_2, the status of satisfactions of goals are the same for the two players: g_a is satisfied but g_b is not. However, if the game is not stopped at q_1 and q_2, then q_2 may not be stable, because player b would be willing to make a move to reach q_3 at which her goal is satisfied. Since the game is played in infinite rounds, if a player's goal is not satisfied, she will look into the future for her possible success.

The point is, in such an extensive game, whether an outcome is stable shall depend on how the game may be played in the possible future. A state is stable only if all players, after looking into all the future possibilities, realize that they cannot raise their interests. Stability is one of the main concerns of a solution concept in game theory, so of course it is also one main concern of our work in this paper.

Our another concern is the formation of coalitions. The term "coalition" could be misleading. Players in a coalitional game are not to be forced into coalitions. They are self-interested and by no means concern the welfare of a group. However, players will form a coalition if they will benefit from doing so; otherwise they will defect from a coalition. The key question for coalitional games is which coalition will be formed. For this, one of the well-known solution concepts, the core, is dominating in literature of one-step coalitional games. The solution concept presented here will depend on players reasoning about other players and the future possible moves.

2. The Model

Definition 2.1. A *goal-directed unbounded coalitional game (GUCG)* is a tuple $S = (\Sigma, Q, \Pi, \pi, G, d, \delta)$, where

(1) $\Sigma = \{a_1, \ldots, a_k\}$ is a finite set of *players;*

(2) Q is a finite, non-empty set of *states*;
(3) Π is a finite, non-empty set of *propositions;*
(4) For each state $q \in Q$, $\pi(q) \subseteq \Pi$ is a set of propositions true at q;
(5) $G = \{g_1, \ldots, g_k\}$ is a set of goals, one for each player in Σ, where each $g_i \in G$ is a propositional formula built over Π.
(6) For each player $a \in \Sigma$ and each state $q \in Q$, $d_a(q) \geq 1$ is a natural number meaning that moves available to a at q are labeled with the natural numbers $1, \ldots, d_a(q)$. For each state q, a move vector at q is a tuple $(j_1, \ldots, j_k)$ such that $1 \leq j_a \leq d_a(q)$.

The transition function δ maps a state q and a move vector at q to a state, so that $\delta(q, j_1, \ldots, j_k) = q'$ means that q' is the resultant state if at state q every player a chooses move j_a in the move vector.

A coalition is a subset A of Σ. For a coalition A, let $G_A = \{g_i \mid a_i \in A\}$. Let $\models$ denote the propositionally logical consequence relation. We say that a state q satisfies a goal g if $q \models g$. Let $A_q = \{a_i \mid q \models g_i\}$, which is the set of players whose goals are satisfies at q.

For two states q and q', we say that q' is a successor of q if there is a move vector $(j_1, \ldots, j_k)$ such that $\delta(q, j_1, \ldots, j_k) = q'$. Intuitively, if the game is at state q, players can choose their moves so that q' is the next state. We denote by $succ(q)$ the set of successors of q.

Definition 2.2. A *computation* is an infinite sequence of states $\lambda = q_0, q_1, \ldots$ such that for all $i \geq 0$, the state q_{i+1} is a successor of q_i. A computation is one possible run of a game. A computation starting at state q is referred to as a *q-computation*. $\lambda[i]$ denotes q_i in the sequence λ; $\lambda[i, j]$ denotes the segment $q_i, \ldots, q_j$; $\lambda[i, \infty]$ denotes the infinite suffix $q_i, q_{i+1}, \ldots$.

Definition 2.3. Let Q^+ be the set of all non-empty finite sequences over Q. A strategy f_a for a player a is a function $f_a : Q^+ \to \mathbb{N}$ such that if the last state of a sequence λ is q, then $f_a(\lambda) \leq d_a(q)$. Thus, a strategy f_a determines for each finite sequence an a's next move. Each strategy f_a induces a set of computations that player a can enforce. Given a coalition A, a collective strategy $F_A = \{f_a \mid a \in A\}$ is a set of strategies, one for each player in A. For a state q, we define the *outcomes* of F_A from q to be the set $out(q, F_A)$ of the q-computations that the players in A enforce when they follow the strategies in F_A. Formally, if $A = \emptyset$ then $out(q, F_A)$ is the set of all q-computations; otherwise, a computation $\lambda = q_0, q_1, \ldots$ is in $out(q, F_A)$ if $q_0 = q$ and for all positions $i \geq 0$, there is a move vector $(j_1, \ldots, j_k)$ at q_i

such that $j_a = f_a(\lambda[0, i])$ for all $a \in A$, and $\delta(q_i, j_1, \ldots, j_k) = q_{i+1}$.

3. Solution Concepts

3.1. *Ending States*

An ending state is a state at which all players are not willing to make further moves. At which states players will stop their actions? It is natural to assume that players whose goals are reached at a state will be happy to stop at the state, while other players tend to make further moves in order to reach the states that satisfy their goals. Thus, the key question is, under which conditions those players will stop acting and accept their failures. The answer to this question depends on certain hypothesis of rational players.

The definition of ending states is from the following argument. Fix a state q.

(1) Players, say A, whose goals are satisfied at q have the same incentive to stop the game at q. They will form a coalition for this purpose.
(2) For players whose goals are not satisfied at q to stop making further moves, players in A should have a collective strategy F_A to make sure that the goals of players not in A will always be false, i.e., the coalition A can prevent other players to reach their goals. In this case, players not in A, as rational players, should admit the fact that it is not possible for them to succeed in the game.
(3) Also, the goals of players in A should always be satisfied if A takes the strategy F_A. If we do not have this condition, players not in A can make moves to fail some players in A, say a. Player a, as a self-interest player, may deviate from the coalition A and break her strategy in F_A for her own interest. In this case, players not in A may still have chance to win the game, and they will not stop their moves.

The point of the argument is: if such a strategy F_A exists, it will be useless for players not in A to make further moves. The exception of the argument is when $A = \Sigma$. In this case the answer is simple: Since every player is willing to stay at q, q is an end state. Formally, we have:

Definition 3.1. Given a GUCG $S = (\Sigma, Q, \Pi, \pi, G, d, \delta)$, a state q in S is an *ending state*, if there is a set of players A such that $q \models \bigwedge G_A$, and if

$A \neq \Sigma$ then there is a collective strategy F_A such that for all computations $\lambda \in out(q, F_A)$, $\lambda[j] \models \bigwedge G_A \wedge \bigwedge_{i\in\Sigma\setminus A} \neg g_i$, for every natural number j.

In this case, we call the ending state q a *winning state* for the coalition A, denoted as an *A-winning state.*

Lemma 3.1. *If q is a winning state for A, then it is not a winning state for any other coalition $A' \neq A$.*

Proof. Straightforward by Definition 3.1. □

By Lemma 3.1, it is safe to say that a state is a winning state without referring to a specified coalition. The coalition must be $A_q = \{a_i \mid q \models g_i\}$, which is totally determined by the given state q.

GUCG is an ideal model that assumes that players do not modify their goals. In real world games, for example, in an oligopolistic competition, an oligopolist is unlikely to stop acting when it is impossible to reach its current goal. Instead, it will change its goal and play another game on the new goal. Reasoning about goals is not discussed in this paper. (See Ref. 18 for a collection of papers on agents who can reason about their goals.)

We consider decision problems of winning states.

A-WINNING STATE: Given a GUCG, a coalition A and a state q, is q a winning state for A?

WINNING STATE: Given a GUCG and a state q, is q a winning state?

A GUCG $S = (\Sigma, Q, \Pi, \pi, G, d, \delta)$ is *turn-based* if at each $q \in Q$, there is a player a such that $d_b(q) = 1$ for all players $b \in \Sigma\setminus\{a\}$. In this case, we say that at state q, it is the turn of player a. A turn-based game is such that at each state, only one player can choose her moves.

Theorem 3.1. *The complexity of A-WINNING STATE is PTIME-complete.*

Proof. The case that $A = \Sigma$ is trivial. When $A \neq \Sigma$, determining whether q is a winning state for A is equivalent to checking whether the ATL formula $\langle\langle A\rangle\rangle\Box(\bigwedge G_A \wedge \bigwedge_{i\in\Sigma\setminus A} \neg g_i)$ is true at q. The model checking problem for ATL is PTIME-complete[11]. It follows that A-WINNING STATE is in PTIME.

The reachability in and/or graphs is PTIME-complete[19]. Given an and/or graph G and two vertices x and y in the graph, we construct a turn-based GUCG with the same vertices (states) and edges (transitions) as G's, except that we change each dead ends in G to be a reflexive state in the GUCG so that the GUCG is well defined. The GUCG has two players, a and b, with respective goals g_a and g_b. Let g_a be true at all states and g_b false at all states except for y. Furthermore, let any "and" vertice in G be the turn of a and any "or" vertice in G be the turn of b. Then y is reachable from x in the and\or graph iff a does not have a strategy to avoid y, which is equivalent to that x is not an $\{a\}$-winning state in the turn-based GUCG. This provides a logspace reduction from the reachability in and/or graphs to A-WINNING STATE. The lower bound is immediate. □

Because the size of the GUCG is the only input parameter, by Lemma 3.1, WINNING STATE shares the same complexity with A-WINNING STATE.

However, if the number of players is also an input parameter, the complexity of the two problems will be much different: A-WINNING STATE is still PTIME-complete, while WINNING STATE is $\Sigma_2 P$-complete! (See Section 3.4 for more discussion.)

3.2. *Successful Coalition*

A-WINNING STATE does not answer the question of coalition formation. Players will form a coalition if doing so can let them win a game. In other words, if players in A can cooperate to guarantee that the game ends at some A-winning state, no matter what other players do, then they will form coalition A. Fix a starting state s.

Definition 3.2. We say that a coalition A is *successful* (at s), if there is a collective strategy F_A such that for any computation $\lambda \in out(s, F_A)$, there is a winning state $q \in \lambda$ for A.

A successful coalition is a coalition that may actually be formed.

SUCCESSFUL COALITION: Given a GUCG, a coalition A, and a starting state s, is A successful at s?

SUCCESSFUL COALITION EXIST: Given a GUCG and a starting state s, is there a successful coalition at s?

Theorem 3.2. *The complexity of SUCCESSFUL COALITION and SUCCESSFUL COALITION EXIST is PTIME-complete.*

Proof. Because the number of players is not an input parameter, the two problems share the same complexity. Given a coalition $A \neq \Sigma$, the problem of whether it is successful at s is equivalent to whether the ATL* formula

$\langle\langle A\rangle\rangle \diamond \langle\langle A\rangle\rangle\Box(\bigwedge G_{A_q} \wedge \bigwedge_{i\in\Sigma\setminus A_q} \neg g_i)$ is true at the starting state s. The model checking problem for ATL* is 2EXPTIME-complete. However, it is PTIME-complete if formulas being checked are bounded by size[11]. Thus, SUCCESSFUL COALITION is in PTIME.

For hardness, we again use and/or graphs. Given an and\or graph G and two vertices x and y. First, consider the and/or graph as a usual graph, that is, reset all its vertices to be marked "or". Then we obtain a graph by eliminating all vertices that are reachable from y. After that we recover an and/or graph from it so that all vertice (not eliminated) have their marks as in G, either "and" or "or". Call the new and/or graph G'. Because an and\or graph is a directed and acyclic graph, the procedure will not change the status of the reachability from x to y in the original and/or graph, that is, y is reachable from x in G iff it is in G'.

We construct a turn-based GUCG in exactly the same way as in the proof of Theorem 3.1. Then y is reachable from x in G' (and G) iff $\{a\}$ is not a successful coalition in the turn-based game. The lower bound is immediate. □

The most studied solution concept in coalitional games is so-called the *core* (pp. 257-274,[1]). Successful coalition and the core are closely related concepts. Intuitively, the core of a coalition is the set of winning strategies for that coalition. In a GUCG, given a coalition A and a state s, we say that a collective strategy F_A is in the *core* of A at s, if for all computation $\lambda \in out(s, F_A)$, there is a winning state $q \in \lambda$ for A.

The problem of CORE MEMBERSHIP is to determine whether a given collective strategy is in the core. However, a strategy in a GUCG is potentially infinite without any finite representation. Thus, in general the recognition time is unbounded for such an input.

CORE NON-EMPTY: Given a GUCG, a coalition A and a state s, is there a collective strategy that is in the core of A.

Apparently, CORE NON-EMPTY is the same problem as SUCCESSFUL COALITION.

A GUCG is a potentially infinite game. However, it will be bounded by wining states if they exist: A winning state is an end state where no player will make further moves. We say that a GUCG with a starting state s is *strongly bounded* if for any s-computation λ in the game, there is a winning state $\lambda[i]$ for some i. A strongly bounded GUCG is guaranteed to terminate in finite steps. A *cycle* in a computation λ is a segment $\lambda[i,j]$ of λ such that $\lambda[i] = \lambda[j]$. It is easy to check that a GUCG with a starting state s is strongly bounded if and only if for any s-computation λ in it, there is no cycle $\lambda[i,j]$ such that $\lambda[k]$ is not a winning state for any $i \leq k \leq j$.

Strongly bounded, as the name suggests, is not a necessary condition for a game to be bounded. Even though there exist unbounded computations, the game may never enter into those computations because they may be avoided by a successful coalition taking its winning strategy. We say that a GUCG with a starting state s is *bounded* if there exists some successful coalition at s in the GUCG. To check the boundedness is the same problem as SUCCESSFUL COALITION EXIST.

3.3. *Dependencies among Players*

Lemma 3.2. *Given a GUCG and a state s, if a coalition A is successful at s, then any coalition A' with $A \cap A' = \emptyset$ is not successful at s.*

Proof. Straightforward and Omitted. □

It is possible that there are more than one successful coalitions at a given state. By Lemma 3.2, any pair of successful coalitions intersects, e.g. both $\{a, b\}$ and $\{a, c\}$ being successful coalitions. In this case, players may negotiate to form an actual coalition. Negotiation is not the topic of the present paper. In the current framework, we can tell that player a is the key player in the game, because she is guaranteed in a successful coalition, and whether b wins or c wins depends on which coalition player a will join.

Let $A_1, \ldots, A_m$ be the set of successful coalitions at a given state. We call the set $K = \bigcap\{A_1, \ldots, A_m\}$ the *key*, and players in K *key players.*

KEY MEMBERSHIP: Given a GUCG, a player, and a state, is the player a key player at the state?

KEY NON-EMPTY: Given a GUCG and a state, is there some key player at the state?

Theorem 3.3. *The complexity of KEY MEMBERSHIP and KEY NON-EMPTY are PTIME-complete.*

Proof. Since the number of players is not an input parameter, these two problems are in the same complexity class as SUCCESSFUL COALITION. □

A player a is a *veto player* of another player b at state q, if $a \in A$ for any successful coalition A at q with $b \in A$. Players a and b are *mutually dependent* if they are veto players of each other. Also because the number of players is not an input parameter, the following decision problems are PTIME-complete.

VETO PLAYER: Given a GUCG, a state, and two players a and b, is a a veto player of b at the state?

MUTUALLY DEPENDENT: Given a GUCG, a state, and two players, are the two players mutually dependent at the state?

3.4. *Player-complexity*

All the decision problems so far in this paper are PTIME-complete. We have seen that one of the reasons of the tractability is that we do not take the number of players into account. If the number of players is not fixed, the decision problems of GUCG will be intractable. The *player-complexity* of a decision problem (in this paper) is its complexity with the number of players also as an input parameter.

Theorem 3.4. *The player-complexity of WINNING STATE is $\Sigma_2 P$-complete.*

Proof. That a state q is a winning state for A is equivalent to that the ATL formula $\langle\langle A\rangle\rangle\Box(\bigwedge G_A \wedge \bigwedge_{i\in\Sigma\backslash A}\neg g_i)$ is true at q. It was shown in Ref. 20 that the model checking problem of ATL, if the number of players is also an input parameter, is $\Sigma_2 P$-complete.

The hardness is through a polynomial reduction of $QSAT_2$ to WINNING STATE. $QSAT_2$ is a well-known $\Sigma_2 P$-complete problem. For each quantified Boolean formula $\exists P_1 \forall P_2 \theta$, where $P_1 \cup P_2 = \{p_1, \ldots, p_k\}$, we construct a GUCG consists of three states q_0, $q_\bot$, and $q_\top$. The set of players is $\Sigma = \{a_1, \ldots, a_k\}$, one for each propositions, with the intended meaning that agent a_i controls the value of proposition p_i. Thus, move vectors are exactly propositional valuations to $\{p_1, \ldots, p_k\}$. We define $\delta(q_0, j_1, \ldots, j_k) = q_\top$ iff $j_1, \ldots, j_k \models \theta$; otherwise, $\delta(q_0, j_1, \ldots, j_k) = q_\bot$. Let A be the set of players that control propositions in P_1. Then $\Sigma\backslash A$ consists of players that control propositions in P_2. Let G_A be true at q_0 and $q_\top$, and false at $q_\bot$; g_b is false at all states for all $b \in \Sigma\backslash A$. It is not hard to check that $\exists P_1 \forall P_2 \theta$ iff q_0 is a winning state. $\Box$

We will not proceed with player-complexity of other decision problems. We conclude that GUCG is only suitable for characterizing games played by a relatively small group.

4. Conclusion

We presented goal-directed unbounded coalitional games in this paper, and study various solution concepts of these games, and give complexity of their decision problems. GUCG is an ideal game model with certain assumptions. We may investigate more practical games by losing these assumptions.

One of the bases of the framework of GUCG is that players reason about each other, and reason about the future possibilities. Resource-bounded agents is an active research area of AI. In GUCG, a resource-bounded player may have only limited reasoning power. For example, we could limit the reasoning power of a player by assigning a natural number n to it so that she can only reason about the future for at most n steps. Another interesting topic would be players with complex and changing goals.

Complexity in this paper is measured with respect to the size of a game model (except for Section 3.4). As shown in Ref. 21, the basic structure can be built from a given set of propositions. The size of the structure is exponentially larger than the number of propositions. Though most complexity measurements in the literature are against the size of a model, there are many investigations on measurements against the number of propositions[22]. Usually, the latter is in higher complexity classes than the former. It would be interesting to investigate it in the context of GUCG.

References

1. M. J. Osborne and A. Rubinstein, *A Course in Game Theory* (The MIT press, 1994).
2. P. Dunne, W. van der Hoek, S. Kraus and M. Wooldridge, Cooperative boolean games, in *Proceedings of the 7th international joint conference on Autonomous agents and multiagent systems*, eds. L. Padgham and D. Paekes (Estoril, 2008).
3. L. Sauro, L. van der Torre and S. Villata, Dependency in cooperative boolean games, in *Proceedings of the Third KES International Symposium on Agent and Multi-agent Systems: Technologies and Applications*, (Springer, 2009).
4. L. De Alfaro, T. Henzinger and O. Kupferman, Concurrent reachability games, *Theoretical Computer Science* **386**, 188 (2007).
5. M. Wooldridge and P. E. Dunne, On the computational complexity of qualitative coalitional games, *Artificial Intelligence* **158**, 27 (2004).
6. M. Wooldridge and P. E. Dunne, On the computational complexity of coalitional resource games, *Artificial Intelligence* **170**, 835 (2006).

7. M. Wooldridge, *Reasoning about rational agents* (MIT press, Cambridge, MA, 2000).
8. M. Ghallab, D. Nau and P. Traverso, *Automated Planning: Theory and Practice* (Morgan Kaufmann, 2004).
9. P. Haddawy and S. Hanks, Utility models for goal-directed, decision-theoretic planners, *Computational Intelligence* **14**, 392 (1998).
10. R. Alur, T. Henzinger and O. Kupferman, Alternating-time temporal logic, *Proceedings of the 38th Annual Symposium on Foundations of Computer Science* , 100 (1997).
11. R. Alur, T. Henzinger and O. Kupferman, Alternating-time temporal logic, *Journal of the ACM* **49**, 672 (2002).
12. M. Pauly, Logic for social software, PhD thesis, Institute for Logic, Language and Computation, University of Amsterdam, (Amsterdam University Press, 2001).
13. M. Pauly, A modal logic for coalitional power in games, *Journal of Logic and Computation* **12**, 149 (2002).
14. V. Goranko and W. Jamroga, Comparing semantics of logics for multi-agent systems, *Synthese* **139**, 241 (2004).
15. D. G. Pearce, Repeated games: Cooperation and rationality, in *Advances in Economic Theory: Sixth World Congress*, ed. J. J. Laffont (Cambridge University Press, 1992).
16. T. Ågotnes, W. Van Der Hoek and M. Wooldridge, Logics for qualitative coalitional games, *Logic Journal of IGPL* **17**, 299 (2009).
17. L. De Alfaro and T. A. Henzinger, Concurrent omega-regular games, *Proceedings of 15th Annual IEEE Symposium on Logic in Computer Science* , 141 (2000).
18. D. W. Aha, M. Klenk, H. Munoz-Avila, A. Ram and D. Shapiro, *Proceedings of AAAI-10 Workshop on Goal Directed Autonomy* (AAAI, 2010).
19. N. Immerman, Number of quantifiers is better than number of tape cells, *Journal of Computer and System Sciences* **22**, 384 (1981).
20. W. Jamroga and J. Dix, Do agents make model checking explode (computationally)?, in *Proceedings of the 4th International Central and Eastern European Conference on Multi-Agent Systems and Applications*, eds. M. Pechoucek, P. Petta and L. Z. Varga (Springer-Verlag, 2005).
21. E. M. Clarke, O. Grumberg and D. A. Peled, *Model checking* (MIT press, Cambridge, MA, 2000).
22. P. Schnoebelen, The complexity of temporal logic model checking, in

Proceedings of the 4th Conference Advances in Modal Logic (AiML 2002), volume 4, (King's College Publications, 2003).

On Extensions of Basic Propositional Logic

Minghui Ma

Institute for Logic and Intelligence, Southwest University
Chongqing, 400715/Beibei, P.R.China
E-mail: mmh.thu@gmail.com

Katsuhiko Sano

School of Information Science,
Japan Advanced Institute of Science and Technology
Nomi, Ishikawa, 923-1292, Japan
E-mail: katsuhiko.sano@gmail.com

Extensions of Visser's basic propositional logic (**BPL**) are explored. We present the lattice of extensions of **BPL**: the lattice of superbasic logics. From a uniform perspective, we establish logical properties of superbasic logics, including strong completeness, finite model property, disjunction property, Post completeness, and two constant property. Moreover, the embedding of superbasic logics into normal modal logics by Gödel-McKinsey-Tarski style translations is revisited.

Keywords: Basic propositional logic; Superbasic logics; Two constant property; Post-completeness; Embedding.

1. Introduction

Visser introduced basic propositional logic (**BPL**) and formal provability logic **FPL**. The logic **FPL** is an extension of **BPL** and can be used to interpret the notion of formal provability in Peano Arithmetic (see Ref. 1). Visser proved that Gödel-McKinsey-Tarski style translations, which embed the intuitionistic logic (**Int**) into modal logics, can also embed **FPL** into the modal provability (Gödel-Löb) logic. Note that **Int** is not comparable with **FPL**, but both of them are extensions of **BPL**.

Visser's logic **BPL** was originally presented by a natural deduction system. However, sequent calculi for **BPL** were discussed in several papers (Refs. 2–4). The Gentzen-style calculus for the predicate extension of basic propositional logic is also given in Ref. 5. Suzuki and Ono presented a Hilbert-style axiomatization for **BPL** in Ref. 6. Ishigaki and Kikuchi presented tree sequent calculi and Hilbert-style systems for subintuitionistic predicate logics in Ref. 7. The propositional parts of those predicate logics include **BPL**, **Int** and a sublogic of **BPL**. Moreover, the algebraic semantics for **BPL** is also presented in Refs. 8,9.

It is already known from Ref. 10 that **BPL** is embedded into the modal logic **K4**. The lattice `NExt`(**K4**) of all normal extensions of the modal logic **K4** is well studied and many important results are obtained (cf. Ref. 11). It is interesting to look at extensions of **BPL** via Gödel-McKincy-Tarski style translations, i.e., to seek for the propositional companion of `NExt`(**K4**).

The aim of this paper is to investigate superbasic logics, which enable us to capture extensions of **BPL**. The results we obtained are the following:

- We construct canonical model for every superbasic logic which is a variant of the canonical general model in Ref. 6, and we show strong completeness of **BPL** (Theorem 3.2).
- By model constructions, we show the finite model property and the disjunction property of **BPL** (first shown by Ardeshir in Ref. 12, Theorem 4.1) and **FPL** (Theorem 6.4). These properties can also be proved for some other superbasic logics (Theorem 5.4).
- We define the concept of two constant property (every constant formula is equivalent to $\top$ or $\bot$), and show that **BPL** has no two constant property (Theorem 5.1), but the superbasic logic **DNT**, which is a proper subintuitionistic logic, has this property (Theorem 5.2).
- We construct a sound and complete natural deduction system for the logic **Log**($\bullet$) of the single irreflexive point, which can also be characterized as a set of formulas inductively constructed from implicational formulas (Theorem 7.4).
- Using the construction of skeleton frames, we show that some superbasic logics can be embedded into modal logics via Gödel-McKinsey-Tarski style translations (Theorems 8.1, 8.2, 8.3 and 8.4).

The outline of this paper is as follows. We first give some basic concepts of **BPL** and its extensions in Section 2. The canonical model for a superbasic logic is discussed in Section 3. Then in Section 4, we discuss some model and frame constructions which can be used to show the disjunction property and the finite model property and of **BPL**. Section 5 introduces the concept of two constant property for superbasic logics. Section 6 compares intuitionistic logic with formal provability logic. Section 7 discusses Post complete logics and sets. Section 8 discusses some results on the embedding of superbasic logics into normal modal logics.

2. Basic Concepts of BPL and its Extensions

The language $\mathcal{L}$ of **BPL** is the same as the language of intuitionistic logic (**Int**) and classical propositional logic (**CL**) in Ref. 11 (p.26), i.e., the language consisting of a denumerable set Prop of propositional variables, and propositional connectives $\bot, \wedge, \vee$ and $\rightarrow$. The set For of all formulas is given by the following rule:

$$\alpha ::= p \mid \bot \mid (\alpha \rightarrow \alpha) \mid (\alpha \wedge \alpha) \mid (\alpha \vee \alpha), \text{ where } p \in \mathsf{Prop}.$$

Define $\top := \bot \rightarrow \bot$ and $\neg\alpha := \alpha \rightarrow \bot$. Let $\mathsf{Var}(\alpha)$ be the set of propositional variables occurring in α. We say that α is a *constant formula*, if $\mathsf{Var}(\alpha) = \emptyset$. Let CFor be the set of all constant formulas. Given any formula α, we denote by $\mathsf{Sub}(\alpha)$ the set of all subformulas of α. We also extend this notation also for a set Σ of formulas, i.e., $\mathsf{Sub}(\Sigma) := \bigcup_{\alpha\in\Sigma} \mathsf{Sub}(\alpha)$.

Now we introduce the Kripke semantics for **BPL** (Ref. 1).

Definition 2.1. A (*Kripke*) *frame* is a pair $\mathfrak{F} = (W, R)$ where W is a non-empty set and $R \subseteq W \times W$ is a transitive relation. Given a frame $\mathfrak{F} = (W, R)$, a valuation in $\mathfrak{F}$ is a function $V : \mathsf{Prop} \rightarrow \mathcal{P}(W)$ from the set Prop to the power set of W satisfying the following persistency condition: if $w \in V(p)$ and wRu imply $u \in V(p)$ for all $w, u \in W$ and all $p \in \mathsf{Prop}$. A (*Kripke*) *model* $\mathfrak{M} = (W, R, V)$ is a pair of a Kripke frame (W, R) and a valuation V in (W, R).

Throughout this paper, we assume that all (Kripke) frames are transitive and that all models are persistent. Given a model $\mathfrak{M} = (W, R, V)$, the satisfaction relation $\mathfrak{M}, w \models \alpha$ is defined as follows:

$$\begin{array}{ll}
\mathfrak{M}, w \models p & \text{iff } w \in V(p), \\
\mathfrak{M}, w \models \bot & \quad\text{Never}, \\
\mathfrak{M}, w \models \alpha \wedge \beta & \text{iff } \mathfrak{M}, w \models \alpha \text{ and } \mathfrak{M}, w \models \beta, \\
\mathfrak{M}, w \models \alpha \vee \beta & \text{iff } \mathfrak{M}, w \models \alpha \text{ or } \mathfrak{M}, w \models \beta, \\
\mathfrak{M}, w \models \alpha \rightarrow \beta & \text{iff } \forall u \in W((wRu \text{ and } \mathfrak{M}, u \models \alpha) \text{ implies } \mathfrak{M}, u \models \beta).
\end{array}$$

For every formula α, define $[\![\alpha]\!]_{\mathfrak{M}} = \{w \in W \mid \mathfrak{M}, w \models \alpha\}$, the *truth set* of α in $\mathfrak{M}$. For $w \in W$ and $A \subseteq W$, define $R(w) = \{u \in W \mid Rwu\}$ and $R[A] = \bigcup_{w\in A} R(w)$. Define the binary operation $\Rightarrow_R$: $\mathcal{P}(W)^2 \rightarrow \mathcal{P}(W)$ over the power set of W by putting $A \Rightarrow_R B = \{w \in W \mid R(w) \cap A \subseteq B\}$. It is easy to see that $[\![\alpha \rightarrow \beta]\!]_{\mathfrak{M}} = [\![\alpha]\!]_{\mathfrak{M}} \Rightarrow_R [\![\beta]\!]_{\mathfrak{M}}$.

Definition 2.2. A formula α is *globally true* or *valid* in $\mathfrak{M} = (W, R, V)$ (notation: $\mathfrak{M} \models \alpha$), if $[\![\alpha]\!]_{\mathfrak{M}} = W$. A formula α is *locally valid* at state w in a frame $\mathfrak{F}$ (notation: $\mathfrak{F}, w \models \alpha$), if α is true at w in $(\mathfrak{F}, V)$ for any valuation V in $\mathfrak{F}$. A formula α is *valid* in $\mathfrak{F}$ (notation: $\mathfrak{F} \models \alpha$), if $\mathfrak{F}, w \models \alpha$ for all states w in $\mathfrak{F}$. We also naturally extend these notions also for a set of formulas. For example, $\mathfrak{F} \models \Gamma$ means that $\mathfrak{F} \models \alpha$ for all $\alpha \in \Gamma$.

Proposition 2.1. *Given a model* $\mathfrak{M} = (W, R, V)$ *and a formula* α, $[\![\alpha]\!]_{\mathfrak{M}}$ *is persistent with respect to* R, *i.e., for all* $w, v \in W$, $\mathfrak{M}, w \models \varphi$ *and* wRv *jointly imply* $\mathfrak{M}, v \models \varphi$ *(in other words,* $R[[\![\alpha]\!]_{\mathfrak{M}}] \subseteq [\![\alpha]\!]_{\mathfrak{M}}$*).*

Proof. By induction on the length of α. The cases for propositional variables, conjunction and disjunction are obvious. Suppose $\alpha = \beta \to \gamma$. Assume $\mathfrak{M}, w \models \beta \to \gamma$ and wRu. Our goal is to show $\mathfrak{M}, u \models \beta \to \gamma$. Assume uRv and $\mathfrak{M}, v \models \beta$. By the transitivity of R, we get wRv. By the assumption, $\mathfrak{M}, v \models \gamma$. Hence $\mathfrak{M}, u \models \beta \to \gamma$. □

Given a class $\mathbb{F}$ of frames, define the logic of $\mathbb{F}$ as the set:

$$\mathbf{Log}(\mathbb{F}) = \{\alpha \in \mathsf{For} \mid \mathfrak{F} \models \alpha \text{ for all } \mathfrak{F} \in \mathbb{F}\}.$$

Conversely, given a set Γ of formulas, define the frame class for Γ as:

$$\mathbb{FR}(\Gamma) = \{\mathfrak{F} \mid \mathfrak{F} \models \gamma \text{ for all } \gamma \in \Gamma\}.$$

The following proposition is obvious:

Proposition 2.2. *For any frame classes* $\mathbb{F}_1$ *and* $\mathbb{F}_2$, *and any sets* Γ_1 *and* Γ_2 *of formulas,* (i) *if* $\mathbb{F}_1 \subseteq \mathbb{F}_2$, *then* $\mathbf{Log}(\mathbb{F}_2) \subseteq \mathbf{Log}(\mathbb{F}_1)$*; and* (ii) *if* $\Gamma_1 \subseteq \Gamma_2$, *then* $\mathbb{FR}(\Gamma_2) \subseteq \mathbb{FR}(\Gamma_1)$.

Table 1. Classes of transitive Kripke frames

$\mathbb{TR}$	the class of all the transitive frames
$\mathbb{ITR}_{\mathrm{fin}}$	the class of all the irreflexive and transitive finite frames
$\mathbb{ISR}$	the class of all the transitive frame satisfying $\forall x.\exists y.yRx$.
$\mathbb{SR}$	the class of all the transitive frame satisfying $\forall x.\exists y.xRy$.
$\mathbb{PRE}$	the class of all the transitive and reflexive frames
$\circ$	the singleton reflexive Kripke frame
$\bullet$	the singleton irreflexive Kripke frame

Table 1 lists classes of transitive Kripke frames considered in this paper, where $\mathbb{SR}$ is the class of all transitive and serial frames and $\mathbb{ISR}$ is the class of all transitive and inverse-serial (i.e., the inverse of R is serial) frames.

Before introducing Visser's **BPL**, we introduce the construction of lifting which allows us to show the admissibility of modus ponens, i.e., from α and $\alpha \to \beta$, we may infer β.

Definition 2.3. A frame class $\mathbb{F}$ *admits modus ponens* if $\mathbb{F} \models \alpha$ and $\mathbb{F} \models \alpha \to \beta$ jointly imply $\mathbb{F} \models \beta$, for all formulas α and β.

In other words, $\mathbb{F}$ admits modus ponens, if for formulas α and β, $\alpha \in \mathbf{Log}(\mathbb{F})$ and $\alpha \to \beta \in \mathbf{Log}(\mathbb{F})$ imply $\beta \in \mathbf{Log}(\mathbb{F})$.

Definition 2.4 (Lifting). *Let $\mathfrak{F} = (W, R)$ a frame and $\circ \notin W$ a point. The frame $\mathfrak{F}^\circ = (W \cup \{\circ\}, R \cup \{\langle \circ, \circ \rangle, \langle \circ, u \rangle \mid u \in W\})$ is called the* lifting *of $\mathfrak{F}$.*[a] *We say that a class of frames $\mathbb{F}$ is* closed under liftings, *if $\mathfrak{F}^\circ \in \mathbb{F}$ for all $\mathfrak{F} \in \mathbb{F}$. Let V be a valuation. The model $\mathfrak{M}^\circ = (\mathfrak{F}^\circ, V^\circ)$ is called the* lifting *of $\mathfrak{M} = (\mathfrak{F}, V)$ where $V^\circ : \mathsf{Prop} \to \mathcal{P}(W \cup \{\circ\})$ is defined by $V^\circ(p) = V(p)$ for every $p \in \mathsf{Prop}$ (no propositional variable is true at $\circ$).*

Proposition 2.3. *Let $\mathfrak{M} = (W, R, V)$. Then, for all formula α and $w \in W$, $\mathfrak{M}, w \models \alpha$ iff $\mathfrak{M}^\circ, w \models \alpha$.*

Proposition 2.4. *Let $\mathbb{F}$ be a class of frames closed under liftings. Then $\mathbb{F}$ admits modus ponens.*

Proof. Let $\mathbb{F}$ be closed under liftings. Assume that $\mathbb{F} \models \alpha$ and $\mathbb{F} \models \alpha \to \beta$. In order to show $\mathbb{F} \models \beta$, fix a frame $\mathfrak{F} = (W, R) \in \mathbb{F}$, a state $w \in W$ and a persistent valuation $V : \mathsf{Prop} \to \mathcal{P}(W)$. Write $\mathfrak{M} = (\mathfrak{F}, V)$. Since $\mathbb{F}$ is closed under liftings, consider the lifting $\mathfrak{F}^\circ \in \mathbb{F}$ of $\mathfrak{F}$. By the assumption, $\mathfrak{F}^\circ \models \alpha$ and $\mathfrak{F}^\circ \models \alpha \to \beta$. It follows that $\mathfrak{M}^\circ, \circ \models \alpha$ and $\mathfrak{M}^\circ, \circ \models \alpha \to \beta$, which implies $\mathfrak{M}^\circ, w \models \beta$ since $\circ$ can see w in the lifting $\mathfrak{M}^\circ$. It follows from Proposition 2.3 that $\mathfrak{M}, w \models \beta$. □

Note that the lifting $\mathfrak{F}^\circ$ is transitive since $\mathfrak{F}$ is transitive. In Table 1, all classes except $\circ$ and $\bullet$ are closed under liftings. So they admit modus ponens in the sense of Proposition 2.4. However, it is still easy to show that $\{\circ\}$ admits modus ponens by the following proposition (because $\circ \in \mathbb{ISR}$):

Proposition 2.5. *If $\mathbb{F} \subseteq \mathbb{ISR}$, then $\mathbb{F}$ admits modus ponens.*

Proof. Assume that $\mathbb{F} \models \alpha$ and $\mathbb{F} \models \alpha \to \beta$. For showing $\mathbb{F} \models \beta$, fix a frame $\mathfrak{F} = (W, R) \in \mathbb{F}$, $w \in W$ and a valuation V in $\mathfrak{F}$. Let $\mathfrak{M} = (W, R, V)$.

[a]The notion of lifting is from order theory in Ref. 13 (p.15). The lifting of an partial order is to add an element as the bottom of the new expanded partial order.

Since $w \in W$ has a predecessor w' such that $w'Rw$, we have $\mathfrak{M}, w' \models \alpha$ and $\mathfrak{M}, w' \models \alpha \to \beta$. By $w'Rw$, we have $\mathfrak{M}, w \models \beta$. □

Since $\mathbb{ISR} \subseteq \mathbb{PRE}$, we also obtain the following:

Corollary 2.1. *For any class $\mathbb{F} \subseteq \mathbb{PRE}$, $\mathbb{F}$ admits modus ponens.*

Now we recall some proof systems for **BPL**. First we present the natural deduction system of **BPL** given by Visser.

Definition 2.5 (Ref. 1). *The natural deduction system $\mathcal{N}$**BPL** consists of the following rules:*

$$\frac{\alpha \quad \beta}{\alpha \wedge \beta}(\wedge I) \quad \frac{\alpha_1 \wedge \alpha_2}{\alpha_i}(\wedge E_i) \quad \frac{\alpha_i}{\alpha_1 \vee \alpha_2}(\vee I_i) \quad \frac{\alpha \vee \beta \quad \begin{array}{c}[\alpha]\\ \vdots\\ \gamma\end{array} \quad \begin{array}{c}[\beta]\\ \vdots\\ \gamma\end{array}}{\gamma}(\vee E) \quad \frac{\begin{array}{c}[\alpha]\\ \vdots\\ \beta\end{array}}{\alpha \to \beta}(\to I)$$

$$\frac{\bot}{\alpha}(\bot E) \quad \frac{\alpha \to \beta \quad \beta \to \gamma}{\alpha \to \gamma}(Tr) \quad \frac{\alpha \to \beta \quad \alpha \to \gamma}{\alpha \to (\beta \wedge \gamma)}(\wedge I_f) \quad \frac{\alpha \to \gamma \quad \beta \to \gamma}{(\alpha \vee \beta) \to \gamma}(\wedge E_f)$$

Next we recall Suzuki and Ono's Hilbert-style axiomatization of **BPL** in Ref. 6, which is an extension of Corsi's axiomatization of the logic **F** of strict implication in Ref. 10.

Definition 2.6 (Ref. 6). *Hilbert-style axiomatization $\mathcal{H}$**BPL** consists of the following axioms and inference rules:*

(A1) $p \to p$	(A7) $p \to (q \to p \wedge q)$
(A2) $p \to (q \to p)$	(A8) $p \to p \vee q$
(A3) $(p \to q) \wedge (q \to r) \to (p \to r)$	(A9) $q \to p \vee q$
(A4) $p \wedge q \to p$	(A10) $(p \to r) \wedge (q \to r) \to (p \vee q \to r)$
(A5) $p \wedge q \to q$	(A11) $p \wedge (q \vee r) \to (p \wedge q) \vee (p \wedge r)$
(A6) $(p \to q) \wedge (p \to r) \to (p \to q \wedge r)$	(A12) $\bot \to p$

(MP) *From α and $\alpha \to \beta$, infer β*

(US) *From α, infer any uniform substitution instance of α.*

In what follows, we also use **BPL** to denote the set of all the theorems of $\mathcal{H}$**BPL**. It is easy to show that $\mathcal{H}$**BPL** is sound with respect to $\mathbb{TR}$, i.e., **BPL** $\subseteq$ **Log**$(\mathbb{TR})$.

Proposition 2.6. **BPL** $\subsetneq$ **Int**.

Proof. Note that **Int** = **Log**($\mathbb{PRE}$) and $\mathbb{PRE} \subseteq \mathbb{TR}$. Hence **BPL** $\subseteq$ **Int**. For **Int** $\neq$ **BPL**, we show that both $(p \wedge (p \to q)) \to q$ and $(p \to (p \to q)) \to (p \to q)$ are theorems of **Int** but they are not in **BPL**. It is easy to check that they are valid in $\mathbb{TR}$. Now consider the model (W, R, V) where $W = \{0, 1\}$, $R = \{(0, 1)\}$ and $V(p) = \{0, 1\}$, $V(q) = \{1\}$. Then, the formula $(p \wedge (p \to q)) \to q$ is false at 0. For the formula $(p \to (p \to q)) \to (p \to q)$, we change the valuation into V' such that $V'(p) = \{1\}$ and $V'(q) = \emptyset$. Then this formula is false at 0. □

In Ref. 6, Suzuki and Ono defined the canonical model for $\mathcal{H}$**BPL** based on general frames and proved the weak completeness of $\mathcal{H}$**BPL**. In the next section, we will generalize their completeness result to obtain the strong completeness of **BPL** with respect to the class of all transitive frames. In order to generalize their completeness result as much as possible, here we introduce *superbasic logics* as follows:

Definition 2.7. A *superbasic logic* is a set Λ of formulas such that **BPL** $\subseteq \Lambda$ and Λ is closed under both modus ponens (MP) and uniform substitutions (US) of Definition 2.6.

Definition 2.8. Given any superbasic logic Λ, define the class $\mathtt{Ext}(\Lambda)$ of all superbasic logics extending Λ by:

$$\mathtt{Ext}(\Lambda) := \{\Lambda' \mid \Lambda \subseteq \Lambda' \text{ and } \Lambda' \text{ is a superbasic logic}\}.$$

By the definition of superbasic logics, we have the following:

Proposition 2.7. *For* $\{\Lambda_i : i \in I\} \subseteq \mathtt{Ext}(\mathbf{BPL})$, $\bigcap_{i \in I} \Lambda_i \in \mathtt{Ext}(\mathbf{BPL})$.

Given a set Γ of formulas, define the *superbasic logic generated by* Γ as:

$$\mathbf{BPL} \oplus \Gamma = \bigcap \{\Lambda : \mathbf{BPL} \cup \Gamma \subseteq \Lambda \text{ and } \Lambda \text{ is a superbasic logic}\}.$$

Given a family $\{\Lambda_i : i \in I\} \subseteq \mathtt{Ext}(\mathbf{BPL})$, define $\bigoplus_{i \in I} \Lambda_i = \mathbf{BPL} \oplus \bigcup_{i \in I} \Lambda_i$, i.e., the closure of $\bigcup_{i \in I} \Lambda_i$ under MP and US. It is clear that $\bigoplus_{i \in I} \Lambda_i \in \mathtt{Ext}(\mathbf{BPL})$. In the case of two superbasic logics Λ_1 and Λ_2, we use notations such as $\Lambda_1 \cap \Lambda_2$ and $\Lambda_1 \oplus \Lambda_2$. Then we get the following propositions immediately.

Proposition 2.8. *The class of all super basic logics* $\mathtt{Ext}(\mathbf{BPL})$ *forms a bounded lattice with respect to binary operations* $\cap$ *and* $\oplus$, *where* **BPL** *is the bottom element, and the set* For *of all formulas is the top element.*

Proposition 2.9. $\mathbf{BPL} \oplus \Gamma = \mathbf{BPL} \oplus \alpha_1 \wedge \ldots \wedge \alpha_n$ *for any finite set* $\Gamma = \{\alpha_1, \ldots, \alpha_n\}$ *of formulas.*

Here we present several superbasic logics which will be studied in the following sections:

$$\mathbf{DNT} = \mathbf{BPL} \oplus \neg\neg\top \qquad \mathbf{Int} = \mathbf{BPL} \oplus (p \wedge (p \to q)) \to q$$
$$\mathbf{CL} = \mathbf{Int} \oplus p \vee \neg p \qquad \mathbf{FPL} = \mathbf{BPL} \oplus ((\top \to p) \to p) \to (\top \to p)$$

Recall that we have defined the logics of classes of frames. Which of them are superbasic? Since every logic $\mathbf{Log}(\mathbb{F})$ is closed under US, then the question is converted into: for which frame classes $\mathbb{F}$ is the logic $\mathbf{Log}(\mathbb{F})$ closed under modus ponens? By Propositions 2.5 and 2.4, we obtain the following corollary:

Corollary 2.2. (i) *If* $\mathbb{F}$ *is a class of frames closed under liftings,* $\mathbf{Log}(\mathbb{F})$ *is a superbasic logic.* (ii) *If* $\mathbb{F} \subseteq \mathbb{ISR}$, *then* $\mathbf{Log}(\mathbb{F})$ *is a superbasic logic.*

3. Canonical Models for Superbasic Logics

The general frame semantics and the canonical general model for **BPL** can be found in Ref. 6 by Suzuki and Ono, while Ardeshir and Ruitenburg (see Ref. 14) also considered the notion of canonical model in a slightly different setting based on a Gentzen-style sequent calculus. In this section, we generalize the canonical model construction by Suzuki and Ono to every superbasic logic $\Lambda \in \mathtt{Ext}(\mathbf{BPL})$, and then we prove the completeness of some superbasic logics. First we list some theorems of **BPL** which will be used in this section.

Proposition 3.1 (Refs. 6,10). *The following are theorems of* **BPL**:

(1) $(\alpha \to \beta) \to (\top \to (\alpha \to \beta))$.
(2) $((\alpha \wedge \beta) \to \gamma) \to (\alpha \to (\beta \to \gamma))$.
(3) $\neg\beta \to (\beta \to \alpha)$.
(4) $((\alpha \to \beta) \wedge (\alpha \wedge \beta \to \gamma)) \to (\alpha \to \gamma)$.
(5) $\bigwedge_{1 \leq i \leq n}(\alpha_i \to \beta) \leftrightarrow ((\bigvee_{1 \leq i \leq n} \alpha_i) \to \beta)$.

Moreover, we will introduce some basic concepts. For any sets X and Y, let $X \subseteq_\omega Y$ denote that X is a finite subset of Y. For a finite set Σ of formulas, let $\bigwedge \Sigma$ be the conjunction of all formulas in Σ, and $\bigvee \Sigma$ be the disjunction of all formulas in Σ (when $\Sigma = \emptyset$, we understand $\bigwedge \Sigma := \top$ and $\bigvee \Sigma := \bot$).

Definition 3.1. Let $\mathbb{F}$ be a class of transitive frames. For a set $\Gamma \cup \{\varphi\}$ of formulas, we say that φ is a *local* $\mathbb{F}$*-consequence* of Γ (notation: $\Gamma \models_{\mathbb{F}} \varphi$), if for every frame $\mathfrak{F} = (W, R) \in \mathbb{F}$, valuation V and $w \in W$, if $(\mathfrak{F}, V), w \models \gamma$ for all $\gamma \in \Gamma$, then $(\mathfrak{F}, V), w \models \varphi$.

We observe from Ref. 15 that $\{\varphi_1, \ldots, \varphi_{n-1}, \varphi_n\} \models_{\mathbb{TR}} \psi$ if and only if $\bigwedge_{1 \leq i \leq n} \varphi_i \models_{\mathbb{TR}} \psi$, where $\mathbb{TR}$ is the class of all transitive frames. Based on this idea, we define the concept of Λ-derivation as follows:

Definition 3.2. Given a superbasic logic Λ, and sets Γ, Δ of formulas, we say that Δ is Λ*-derivable* from Γ (notation: $\Gamma \vdash_\Lambda \Delta$), if there exists $\Gamma' \subseteq_\omega \Gamma$ and $\Delta' \subseteq_\omega \Delta$ such that $\bigwedge \Gamma' \to \bigvee \Delta' \in \Lambda$. When Δ is a singleton $\{\alpha\}$, we write $\Gamma \vdash_\Lambda \alpha$ instead of $\Gamma \vdash_\Lambda \{\alpha\}$.

Proposition 3.2. *Given a superbasic logic* Λ *and a set of formulas* $\Gamma \cup \{\alpha\}$, $\Gamma \vdash_\Lambda \alpha$ *iff there exists a set* $\Gamma' \subseteq_\omega \Gamma$ *such that* $\bigwedge \Gamma' \to \alpha \in \Lambda$.

Proof. The right-to-left direction is immediate. For the other direction, it suffices to note that $\neg\beta \to (\beta \to \alpha) \in \Lambda$ for all superbasic logics Λ. □

Definition 3.3 (Strong and weak completeness). *Let* Λ *be a superbasic logic and* $\mathbb{F}$ *a class of transitive frames. We say that* Λ *is* strongly complete *with respect to* $\mathbb{F}$, *if for every set* $\Gamma \cup \{\alpha\}$ *of formulas,* $\Gamma \models_{\mathbb{F}} \alpha$ *implies* $\Gamma \vdash_\Lambda \alpha$. *We say that* Λ *is* weakly complete *with respect to* $\mathbb{F}$, *if for every formula* α, $\mathbb{F} \models \alpha$ *implies* $\alpha \in \Lambda$.

Definition 3.4 (Λ-consistency). *Let* Λ *be a superbasic logic, and* Γ *and* Σ *be non-empty sets of formulas. We say that the pair* (Γ, Σ) *is* Λ-consistent, *if* $\Gamma \nvdash_\Lambda \Sigma$. *We say that* (Γ, Σ) *is* maximally Λ-consistent, *if* (Γ, Σ) *is* Λ*-consistent and* $\Gamma \cup \{\alpha\} \vdash_\Lambda \Sigma$ *for every* $\alpha \notin \Gamma$.

Lemma 3.1. *Let* (Γ, Σ) *be* Λ-consistent *for a superbasic logic* Λ. *Then, there exists a set* Γ^+ *such that* $\Gamma \subseteq \Gamma^+$ *and* (Γ^+, Σ) *is maximally* Λ*-consistent.*

Proof. Let $(\alpha_n)_{n\in\omega}$ be an enumeration of all formulas. Define a sequence $(\Gamma_n)_{n\in\omega}$ by induction on $n \in \omega$ as follows: $\Gamma_0 = \Gamma$ and

$$\Gamma_{n+1} = \begin{cases} \Gamma_n \cup \{\alpha_n |\} & \text{if } (\Gamma_n \cup \{\alpha_n\}, \Sigma) \text{ is } \Lambda\text{-consistent,} \\ \Gamma_n & \text{otherwise.} \end{cases}$$

Let $\Gamma^+ = \bigcup_{n\in\omega} \Gamma_n$. Note that every (Γ_n, Σ) is Λ-consistent. So is (Γ^+, Σ). Then it is easy to verify that (Γ^+, Σ) is maximally Λ-consistent. □

Proposition 3.3. *If* $\Gamma, \alpha \vdash_\Lambda \beta$, *then* $\Gamma \vdash_\Lambda \alpha \to \beta$.

Proof. Assume $\Gamma, \alpha \vdash_\Lambda \beta$. Then there exists $\Gamma' \subseteq_\omega \Gamma$ such that $\bigwedge \Gamma' \wedge \alpha \to \beta \in \Lambda$. By $(p \wedge q \to r) \to (p \to (q \to r)) \in$ **BPL**, $\bigwedge \Gamma' \to (\alpha \to \beta) \in \Lambda$. Hence $\Gamma \vdash_\Lambda \alpha \to \beta$. □

The converse of above deduction theorem does not hold generally. For instance, it is easy to see that $p \to q \vdash_\Lambda p \to q$, but $p \to q, p \nvdash_\Lambda q$ since $((p \to q) \wedge p) \to q$ may not be a theorem of Λ. Actually, $((p \to q) \wedge p) \to q$ is not an theorem of **BPL** (recall the proof of Proposition 2.6).

Proposition 3.4. *For every superbasic logic* Λ, *if* $((p \to q) \wedge p) \to q \in \Lambda$, *then* $\Gamma, \alpha \vdash_\Lambda \beta$ *iff* $\Gamma \vdash_\Lambda \alpha \to \beta$.

Proof. Assume $((p \to q) \wedge p) \to q \in \Lambda$. It suffices to show the converse of Proposition 3.3. Assume $\Gamma \vdash_\Lambda \alpha \to \beta$. Then $\Gamma, \alpha \vdash_\Lambda \alpha \to \beta$ and $\Gamma, \alpha \vdash_\Lambda \alpha$. Let $\bigwedge \Gamma_1 \to (\alpha \to \beta) \in \Lambda$ and $\bigwedge \Gamma_2 \to \alpha$ for $\Gamma_1, \Gamma_2 \subseteq_\omega \Gamma \cup \{\alpha \mid \}$. So $\bigwedge(\Gamma_1 \cup \Gamma_2) \to (\alpha \wedge (\alpha \to \beta)) \in \Lambda$. Since $((p \to q) \wedge p) \to q \in \Lambda$, $\bigwedge(\Gamma_1 \cup \Gamma_2) \to \beta$. Hence $\Gamma, \alpha \vdash_\Lambda \beta$. □

It is already known that **BPL** $\oplus (p \to q) \wedge p \to q =$ **Int** (see Ref. 2). If $(p \to q) \wedge p \to q \in \Lambda$ for a superbasic logic Λ, then Λ is an extension of **Int** and so it becomes an intermediate logic.

Lemma 3.2. *Let* (Γ, Σ) *be maximally* Λ*-consistent. Then*

(i) *if* $\Gamma \vdash_\Lambda \alpha$, *then* $\alpha \in \Gamma$.
(ii) $\Lambda \subseteq \Gamma$.
(iii) *if* $\alpha \in \Gamma$ *and* $\alpha \to \beta \in \Lambda$, *then* $\beta \in \Gamma$.
(iv) $\alpha \wedge \beta \in \Gamma$ *iff* $\alpha \in \Gamma$ *and* $\beta \in \Gamma$.
(v) $\alpha \vee \beta \in \Gamma$ *iff* $\alpha \in \Gamma$ *or* $\beta \in \Gamma$.
(vi) *if* $\alpha \to \beta \in \Gamma$ *and* $\beta \to \gamma \in \Gamma$, *then* $\alpha \to \gamma \in \Gamma$.
(vii) *if* $\alpha \wedge \beta \to \gamma \in \Gamma$ *and* $\beta \in \Lambda$, *then* $\alpha \to \gamma \in \Gamma$.
(viii) *if* $\alpha \wedge \eta \to \beta \vee \gamma \in \Lambda$, $\delta \to \eta \in \Gamma$, $\gamma \to \theta \in \Gamma$, *then* $\alpha \wedge \delta \to \beta \vee \theta \in \Gamma$.
(ix) *if* $\alpha \to \delta \in \Gamma$ *and* $\alpha \wedge \delta \to \beta \in \Gamma$, *then* $\alpha \to \beta \in \Gamma$.

Proof. Proofs for items from (ii) to (ix) are the same as those for $\Lambda =$ **BPL** in Refs. 6,10. For (i), assume that $\Gamma \vdash_\Lambda \alpha$. Then $\bigwedge \Gamma' \to \alpha \in \Lambda$ for some $\Gamma' \subseteq_\omega \Gamma$. Assume for contradiction that $\alpha \notin \Gamma$. Because (Γ, Σ) is maximally Λ-consistent, $\Gamma \cup \{\alpha\} \vdash_\Lambda \Sigma$. Then there exist $\Gamma'' \subseteq_\omega \Gamma$ and $\Sigma' \subseteq_\omega \Sigma$ such that $\bigwedge \Gamma'' \wedge \alpha \to \bigvee \Sigma' \in \Lambda$. Then, by the assumption, $\bigwedge(\Gamma' \cup \Gamma'') \to \Sigma' \in \Lambda$, a contradiction with the Λ-consistency of (Γ, Σ). □

Definition 3.5 (Canonical Model). *For every superbasic logic* Λ, *define the* Λ-canonical model $\mathfrak{M}^\Lambda = (W^\Lambda, R^\Lambda, V^\Lambda)$ *as follows:*

(i) $W^\Lambda = \{w \mid (w, \Sigma)$ *is maximally* Λ*-consistent for some* $\Sigma \neq \emptyset\}$.
(ii) $wR^\Lambda u$ *iff* $\alpha \to \beta \in w$ *and* $\alpha \in u$ *imply* $\beta \in u$.
(iii) $V^\Lambda(p) = \{w \mid p \in w\}$ *for each* $p \in \mathsf{Prop}$.

The frame $\mathfrak{F}^\Lambda = (W^\Lambda, R^\Lambda)$ *is called the* Λ-canonical frame.

Proposition 3.5. *Let* $\mathfrak{M}^\Lambda = (W^\Lambda, R^\Lambda, V^\Lambda)$ *be the* Λ*-canonical model. Then* (i) R^Λ *is transitive and* (ii) V^Λ *is persistent.*

Proof. For (i), assume $wR^\Lambda u$ and $uR^\Lambda v$. Suppose $\alpha \to \beta \in w$ and $\alpha \in v$. We show $\beta \in v$. Since $(\alpha \to \beta) \to (\top \to (\alpha \to \beta)) \in \Lambda$ and $\alpha \to \beta \in w$, we have $\top \to (\alpha \to \beta) \in w$. By $\top \in u$ and $wR^\Lambda u$, we obtain $\alpha \to \beta \in u$. Since $\alpha \in v$ and $uR^\Lambda v$, we have $\beta \in v$. For (ii), let $p \in w$ and $wR^\Lambda u$. By $p \to (\top \to p) \in \Lambda$, we get $\top \to p \in w$. By $\top \in u$, we get $p \in u$. □

Consider the Λ-canonical model $\mathfrak{M}^\Lambda$, for each formula β and $w \in W^\Lambda$, define $\mathsf{Ant}(\beta, w) = \{\varphi \mid \varphi \to \beta \in w\}$, i.e., the set of antecedents of β in w. Note that $\mathsf{Ant}(\beta, w) \neq \emptyset$, since $\beta \to \beta \in \Lambda$ and so $\beta \in \mathsf{Ant}(\beta, w)$.

Lemma 3.3. *Let* (w, Σ) *be maximally* Λ*-consistent for a superbasic logic* Λ *and* $\alpha \to \beta \notin w$. *Then* $(\{\alpha\}, \mathsf{Ant}(\beta, w))$ *is* Λ*-consistent.*

Proof. Assume that (w, Σ) is maximally Λ-consistent and $\alpha \to \beta \notin w$. For a contradiction, suppose that $(\{\alpha\}, \mathsf{Ant}(\beta, w))$ is Λ-inconsistent. Let $\Sigma' = \{\varphi_1, \ldots, \varphi_n\} \subseteq_\omega \mathsf{Ant}(\beta, w)$ such that $\alpha \to \varphi_1 \vee \cdots \vee \varphi_n \in \Lambda$, where $\varphi_i \to \beta \in w$ for each $1 \leq i \leq n$. By Lemma 3.2 and Proposition 3.1, $\bigwedge_{1 \leq i \leq n}(\varphi_i \to \beta) \in w$. Since $\bigwedge_{1 \leq i \leq n}(\varphi_i \to \beta) \to ((\bigvee_{1 \leq i \leq n} \varphi_i) \to \beta) \in \Lambda$, $(\bigvee_{1 \leq i \leq n} \varphi_i) \to \beta \in w$ again by Lemma 3.2. By $\alpha \to \bigvee_{1 \leq i \leq n} \varphi_i \in \Lambda$, $\alpha \to \beta \in w$, a contradiction. □

Lemma 3.4 (Existence Lemma). *Let* (w, Σ) *be maximally* Λ*-consistent for a superbasic logic* Λ, *and* $\alpha \to \beta \notin w$. *Then, there exists a set* w' *such that* $(w', \mathsf{Ant}(\beta, w))$ *is maximally* Λ*-consistent,* $\alpha \in w'$, $\beta \notin w'$ *and* $wR^\Lambda w'$.

Proof. Recall that $\mathsf{Ant}(\beta, w) \neq \emptyset$, since $\beta \in \mathsf{Ant}(\beta, w)$. By the assumption and Lemma 3.3, $(\{\alpha\}, \mathsf{Ant}(\beta, w))$ is Λ-consistent. Let $(w', \mathsf{Ant}(\beta, w))$ be maximally Λ-consistent and $\alpha \in w'$ by Lemma 3.1. Since $\beta \in \mathsf{Ant}(\beta, w)$, so $\beta \notin w'$. For showing $wR^\Lambda w'$, assume $\delta \to \eta \in w$, $\delta \in w'$ but $\eta \notin w'$. Since w' is maximal and $\eta \notin w'$, there exist $\Sigma' = \{\sigma'_1, \ldots, \sigma'_n\} \subseteq_\omega \mathsf{Ant}(\beta, w)$ and $\Gamma' \subseteq_\omega w'$ such that $\bigwedge \Gamma' \wedge \eta \to \bigvee_{1 \leq i \leq n} \sigma'_i \in \Lambda$, where $\sigma'_i \to \beta \in w$

for $1 \leq i \leq n$. So $(\bigvee_{1\leq i\leq n} \sigma_i') \to \beta \in w$. Then $\bigwedge \Gamma' \wedge \eta \to \beta \in w$. Thus $(\delta \to \eta) \wedge (\bigwedge \Gamma' \wedge \eta \to \beta) \in w$, which implies $\delta \wedge \bigwedge \Gamma' \to \beta \in w$. It follows that $\delta \wedge \bigwedge \Gamma' \in \mathsf{Ant}(\beta, w)$. But then, $\delta, \bigwedge \Gamma' \in w'$ implies that $(w', \mathsf{Ant}(\beta, w))$ is not Λ-consistent, a contradiction. □

Lemma 3.5 (Truth Lemma). *For every superbasic logic Λ, formula α and $w \in W^\Lambda$, $\mathfrak{M}^\Lambda, w \models \alpha$ iff $\alpha \in w$.*

Proof. By induction on the length of α. For the case of implication, use the existence lemma (Lemma 3.4) and the required result follows. □

Theorem 3.1. *For every superbasic logic Λ, Λ is strongly complete with respect to the class $\{ \mathfrak{F}_\Lambda \}$.*

Proof. Let $\Gamma \cup \{ \alpha \}$ be a set of formulas. Suppose $\Gamma \nvdash_\Lambda \alpha$. Then $(\Gamma, \{ \alpha \})$ is Λ-consistent. By Lemma 3.1, there exists $\Gamma^+ \supseteq \Gamma$ such that $(\Gamma^+, \{ \alpha \})$ is maximally Λ-consistent. Then $\alpha \notin \Gamma^+$. By $\Gamma^+ \in W_\Lambda$ and Truth Lemma, $\mathfrak{M}^\Lambda, \Gamma^+ \models \gamma$ for all $\gamma \in \Gamma^+$ and $\mathfrak{M}^\Lambda, \Gamma^+ \not\models \alpha$. Hence $\Gamma \not\models_{\{ \mathfrak{F}_\Lambda \}} \alpha$. □

Theorem 3.2. **BPL** *is strongly complete with respect to the class $\mathbb{TR}$ of all transitive frames.*

Proof. By Theorem 3.1 and Proposition 3.5. □

Theorem 3.3. **BPL** *is strongly complete with respect to the class $\mathbb{ISR}$ of all transitive frames satisfying $\forall x \exists y (yRx)$, i.e., every state has a predecessor.*

Proof. Let Λ = **BPL**. Suppose that $\Gamma \nvdash_\Lambda \alpha$. By the proof of Theorem 3.1, $\mathfrak{M}^\Lambda, \Gamma^+ \models \gamma$ for all $\gamma \in \Gamma^+$ and $\mathfrak{M}^\Lambda, \Gamma^+ \not\models \alpha$. By Proposition 3.5, $\mathfrak{F}^\Lambda$ is transitive. By taking the lifting $(\mathfrak{M}^\Lambda)^\circ$ of $\mathfrak{M}^\Lambda$, Proposition 2.3 implies that $(\mathfrak{M}^\Lambda)^\circ, \Gamma^+ \models \gamma$ for all $\gamma \in \Gamma^+$ and $(\mathfrak{M}^\Lambda)^\circ, \Gamma^+ \not\models \alpha$. The frame $(\mathfrak{F}^\Lambda)^\circ$ is still transitive but it satisfies $\forall x \exists y (yRx)$. Therefore, $\Gamma \not\models_{\mathbb{ISR}} \alpha$. □

4. Model and Frame Constructions

We discuss some model and frame constructions for **BPL** in this section. First, we introduce the notion of bisimulation, and then, based on this concept, we introduce the notions of disjoint union, generated subframe and reduction for both frames and models. By using these constructions, we show that **BPL** has the disjunction property. We also introduce Lemmon-filtration to show that **BPL** enjoys the finite model property.

Definition 4.1. Let $\mathfrak{M} = (W, R, V)$ and $\mathfrak{M}' = (W', R', V')$ be models. A non-empty relation $Z \subseteq W \times W'$ is called a *bisimulation* between $\mathfrak{M}$ and $\mathfrak{M}'$ (notation: $Z : \mathfrak{M} \leftrightarrows \mathfrak{M}'$), if the following hold: for each pair $(x, x') \in Z$,

(i) Atomic: x and x' satisfy the same propositional variables.
(ii) Forth: if xRy, then there exists $y' \in W'$ such that $x'R'y'$ and yZy'.
(iii) Back: if $x'R'y'$, then there exists $y \in W$ such that xRy and yZy'.

Bisimulation between frames is defined without the atomic condition. We write $Z : \mathfrak{M}, x \leftrightarrows \mathfrak{M}', x'$ to denote that Z is a bisimulation between $\mathfrak{M}$ and $\mathfrak{M}'$ such that xZx', and write $\mathfrak{M}, x \leftrightarrows \mathfrak{M}', x'$ to mean that $Z : \mathfrak{M}, x \leftrightarrows \mathfrak{M}', x'$ for some bisimulation Z. We say that Z is a *global bisimulation*, if for all $x' \in W'$ there exists $x \in W$ such that xZx', and vice versa.

We note that this notion of bisimulation can be also studied in the previous works (see Refs. 14,16) and also that Bou studied a weaker notion of bisimulation, called *quasi-bisimulation* in Ref. 17.

Proposition 4.1. *Let $\mathfrak{M} = (\mathfrak{F}, V)$ and $\mathfrak{M}' = (\mathfrak{F}', V')$ be models. Assume that $Z : \mathfrak{M}, x \leftrightarrows \mathfrak{M}', x'$ and so $Z : \mathfrak{F}, x \leftrightarrows \mathfrak{F}', x'$. Then for any formula α,*

(i) $\mathfrak{M}, x \models \alpha$ *iff* $\mathfrak{M}', x' \models \alpha$.
(ii) *If Z is global, then $\mathfrak{M} \models \alpha$ iff $\mathfrak{M}' \models \alpha$.*
(iii) *If α is constant, then $\mathfrak{F}, x \models \alpha$ iff $\mathfrak{F}', x' \models \alpha$.*
(iv) *If Z is global and α is constant, then $\mathfrak{F} \models \alpha$ iff $\mathfrak{F}' \models \alpha$.*

Proof. The (i) is shown by induction on α and others follow. □

Let $(\mathfrak{M}_i)_{i \in I}$ be a family of (pairwise disjoint) models where $\mathfrak{M}_i = (\mathfrak{F}_i, V_i)$ and $\mathfrak{F}_i = (W_i, R_i)$. Define the *disjoint union* $\biguplus_{i \in I} \mathfrak{M}_i = (W, R, V)$ of $(\mathfrak{M}_i)_{i \in I}$ by putting: $W = \bigcup_{i \in I} W_i$, $R = \bigcup_{i \in I} R_i$ and $V(p) = \bigcup_{i \in I} V_i(p)$ for each $p \in \mathsf{Prop}$. Let $\biguplus_{i \in I} \mathfrak{M}_i = (W, R)$ be the disjoint union of frames. For the disjoint union of two structures, we use the notation $(.) \uplus (.)$.

Proposition 4.2. *Let $(\mathfrak{M}_i)_{i \in I}$ be a family of (pairwise disjoint) models, where $\mathfrak{M}_i = (\mathfrak{F}_i, V_i)$. For every formula α, the following hold:*

(i) $\biguplus_{i \in I} \mathfrak{M}_i, w \models \alpha$ *iff* $\mathfrak{M}_i, w \models \alpha$.
(ii) $\biguplus_{i \in I} \mathfrak{M}_i \models \alpha$ *iff* $\mathfrak{M}_i \models \alpha$ *for all* $i \in I$.
(iii) $\biguplus_{i \in I} \mathfrak{F}_i, w \models \alpha$ *iff* $\mathfrak{F}_i, w \models \alpha$.
(iv) $\biguplus_{i \in I} \mathfrak{F}_i \models \alpha$ *iff* $\mathfrak{F}_i \models \alpha$ *for all* $i \in I$.

Proof. For (i), it suffices to note that $\biguplus_{i\in I}\mathfrak{M}_i, w \leftrightarroows \mathfrak{M}_i, w$ for all $i \in I$ by Proposition 4.1. Other items follow from (i). □

Now we consider (generated) subframe and submodels. A frame $\mathfrak{G} = (G, S)$ is a *subframe* of $\mathfrak{F} = (W, R)$, if $G \subseteq W$ and $S = R \cap G^2$. A model $\mathfrak{N} = (G, S, U)$ is a *submodel* of $\mathfrak{M} = (W, R, V)$, if (G, S) is a subframe of (W, R) and $U(p) = V(p) \cap G$ for all $p \in \mathsf{Prop}$.

Given any frame $\mathfrak{F} = (W, R)$ and non-empty subset $X \subseteq W$, the *generated subframe* of $\mathfrak{F}$ induced by X is defined as the subframe $\mathfrak{F}_X = (R[X], R_X, V_X)$ of $\mathfrak{F}$. A frame $\mathfrak{G} = (G, S)$ is a generated subframe of $\mathfrak{F} = (W, R)$, if $\mathfrak{G} = \mathfrak{F}_X$ for some $X \subseteq W$. The notions of generated submodels are defined naturally. For any $w \in W$, we say that $\mathfrak{F}_w$ is the *point generated subframe* of $\mathfrak{F}$ induced by $\{w\}$. A frame $\mathfrak{F}$ is *pointed generated*, if $\mathfrak{F} = \mathfrak{F}^w$ for some w in $\mathfrak{F}$. Similar notions for models are defined naturally.

Proposition 4.3. *Let $\mathfrak{N} = (\mathfrak{G}, U)$ be a generated submodel of $\mathfrak{M} = (\mathfrak{F}, V)$. For any formula α and w in $\mathfrak{N}$, the following hold:*

(i) $\mathfrak{N}, w \models \alpha$ *iff* $\mathfrak{M}, w \models \alpha$.
(ii) $\mathfrak{M} \models \alpha$ *implies* $\mathfrak{N} \models \alpha$.
(iii) $\mathfrak{F}, w \models \alpha$ *iff* $\mathfrak{G}, w \models \alpha$.
(iv) $\mathfrak{F} \models \alpha$ *implies* $\mathfrak{G} \models \alpha$.

Proof. For (i), it suffices to note that $\mathfrak{N}, w \leftrightarroows \mathfrak{M}, w$ by Proposition 4.1. Other items follow from (i). □

The next model and frame construction is *reduction*. For any frames $\mathfrak{F} = (W, R)$ and $\mathfrak{G} = (G, S)$, a *surjective* function $f : W \to G$ is called a *reduction* from $\mathfrak{F}$ to $\mathfrak{G}$ (notation: $f : \mathfrak{F} \twoheadrightarrow \mathfrak{G}$), if for all $x, y \in W$, the following hold:

(i) (*Forth*) if xRy, then $f(x)Sf(y)$.
(ii) (*Back*) if $f(x)Sf(y)$, then $\exists z \in W(xRz \ \&\ f(z) = f(y))$.

We say that $\mathfrak{F}$ is *reducible* to $\mathfrak{G}$ (notation: $\mathfrak{F} \twoheadrightarrow \mathfrak{G}$), if there exists a reduction from $\mathfrak{F}$ to $\mathfrak{G}$. The notion of reduction over models is a frame reduction plus the atomic condition: x and $f(x)$ satisfy the same propositional variables.

Proposition 4.4. *Let $\mathfrak{M} = (\mathfrak{F}, V)$ and $\mathfrak{N} = (\mathfrak{G}, U)$ be Kripke models and $f : \mathfrak{M} \twoheadrightarrow \mathfrak{N}$. Then for any state w in $\mathfrak{F}$ and any formula α,*

(i) $\mathfrak{M}, w \models \alpha$ *iff* $\mathfrak{N}, f(w) \models \alpha$.
(ii) $\mathfrak{M} \models \alpha$ *iff* $\mathfrak{N} \models \alpha$.

(iii) $\mathfrak{F}, w \models \alpha$ *implies* $\mathfrak{G}, f(w) \models \alpha$.
(iv) $\mathfrak{F} \models \alpha$ *implies* $\mathfrak{G} \models \alpha$.

Proof. For (i), it suffices to note that $\mathfrak{M}, w \leftrightarrows \mathfrak{N}, f(w)$ by Proposition 4.1. Other items follow from (i). □

Proposition 4.5. *For any frame* $\mathfrak{F} = (W, R)$, *we have* $\biguplus_{w \in W} \mathfrak{F}_w \twoheadrightarrow \mathfrak{F}$. *Hence for every formula* α, $\mathfrak{F} \models \alpha$ *iff* $\biguplus_{w \in W} \mathfrak{F}_w \models \alpha$.

Now we will apply above model constructions to show the disjunction property of **BPL** (first shown by Ardeshir in Ref. 12) by using above model constructions. Note that the disjunction property in strict implicational logics is studied in Ref. 10 and the Section 4.4 of Ref. 17.

Definition 4.2. We say that a superbasic logic Λ has the *disjunction property* if, for all formulas α and β, $\alpha \vee \beta \in \Lambda$ implies $\alpha \in \Lambda$ or $\beta \in \Lambda$.

Theorem 4.1 (Ref. 12). BPL *has the disjunction property.*

Proof. The proof is similar to the proof of the disjunction property of intuitionistic logic in Ref. 11 (pp. 51-52). Assume that $\alpha \vee \beta \in$ **BPL** but $\alpha \notin$ **BPL** and $\beta \notin$ **BPL**. There exist models $\mathfrak{M}_1 = (W_1, R_1, V_1)$, $\mathfrak{M}_2 = (W_2, R_2, V_2)$ and states $x_1 \in W_1$ and $x_2 \in W_2$ such that $\mathfrak{M}_1, x_1 \not\models \alpha$ and $\mathfrak{M}_2, x_2 \not\models \beta$. Since truth is preserved in generated submodels, we may assume that $\mathfrak{M}_1$ and $\mathfrak{M}_2$ are generated from x_1 and x_2 respectively. Take the disjoint union $\mathfrak{M} = \mathfrak{M}_1 \uplus \mathfrak{M}_2 = (W, R, V)$. Let $\mathfrak{M}^\circ$ be the lifting of $\mathfrak{M}$. Then both $\mathfrak{M}_1$ and $\mathfrak{M}_2$ are generated submodels of $\mathfrak{M}'$. Therefore $\mathfrak{M}', \circ \not\models \alpha \vee \beta$ by Proposition 2.3. □

Our last model construction is *Lemmon filtration* which can be used to show the finite model property of **BPL**.

Definition 4.3 (finite model property). *A superbasic logic* Λ *has the* finite model property, *if for every formula* $\alpha \notin \Lambda$ *there exists a finite model* $\mathfrak{M} \models \Lambda$ *such that* $\mathfrak{M} \not\models \alpha$.

Given any model $\mathfrak{M} = (W, R, V)$ and a set Σ of formulas which is closed under taking subformulas, define an equivalence relation $\sim_\Sigma$ over W by

$$x \sim_\Sigma y \text{ iff for all } \alpha \in \Sigma, (\mathfrak{M}, x \models \alpha \text{ iff } \mathfrak{M}, y \models \alpha).$$

Let $[x] = \{y \in W \mid x \sim_\Sigma y\}$ and $W_\Sigma = \{[x] \mid x \in W\}$. It is clear that, if Σ is finite, then W^Σ is finite, the size of which is at most $2^{|\Sigma|}$.

Definition 4.4. Let $\mathfrak{M} = (W, R, V)$ be a Kripke model and Σ a set of formulas closed under taking subformulas. A *filtration* of $\mathfrak{M}$ through Σ is a model $\mathfrak{M}_\Sigma = (W_\Sigma, R_\Sigma, V_\Sigma)$ satisfying the following conditions:

(i) If xRy, then $[x]R_\Sigma[y]$.
(ii) If $[x]R_\Sigma[y]$, then ($\mathfrak{M}, x \models \alpha \to \beta$ and $\mathfrak{M}, y \models \alpha$) imply $\mathfrak{M}, y \models \beta$ for every $\alpha \to \beta \in \Sigma$.
(iii) R_Σ is transitive
(iv) $V_\Sigma(p) = \{[x] \in W_\Sigma \mid x \in V(p)\}$ for each $p \in \Sigma$.

Note that the conditions (i) and (iv) imply that V_Σ is persistent if V is persistent. A filtration satisfying the above conditions exists. Take the *finest filtration* $\mathfrak{M}^s_\Sigma = (W_\Sigma, R^s_\Sigma, V_\Sigma)$ where R^s_Σ is defined by

$$[x]R^s_\Sigma[y] \text{ iff } \exists x' \in [x] \exists y' \in [y] x'Ry'.$$

It is easy to see that R^s_Σ satisfies the conditions (i), (ii) and (iv) of Definition 4.4. However, we cannot assure that R^s_Σ is transitive, and so, we take the transitive closure $(R^s_\Sigma)^+$ of R^s_Σ. Now we need to check the conditions (i) to (iv) for $(R^s_\Sigma)^+$. The most involving part is the condition (ii) for $(R^s_\Sigma)^+$. For this purpose, we first show the following lemma for R^s_Σ.

Lemma 4.1. *If $[x]R^s_\Sigma[y]$, then for any $\alpha \in \Sigma$, $\mathfrak{M}, x \models \alpha$ implies $\mathfrak{M}, y \models \alpha$.*

Proof. Assume that $[x]R^s_\Sigma[y]$ and fix any $\alpha \in \Sigma$ such that $\mathfrak{M}, x \models \alpha$. By $[x]R^s_\Sigma[y]$, $x'Ry'$ holds for some $x' \in [x]$ and $y' \in [y]$. Since $\mathfrak{M}, x \models \alpha$ and $x' \in [x]$, $\mathfrak{M}, x' \models \alpha$. By $x'Ry'$ and persistency, $\mathfrak{M}, y' \models \alpha$. By $y' \in [y]$ and $\alpha \in \Sigma$, $\mathfrak{M}, y \models \alpha$. □

Proposition 4.6. *Given any model $\mathfrak{M}$ and any subformula closed set Σ of formulas, the transitive closure $(\mathfrak{M}^s_\Sigma)^+ = (W_\Sigma, (R^s_\Sigma)^+, V_\Sigma)$ is a filtration of $\mathfrak{M}$ through Σ, and $(R^s_\Sigma)^+$ is antisymmetric, i.e., $[x](R^s_\Sigma)^+[y]$ and $[y](R^s_\Sigma)^+[x]$ imply $[x] = [y]$.*

Proof. The conditions except (ii) in Definition 4.4 are easy to check. For the second condition, assume that $[x](R^s_\Sigma)^+[y]$ and $\mathfrak{M}, x \models \alpha \to \beta$ and $\mathfrak{M}, y \models \alpha$. By $[x](R^s_\Sigma)^+[y]$, we obtain $[x] = [x_0]R^s_\Sigma[x_1]R^s_\Sigma[x_2] \cdots [x_{n-1}]R^s_\Sigma[x_n] = [y]$ for some $x_1, \ldots, x_{n-1} \in W$. By applying Lemma 4.1 repeatedly, we obtain $\mathfrak{M}, x_{n-1} \models \alpha \to \beta$. By $[x_{n-1}]R^s_\Sigma[x_n]$, we can find $u \in [x_{n-1}]$ and $v \in [x_n]$ such that uRv. Since $\alpha \to \beta \in \Sigma$ and $u \in [x_{n-1}]$, $\mathfrak{M}, u \models \alpha \to \beta$. Because $v \in [x_n] = [y]$ and $\mathfrak{M}, y \models \alpha$, we also obtain $\mathfrak{M}, v \models \alpha$. By uRv, $\mathfrak{M}, v \models \beta$. Hence $\mathfrak{M}, y \models \beta$.

For showing the antisymmetry of $(R^s_\Sigma)^+$, assume $[x](R^s_\Sigma)^+[y]$ and $[y](R^s_\Sigma)^+[x]$. By $[x](R^s_\Sigma)^+[y]$, there exist $x_1, \ldots, x_{n-1} \in W$ such that $[x] = [x_0]R^s_\Sigma[x_1] \ldots [x_{n-1}]R^s_\Sigma[x_n] = [y]$. It is easy to check that $\mathfrak{M}, x \models \alpha$ implies $\mathfrak{M}, y \models \alpha$, for all $\alpha \in \Sigma$, by the definition of R^s_Σ and Lemma 4.1. Similarly, from $[y](R^s_\Sigma)^+[x]$, we can show that $\mathfrak{M}, x \models \alpha$ implies $\mathfrak{M}, y \models \alpha$, for every formula $\alpha \in \Sigma$. It follows that $x \sim_\Sigma y$, and hence $[x] = [y]$. □

Lemma 4.2. *Let $\mathfrak{M}_\Sigma = (W_\Sigma, R_\Sigma, V_\Sigma)$ be a filtration of $\mathfrak{M} = (W, R, V)$ through Σ. Then for any $\alpha \in \Sigma$ and $x \in W$, $\mathfrak{M}, x \models \alpha$ iff $\mathfrak{M}_\Sigma, [x] \models \alpha$.*

Proof. The regular proof by induction on the length of α is omitted. □

Theorem 4.2. **BPL** *has the finite model property. Moreover,* **BPL** *is weakly complete with respect to the class of all finite antisymmetric transitive frames.*

Proof. Let $\alpha \notin$ **BPL**. Then by completeness of **BPL** (Theorem 3.2), it is refuted in some Kripke model $\mathfrak{M}$. Consider $\Sigma = \{\beta \mid \beta \text{ is a subformula of } \alpha\}$. Then Σ is finite. By Lemma 4.2 and Proposition 4.6, α is refuted in the finite model $(\mathfrak{M}^s_\Sigma)^+$. Hence **BPL** has the finite model property. Moreover, every formula $\alpha \notin$ **BPL** is refuted in $(\mathfrak{M}^s_\Sigma)^+$, which is antisymmetric by Proposition 4.6. Hence **BPL** is characterized by the class of all finite antisymmetric transitive frames. □

5. Two Constant Property

Recall that CFor is the set of all constant formulas.

Definition 5.1. A superbasic logic Λ has the *two constant property*, if $\alpha \leftrightarrow \top \in \Lambda$ or $\alpha \leftrightarrow \bot \in \Lambda$ for all $\alpha \in \mathsf{CFor}$.

Proposition 5.1. *For any superbasic logics Λ and Λ', if Λ has the two constant property and $\Lambda \subseteq \Lambda'$, then Λ' has the two constant property.*

It is well-known that both **Int** and **CL** have the two constant property (see, e.g., p.35 of Ref. 11). Hence all intermediate logics between **Int** and **CL** have this property. But it is not the case for **BPL**.

Theorem 5.1. **BPL** *has no two constant property.*

Proof. Consider the constant formula $\alpha := \top \to \bot$. Then let us show that $\alpha \leftrightarrow \top \notin$ **BPL** and $\alpha \leftrightarrow \bot \notin$ **BPL**. Consider the models $\mathfrak{M}_1 = (W_1, R_1, V_1)$ and $\mathfrak{M}_2 = (W_2, R_2, V_2)$ where: (i) $W_1 = \{x, y, z\}$; $R_1 =$

$\{(x,y),(y,z)\}$, (ii) $W_2 = \{x,y\}$; $R_2 = \{(x,y)\}$, and (iii) $V_1(p) = V_2(p) = \emptyset$ for each $p \in \mathsf{Prop}$. Then $\mathfrak{M}_1, z \not\models \bot$, and so $\mathfrak{M}_1, y \not\models \top \to \bot$. Hence $\mathfrak{M}_1, x \not\models \top \to (\top \to \bot)$. We obtain $\top \to \alpha \notin \mathbf{BPL}$ hence $\alpha \leftrightarrow \top \notin \mathbf{BPL}$. Now consider $\mathfrak{M}_2$. It is clear that $\mathfrak{M}_2, y \models \top \to \bot$ since there is no successor of y in $\mathfrak{M}_2$, but $\mathfrak{M}_2, y \not\models \bot$. Then $\mathfrak{M}_2, x \not\models (\top \to \bot) \to \bot$. We conclude $\alpha \to \bot$ hence $\alpha \leftrightarrow \bot \notin \mathbf{BPL}$. □

It is known that that **Int** has the two constant property (Proposition 2.26 of Ref. 11) and **Int** is also a superbasic logic. Are there superbasic proper sublogics of **Int** with this property? Here we give an example. Recall $\mathbb{SR} = \{\mathfrak{F} \in \mathbb{TR} \mid \mathfrak{F} \text{ satisfies } \forall x \exists y x R y\}$, i.e., the class of all serial and transitive frames. Consider the logic $\mathbf{Log}(\mathbb{SR})$. Since $\mathbb{SR}$ is closed under liftings, $\mathbf{Log}(\mathbb{SR})$ is a superbasic logic.

Proposition 5.2. $\mathbf{Log}(\mathbb{SR}) \subsetneq \mathbf{Int}$.

Proof. It is easy to see that $(p \wedge (p \to q)) \to q \in \mathbf{Int} \setminus \mathbf{Log}(\mathbb{SR})$, since it is refuted at the following serial frame:

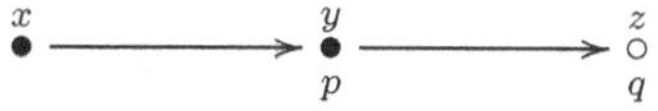

Thus $\mathbf{Log}(\mathbb{SR}) \neq \mathbf{Int}$. Now we show $\mathbf{Log}(\mathbb{SR}) \subseteq \mathbf{Int}$. Suppose $\alpha \notin \mathbf{Int}$. Then $\mathfrak{F} \not\models \alpha$ for some pre-ordered frame $\mathfrak{F}$, since $\mathbf{Int} = \mathbf{Log}(\mathbb{PRE})$. Since $\mathfrak{F} \in \mathbb{SR}$, $\alpha \notin \mathbf{Log}(\mathbb{SR})$. □

Theorem 5.2. *The logic* $\mathbf{Log}(\mathbb{SR})$ *has the two constant property.*

Proof. Let $\alpha \in \mathsf{CFor}$. If $\alpha \in \mathbf{Log}(\mathbb{SR})$, then $\alpha \leftrightarrow \top \in \mathbf{Log}(\mathbb{SR})$. Assume $\alpha \notin \mathbf{Log}(\mathbb{SR})$. Since $\bot \to \alpha \in \mathbf{Log}(\mathbb{SR})$, it suffices to show $\neg\alpha \in \mathbf{Log}(\mathbb{SR})$. Suppose for contradiction that $\neg\alpha \notin \mathbf{Log}(\mathbb{SR})$. Then there are two models $\mathfrak{M} = (W_1, R_1, V_1)$ and $\mathfrak{N} = (W_2, R_2, V_2)$ such that $\mathfrak{M} \not\models \alpha$ and $\mathfrak{N} \not\models \neg\alpha$. Since α is constant, we assume without loss of generality that no variable is satisfiable in both models. Let $\mathfrak{M}, w_1 \not\models \alpha$ and $\mathfrak{N}, w_2 \not\models \neg\alpha$. Then there exists u_2 in $\mathfrak{N}$ such that $w_2 R_2 u_2$ and $\mathfrak{N}, u_2 \models \alpha$. Now consider the generated submodels $\mathfrak{M}_{w_1}$ and $\mathfrak{N}_{u_2}$. Then $\mathfrak{M}_{w_1}, w_1 \not\models \alpha$ and $\mathfrak{N}_{u_2}, u_2 \models \alpha$. Hence $\mathfrak{M}_{w_1} \uplus \mathfrak{N}_{u_2}, w_1 \not\models \alpha$ and $\mathfrak{M}_{w_1} \uplus \mathfrak{N}_{u_2}, u_2 \models \alpha$. Let $\mathfrak{F}_{w_1}$ and $\mathfrak{G}_{u_2}$ be underlying frames of $\mathfrak{M}_{w_1}$ and $\mathfrak{N}_{u_2}$ respectively. Since α is a constant formula, we have $\mathfrak{F}_{w_1} \uplus \mathfrak{G}_{u_2}, w_1 \not\models \alpha$ and $\mathfrak{F}_{w_1} \uplus \mathfrak{G}_{u_2}, u_2 \models \alpha$. Now since the frame $\mathfrak{F}_{w_1} \uplus \mathfrak{G}_{u_2}$ is serial, it can be reduced to the single reflexive point $\circ$. Hence $\circ \models \alpha$, and $\circ \not\models \alpha$, a contradiction. □

As is shown in Ref. 11 (p.28), the classical propositional logic $\mathbf{CL} = \mathbf{Log}(\circ)$. Since $\circ \in \mathbb{SR}$, we get $\mathbf{Log}(\mathbb{SR}) \subseteq \mathbf{CL}$. Then we have the following:

Corollary 5.1. *For every $\alpha \in \mathsf{CFor}$, $\alpha \in \mathbf{Log}(\mathbb{SR})$ iff $\alpha \in \mathbf{CL}$.*

Proof. The left-to-right direction is easy. For the other direction, assume $\alpha \in \mathbf{CL}$. By Theorem 5.2, there are two cases: (i) $\alpha \leftrightarrow \top \in \mathbf{Log}(\mathbb{SR})$, or (ii) $\alpha \leftrightarrow \bot \in \mathbf{Log}(\mathbb{SR})$. For the case (i), $\top \to \alpha \in \mathbf{Log}(\mathbb{SR})$ and so $\alpha \in \mathbf{Log}(\mathbb{SR})$. For the case (ii), $\alpha \leftrightarrow \bot \in \mathbf{Log}(\mathbb{SR})$. Then $\alpha \to \bot \in \mathbf{Log}(\mathbb{SR})$. So $\neg\alpha \in \mathbf{Log}(\mathbb{SR})$, and hence $\neg\alpha \in \mathbf{CL}$, a contradiction with $\alpha \in \mathbf{CL}$. Therefore, $\alpha \in \mathbf{Log}(\mathbb{SR})$. □

A logic Λ is called *0-reducible*, if for all non-theorem $\alpha \notin \Lambda$, there is a constant uniform substitution instance α^σ of α such that $\alpha^\sigma \notin \Lambda$.

Corollary 5.2. *The logic $\mathbf{Log}(\mathbb{SR})$ is not 0-reducible.*

Proof. Assume $\alpha \in \mathbf{CL} \setminus \mathbf{Log}(\mathbb{SR})$. Then any constant substitution instance $\alpha^\sigma \in \mathbf{CL}$ and so by Corollary 5.1, $\alpha^\sigma \in \mathbf{Log}(\mathbb{SR})$. □

By the result that intuitionistic logic has the same constant theorems as classical logic in Ref. 11 (p.35), we get the following result on the relation between $\mathbf{Int}$ and $\mathbf{Log}(\mathbb{SR})$:

Corollary 5.3. *For every $\alpha \in \mathsf{CFor}$, $\alpha \in \mathbf{Int}$ iff $\alpha \in \mathbf{Log}(\mathbb{SR})$.*

It is clear that $\mathbf{Log}(\mathbb{SR}) \cap \mathsf{CFor} = \mathbf{Int} \cap \mathsf{CFor} = \mathbf{CL} \cap \mathsf{CFor}$. However, this is not the case for $\mathbf{BPL}$. Can the superbasic logic $\mathbf{Log}(\mathbb{SR})$ be axiomatizable? The answer is yes. Consider the formula $\neg\neg\top$.

Proposition 5.3. *For every transitive frame $\mathfrak{F}$, $\mathfrak{F} \models \neg\neg\top$ iff $\mathfrak{F}$ satisfies the first-order condition $\forall xy(xRy \to \exists z(yRz))$.*

Proof. Let $\mathfrak{F} = (W, R)$. Assume $\mathfrak{F} \models \neg\neg\top$ and xRy for $x, y \in W$. Let V be an arbitrary valuation in $\mathfrak{F}$ and put $\mathfrak{M} :=(W, R, V)$. Then $\mathfrak{M}, x \models \neg\neg\top$. By xRy, we have $\mathfrak{M}, y \not\models \neg\top$. Then there exists $z \in W$ such that yRz and $\mathfrak{M}, z \models \top$. Conversely, assume $\mathfrak{F}$ satisfies the condition $\forall xy(xRy \to \exists z(yRz))$. Suppose that there is a valuation V and a point $x \in W$ such that $(\mathfrak{F}, V), x \not\models \neg\neg\top$. Then there exists $y \in W$ such that xRy and $(\mathfrak{F}, V), y \models \neg\top$. Hence by the assumption, there exists $z \in W$ such that yRz. Then $(\mathfrak{F}, V), z \not\models \top$, a contradiction. □

Now we show that the logic $\mathbf{Log}(\mathbb{SR})$ of serial frames is equal to $\mathbf{DNT}$ $:= \mathbf{BPL} \oplus \neg\neg\top$ (the proof below is adopted from the proof of the logic $\mathbf{F}$ of strict implication extended with $\neg\neg\top$ in Ref. 10).

Theorem 5.3. *The logic* $\mathbf{DNT} = \mathbf{BPL} \oplus \neg\neg\top$ *is strongly complete with respect to the class* $\mathbb{SR}$ *of serial frames.*

Proof. Let $\Lambda = \mathbf{DNT}$. By Theorem 3.1, it suffices to show the canonical frame $\mathfrak{F}^\Lambda = (W^\Lambda, R^\Lambda)$ is serial. Take any state $w \in W$. We show that $\neg\top \notin w$. Suppose for contradiction that $\neg\top \in w$. Since $\neg\neg\top \to (\neg\top \to \beta) \in \mathbf{BPL}$ and $\neg\neg\top \in \mathbf{DNT}$, we obtain $\neg\top \to \beta \in \mathbf{DNT}$. By assumption, $\beta \in w$ for any formula β, a contradiction. Therefore, $\neg\top \notin w$. By Lemma 3.4, there exists a set w' such that $(w', \mathsf{Ant}(\bot, w))$ is maximally Λ-consistent, $\top \in w'$, $\bot \notin w'$ and $wR^\Lambda w'$. Therefore $R^\Lambda(w) \neq \emptyset$. □

Theorem 5.4. $\mathbf{DNT}$ *has the disjunction property.*

Proof. By the same argument as in the proof of Theorem 4.1. Note that the class of all models for $\mathbf{DNT}$ is closed under taking liftings. □

Theorem 5.5. $\mathbf{DNT}$ *has the finite model property. Moreover,* $\mathbf{DNT}$ *is weakly complete with respect to the class of all finite antisymmetric transitive serial frames.*

Proof. By the similar argument as in the proof of Theorem 4.2, but we need to check that $(R^s_\Sigma)^+$ is serial when R is serial. Assume that R is serial. Fix any point $[x] \in W_\Sigma$. By assumption, we can find some $y \in W$ such that xRy, and so we obtain $[x]R^s_\Sigma[y]$ hence $[x](R^s_\Sigma)^+[y]$. □

6. Intuitionistic Logic and Formal Provability Logic

In this section, we make some observations about intuitionistic logic and formal provability logic. From Ref. 7, we know that $\mathbf{Int}$ is equal to $\mathbf{BPL} \oplus (p \wedge (p \to q)) \to q$. It is also well-known that $\mathbf{Int}$ is sound and complete with respect to the class of all partially ordered frames. However, we can show another completeness theorem for $\mathbf{Int}$.

Proposition 6.1. *For every transitive* $\mathfrak{F}$, $\mathfrak{F} \models (p \wedge (p \to q)) \to q$ *iff* $\mathfrak{F}$ *satisfies the first-order condition* $\forall xy(xRy \to yRy)$.

Proof. The right-to-left direction is easy. For the other direction, assume $\mathfrak{F} \models (p \wedge (p \to q)) \to q$ and xRy. Take a valuation V such that $V(p) =$

$\{y\} \cup R(y)$ and $V(q) = R(y)$. Let $\mathfrak{M} = (W, R, V)$. Then $\mathfrak{M}, y \models p$ and $\mathfrak{M}, y \models p \to q$. Hence $\mathfrak{M}, y \models p \wedge (p \to q)$. By the assumption, $\mathfrak{M}, x \models (p \wedge (p \to q)) \to q$ hence $\mathfrak{M}, y \models q$. Hence $y \in R(y)$, i.e., yRy. □

Notice that the reflexivity $\forall x(xRx)$ implies the first-order condition of weak reflexivity $\forall xy(xRx \to yRy)$, but the converse does not hold.

Theorem 6.1. **Int** *is sound and strongly complete with respect to the class of all transitive frames satisfying the condition* $\forall xy(xRy \to yRy)$.

Proof. Soundness is due to Proposition 6.1. For the strong completeness, it suffices to show that the canonical frame $\mathfrak{F}^{\mathbf{Int}}$ satisfies $\forall xy(xRy \to yRy)$. Let $x, y \in W^{\mathbf{Int}}$ and $xR^{\mathbf{Int}}y$. Assume $\alpha, \alpha \to \beta \in y$. Then $\alpha \wedge (\alpha \to \beta) \in y$. Since $\alpha \wedge (\alpha \to \beta) \to \beta \in \Lambda$, we have $\beta \in y$. Hence $yR^{\mathbf{Int}}y$. □

Another important logic **FPL** in Ref. 1 is obtained from the natural deduction $\mathcal{N}$**BPL** of Definition 2.5 by adding the following Löb's rule:

$$\frac{(\top \to \alpha) \to \alpha}{\top \to \alpha}$$

The Hilbert-style axiomatization of **FPL** can be obtained from **BPL** by adding the following axiom:

$$\text{(Löb)} \quad ((\top \to p) \to p) \to (\top \to p).$$

The following theorem is already known from Ref. 18 (see also Ref. 2):

Theorem 6.2. *The superbasic logic* **FPL** *is sound and weakly complete with respect to the class* $\mathbb{ITR}_{\text{fin}}$ *of all finite irreflexive transitive frames, i.e.,* **FPL** $= \mathbf{Log}(\mathbb{ITR}_{\text{fin}})$.

Corollary 6.1. **FPL** *has the finite model property.*

Here we recall the frame condition for the axiom (Löb). For any frame $\mathfrak{F} = (W, R)$, an infinite chain in $\mathfrak{F}$ is a sequence $(x_i)_{i \in \omega}$ such that $x_i R x_{i+1}$ for all $i \in \omega$. Then it is easy to check that for every transitive frame $\mathfrak{F}$, $\mathfrak{F} \models ((\top \to p) \to p) \to (\top \to p)$ iff for any states x, y in $\mathfrak{F}$, if xRy, then there is no infinite chain starting from y.

Definition 6.1. We say that two superbasic logics Λ_1 and Λ_2 are *comparable*, if $\Lambda_1 \subseteq \Lambda_2$ or $\Lambda_2 \subseteq \Lambda_1$.

Proposition 6.2. **Int** *and* **FPL** *are not comparable.*

Proof. It is easy to see that $((\top \to p) \to p) \to (\top \to p) \notin$ **Int** by the frame conditions for validating the Löb formula and **Int**. Conversely, since $\bullet \not\models p \wedge (p \to q) \to q$ and $\bullet \models$ **FPL**, so $(p \wedge (p \to q)) \to q \notin$ **FPL**. □

Proposition 6.3. **FPL** *has no two constant property.*

Proof. Consider the constant formula $\alpha := \top \to \bot$. Then it is easy to see that $\alpha \leftrightarrow \top \notin$ **FPL** and $\alpha \leftrightarrow \bot \notin$ **FPL**, since $\top \to \alpha$ and $\alpha \to \bot$ are refutable in some finite ascending chain. □

Proposition 6.4. **FPL** *has the disjunction property.*

Proof. The proof is similar to the proof of Theorem 4.1. Assume $\alpha \notin$ **FPL** and $\beta \notin$ **FPL**. By Theorem 6.2, there exist finite irreflexive transitive frames $\mathfrak{F}_1$ and $\mathfrak{F}_2$ such that $\mathfrak{F}_1 \not\models \alpha$ and $\mathfrak{F}_2 \not\models \beta$. Let $\mathfrak{M}_1, x_1 \not\models \alpha$ and $\mathfrak{M}_2, x_2 \not\models \beta$ for some models $(\mathfrak{M}_1, x_1)$ and $(\mathfrak{M}_2, x_2)$ based on frames $\mathfrak{F}_1$ and $\mathfrak{F}_2$ respectively. Assume without loss of generality that $\mathfrak{M}_1$ and $\mathfrak{M}_2$ are generated from x_1 and x_2 respectively. Let $\mathfrak{F} = \mathfrak{F}_1 \uplus \mathfrak{F}_2 = (W, R)$ and $\mathfrak{M} = \mathfrak{M}_1 \uplus \mathfrak{M}_2 = (\mathfrak{F}, V)$. Then $\mathfrak{M}, x_1 \not\models \alpha$ and $\mathfrak{M}, x_2 \not\models \beta$. Let $\bullet \notin W$ be an irreflexive point. Define a model $\mathfrak{M}^\bullet = (\mathfrak{F}^\bullet, V)$ where $\mathfrak{F}^\bullet = (W \cup \{\bullet\}, R \cup \{(\bullet, x) : x \in W\})$. Note that no propositional variable is true at $\bullet$, and $\mathfrak{F}^\bullet \in \mathbb{ITR}_{\mathsf{fin}}$. By induction on χ, it is easy to check that $\mathfrak{M}^\bullet, x \models \chi$ iff $\mathfrak{M}, x \models \chi$, for every $x \in W$. Then $\mathfrak{M}^\bullet, x_1 \not\models \alpha$ and $\mathfrak{M}^\bullet, x_1 \not\models \beta$. Then $\mathfrak{M}^\bullet, \bullet \not\models \alpha \vee \beta$. So $\mathfrak{F}^\bullet \not\models \alpha \vee \beta$. Therefore, $\alpha \vee \beta \notin$ **FPL**. □

7. Post Complete Logics and Post Complete Sets

We say that a superbasic logic Λ is *consistent*, if $\Lambda \neq \mathsf{For}$. Since $\bot \to p \in \Lambda$, a superbasic logic Λ is consistent iff $\bot \notin \Lambda$.

Definition 7.1. A superbasic logic Λ is *Post complete*, if Λ is consistent and there exists no consistent superbasic logic Λ' such that $\Lambda \subsetneq \Lambda'$, i.e., Λ has no proper consistent extensions.

As before, we use $\circ$ to denote a single reflexive point, and $\bullet$ a single irreflexive point. Then **Log**$(\circ)$ = **CL** $\in$ `Ext`(**BPL**). It is well-known that the classical propositional logic **CL** = **Log**$(\circ)$ is the unique Post complete intermediate logic, i.e., in `Ext` (**Int**) (see Ref. 11 [p.50]). Let us consider the sublattice `Ext`(**DNT**) $\subsetneq$ `Ext`(**Int**).

Theorem 7.1. **CL** *is the unique Post complete logic in* `Ext`(**DNT**).

Proof. Suppose for contradiction that $\Lambda \in \texttt{Ext}(\mathbf{DNT})$ is a Post complete superbasic logic and $\Lambda \subsetneq \mathbf{CL}$. Note that Λ is consistent. Then there exists $\alpha \in \Lambda$ and $\alpha \notin \mathbf{CL}$. Let α^σ be a constant substitution instance of α such that $\alpha^\sigma \notin \mathbf{CL}$. So $\neg\alpha^\sigma \in \mathbf{CL}$. By Corollary 5.1, $\neg\alpha^\sigma \in \mathbf{DNT}$. So $\neg\alpha^\sigma \in \Lambda$. Since $\alpha \in \Lambda$ and Λ is closed under uniform substitutions, then $\alpha^\sigma \in \Lambda$ hence $\bot \in \Lambda$, and so Λ is inconsistent. A contradiction. □

More generally, in this section, we consider sets of formulas, because $\mathbf{Log}(\bullet) = \{\alpha : \bullet \models \alpha\}$ is not superbasic as we are going to show in Proposition 7.1. A set Γ of formulas is **BPL**-*consistent*, if $\mathbb{FR}(\Gamma) = \{\mathfrak{F} \mid \mathfrak{F} \models \gamma \text{ for all } \gamma \in \Gamma\} \neq \emptyset$.

Definition 7.2. A set Γ of formulas is *Post complete*, if Γ is **BPL**-consistent and there exists no **BPL**-consistent set Γ' such that $\Gamma \subsetneq \Gamma'$.

Proposition 7.1. $\mathbf{Log}(\bullet)$ *is not a superbasic, i.e.,* $\mathbf{Log}(\bullet) \notin \texttt{Ext}(\mathbf{BPL})$.

Proof. It suffices to show that $\mathbf{Log}(\bullet)$ is not closed under MP. Since $\bullet$ has no successor state, $\{\alpha \to \beta \mid \alpha, \beta \in \mathsf{For}\} \subseteq \mathbf{Log}(\bullet)$. Then $\top \to \bot \in \mathbf{Log}(\bullet)$ and $\top \in \mathbf{Log}(\bullet)$ but $\bot \notin \mathbf{Log}(\bullet)$. □

Makinson's classification of consistent normal modal logic in Ref. 19 can be generalized to our **BPL**-consistent sets.

Lemma 7.1. *For any set* Γ_1 *and* Γ_2 *of formulas, if* $\Gamma_1 = \mathbf{Log}(\mathbb{F})$ *for some class* $\mathbb{F}$ *of transitive frames, and* $\mathbb{F} \models \Gamma_2$*, then* $\Gamma_2 \subseteq \Gamma_1$.

Proof. Assume $\alpha \in \Gamma_2$. Then $\mathbb{F} \models \alpha$. Hence $\alpha \in \mathbf{Log}(\mathbb{F}) = \Gamma_1$. □

Theorem 7.2. *For any* **BPL**-*consistent set* Γ, $\Gamma \subseteq \mathbf{Log}(\bullet)$ *or* $\Gamma \subseteq \mathbf{Log}(\circ)$.

Proof. By Lemma 7.1, it suffices to show that $\bullet \models \Gamma$ or $\circ \models \Gamma$. Since Γ is **BPL**-consistent, there is a frame $\mathfrak{F} \models \Gamma$. If $\mathfrak{F}$ contains a dead end $\bullet$, then $\bullet$ is a generated subframe of $\mathfrak{F}$ and so $\bullet \models \Gamma$. Suppose that $\mathfrak{F}$ contains no dead end, then every state in $\mathfrak{F}$ has a successor. So $\mathfrak{F}$ is reducible to $\circ$. Hence $\circ \models \Gamma$. □

The notion of post completeness in the following is for sets of formulas, since $\mathbf{Log}(\bullet)$ is not superbasic.

Theorem 7.3. $\mathbf{Log}(\bullet)$ *is a Post complete set.*

Proof. It is clear that $\mathbf{Log}(\bullet)$ is **BPL**-consistent. Suppose for contradiction that there exists a **BPL**-consistent Γ such that $\mathbf{Log}(\bullet) \subsetneq \Gamma$. By Theorem 7.1, $\Gamma \subseteq \mathbf{Log}(\circ)$. Then $\mathbf{Log}(\bullet) \subsetneq \mathbf{Log}(\circ)$, which is impossible because $\top \to \bot \in \mathbf{Log}(\bullet)$ as we observed in the proof of Proposition 7.1. □

The set $\mathbf{Log}(\bullet)$ is not a superbasic logic but a Post complete set. How can we give a proof theory to capture $\mathbf{Log}(\bullet)$? Recall that all implication formulas are theorems of this logic. This reminds us to use implication formulas as axioms. Then we need only add rules for other connectives.

Definition 7.3. The natural deduction system $\mathcal{N}_\bullet$ for $\mathbf{Log}(\bullet)$ consists of the following inference rule:

$$\frac{}{\alpha \to \beta}\ (\to)$$

as well as $(\wedge I)$, $(\wedge E_i)$, $(\vee I_i)$, $(\vee E)$ and $(\bot E)$ of Definition 2.5. For any set $\Gamma \cup \{\alpha \mid \}$ of formulas, we say that α is *derivable* from Γ in $\mathcal{N}_\bullet$ (notation: $\Gamma \vdash_{\mathcal{N}_\bullet} \alpha$), if there exists a derivation tree generated by the proof-rules where the root node is α and all non-discharged assumptions are from Γ.

It is clear that the derivability relation $\vdash_{\mathcal{N}_\bullet}$ is monotone: if $\Gamma_1 \subseteq \Gamma_2$ and $\Gamma_1 \vdash_{\mathcal{N}_\bullet} \alpha$, then $\Gamma_2 \vdash_{\mathcal{N}_\bullet} \alpha$.

Definition 7.4. We say that α is a *theorem* of $\mathcal{N}_\bullet$, if $\emptyset \vdash_{\mathcal{N}_\bullet} \alpha$. Let $\mathbf{Thm}(\mathcal{N}_\bullet)$ is the set of all $\mathcal{N}_\bullet$-theorems. A set Γ of formulas is $\mathcal{N}_\bullet$-*consistent*, if $\Gamma \nvdash_{\mathcal{N}_\bullet} \bot$. A set Γ of formulas is a $\mathcal{N}_\bullet$-*theory*, if Γ is closed under $\vdash_{\mathcal{N}_\bullet}$, i.e., $\Gamma \vdash_{\mathcal{N}_\bullet} \alpha$ implies $\alpha \in \Gamma$, for all formulas α. A $\mathcal{N}_\bullet$-theory is *prime*, if $\alpha \vee \beta \in \Gamma$ implies $\alpha \in \Gamma$ or $\beta \in \Gamma$.

Lemma 7.2. *Assume that $\Gamma \nvdash_{\mathcal{N}_\bullet} \alpha$. Then there exists a prime theory Γ^+ such that $\Gamma \subseteq \Gamma^+$ and $\alpha \notin \Gamma^+$.*

Proof. Assume that $\Gamma \nvdash \alpha$. Enumerate all formulas as $(\beta_n)_{n\in\omega}$. Define a sequence $(\Gamma_n)_{n\in\omega}$ as follows:

$$\Gamma_0 = \Gamma; \quad \Gamma_{n+1} = \begin{cases} \Gamma_n \cup \{\beta_n\}, \text{ if } \Gamma_n \cup \{\beta_n\} \nvdash_{\mathcal{N}_\bullet} \alpha \\ \Gamma_n, \text{ otherwise.} \end{cases}$$

Let $\Gamma^+ = \bigcup_{n\in\omega} \Gamma_n$. Then, it is easy to check that Γ^+ is a prime theory. □

Lemma 7.3. *Given a prime $\mathcal{N}_\bullet$-theory Γ, $\mathfrak{M}_\Gamma = (\{\Gamma\}, \emptyset, V)$ is defined as follows: for each $p \in \mathsf{Prop}$, $\Gamma \in V(p)$ iff $p \in \Gamma$. Then, for every formula α, $\mathfrak{M}_\Gamma, \Gamma \models \alpha$ iff $\alpha \in \Gamma$.*

Proof. By induction on α. For the case $\alpha = \beta \to \gamma$, it suffices to note that $\beta \to \gamma$ is valid at the single irreflexive state Γ and it also belongs to Γ, since Γ is an $\mathcal{N}_\bullet$-theory and $\beta \to \gamma \in \mathbf{Thm}(\mathcal{N}_\bullet)$. The other cases are easy. □

Theorem 7.4. *For every set $\Gamma \cup \{\alpha\}$ of formulas, $\Gamma \vdash_{\mathcal{N}_\bullet} \alpha$ iff $\Gamma \models_{\{\bullet\}} \alpha$. Therefore $\mathbf{Thm}(\mathcal{N}_\bullet) = \mathbf{Log}(\bullet)$.*

Proof. The soundness is easy. For the completeness, assume that $\Gamma \nvdash_{\mathcal{N}_\bullet} \alpha$. By Lemma 7.2 and Lemma 7.3, there exists a prime $\mathcal{N}_\bullet$-theory $\Gamma^+ \supseteq \Gamma$ and a model $\mathfrak{M}_{\Gamma^+} = (\{\Gamma^+\}, \emptyset, V)$ such that $\mathfrak{M}_{\Gamma^+}, \Gamma^+ \models \gamma$ for all $\gamma \in \Gamma^+$ but $\mathfrak{M}_{\Gamma^+}, \Gamma^+ \not\models \alpha$. Since $(\{\Gamma^+\}, \emptyset)$ is isomorphic to $\bullet$, $\Gamma \not\models_{\{\bullet\}} \alpha$. □

What is the relationship between $\mathbf{Log}(\bullet)$ and $\mathbf{Log}(\circ)$? Obviously, $\mathbf{Log}(\bullet) \not\subseteq \mathbf{Log}(\circ)$ since there are implication formulas $\alpha \to \beta$ which are not tautologies. Conversely, every tautology is valid in a single state frame, whether it is reflexive or not. Then we have the following proposition:

Proposition 7.2. $\mathbf{Log}(\circ) \subsetneq \mathbf{Log}(\bullet)$. *Therefore, any* **BPL**-*consistent set is a subset of* $\mathbf{Log}(\bullet)$.

Proof. It suffices to show $\mathbf{Log}(\circ) \subseteq \mathbf{Log}(\bullet)$. Fix any formula α such that $\circ \models \alpha$ and any valuation V on $\bullet$. Note that V can be regarded as a valuation on $\circ$. Then, by induction on β, we can show $(\circ, V), \circ \models \beta$ implies $(\bullet, V), \bullet \models \beta$. So, $(\bullet, V), \bullet \models \alpha$. Therefore, $\bullet \models \alpha$. □

Clearly $\mathbf{Log}(\bullet)$ differs from $\mathbf{Log}(\circ)$ on implicational formulas. Define the set Noif of all implication-free formula of For as follows:

$$\mathsf{Noif} \ni \alpha ::= p \mid \bot \mid \alpha \wedge \alpha \mid \alpha \vee \alpha, \text{ where } p \in \mathsf{Prop}.$$

Proposition 7.3. *Let $* \in \{\bullet, \circ\}$ and denote $*$ by $(\{*\}, R_*)$, where $R_* := \emptyset$ (if $* = \bullet$); $\{(\circ, \circ)\}$ (if $*$ is $\circ$). Define a valuation $V_\emptyset : \mathsf{Prop} \to \mathcal{P}(\{*\})$ by $V_\emptyset(p) = \emptyset$ for all $p \in \mathsf{Prop}$. Then, for all $\alpha \in \mathsf{Noif}$, $(\{*\}, R_*, V_\emptyset), * \not\models \alpha$.*

Proposition 7.4. *Let $* \in \{\bullet, \circ\}$. The implication-free fragment of $\mathbf{Log}(*)$ is empty, i.e., $\mathbf{Log}(*) \cap \mathsf{Noif} = \emptyset$.*

Proof. For every $\alpha \in \mathsf{Noif}$, by Proposition 7.3, $\alpha \notin \mathbf{Log}(*)$. □

Corollary 7.1. *Let Λ be a superbasic logic. If $\Lambda \subseteq \mathbf{Log}(\bullet)$, then the implication-free fragment of Λ is empty, i.e., $\Lambda \cap \mathsf{Noif} = \emptyset$.*

Corollary 7.2. *The implication-free fragment of* **BPL** *is empty, i.e.,* $\mathbf{BPL} \cap \mathsf{Noif} = \emptyset$.

Proof. By Corollary 7.1 and $\mathbf{BPL} = \mathbf{Log}(\mathbb{TR})$ and $\bullet \in \mathbb{TR}$. □

8. Some Embedding Theorems

In this section, we will prove some results on the embeddability of superbasic logics into modal logics. Here we first recall some basic concepts of modal logic. The set $\mathsf{For}(\Box)$ of all modal formulas is defined by:

$$\mathsf{For}(\Box) \ni \varphi ::= p \mid \bot \mid (\varphi \to \varphi) \mid (\varphi \wedge \varphi) \mid (\varphi \vee \varphi) \mid \Box\varphi, \text{ where } p \in \mathsf{Prop}.$$

Other propositional connectives are defined as usual. Define $\Diamond\varphi := \neg\Box\neg\varphi$.

A *frame* for the modal language is a pair $\mathfrak{F} = (W, R)$ where W is a non-empty set of states, and R a binary relation over W. Note that the transitivity is not assumed here. A *model* for the modal language is a tuple $\mathfrak{M} = (W, R, V)$ where (W, R) is a modal frame and $V : \mathsf{Prop} \to \mathcal{P}(W)$ is a valuation. Note that we do not assume the persistency condition of the valuation in a model for the modal language. The satisfiability relation $\mathfrak{M}, w \Vdash \alpha$ between $(\mathfrak{M}, w)$ and a modal formula α is defined as usual as in Ref. 11. In particular, we have

(1) $\mathfrak{M}, w \Vdash \Box\varphi$ iff $\mathfrak{M}, u \Vdash \varphi$ for every u such that wRu.
(2) $\mathfrak{M}, w \Vdash \Diamond\varphi$ iff $\mathfrak{M}, u \Vdash \varphi$ for some u such that wRu.

The validity of modal formulas in frames is also defined as usual.

The minimal normal modal logic **K** is the set of all modal formulas valid in all frames. It can be axiomatized by (i) all instances of propositional tautologies, (ii) $\Box(p \to q) \to (\Box p \to \Box q)$, and (iii) the rules (MP), (US) and (Gen): from α infer $\Box\alpha$. A *normal modal logic* is a set Λ of modal formulas such that $\mathbf{K} \subseteq \Lambda$ and Λ is closed under (MP), (US) and (Gen). By $\Lambda \oplus \varphi$ we mean the normal modal logic obtained by adding φ as a new axiom to Λ.

Put **FPL** and **Int** together with other superbasic logics, we have Fig. 1 which is contrasted with the lattice of normal modal logics on the right-side of Fig. 1. We consider the following normal modal logics:

$$\mathbf{Ver} = \mathbf{K} \oplus \Box\bot, \qquad \mathbf{Triv} = \mathbf{K} \oplus p \leftrightarrow \mathcal{B}p$$
$$\mathbf{K4} = \mathbf{K} \oplus \Box p \to \mathcal{B}\mathcal{B}p, \qquad \mathbf{KD4} = \mathbf{K4} \oplus \Diamond\top$$
$$\mathbf{GL} = \mathbf{K} \oplus \Box(\Box p \to p) \to \Box p, \qquad \mathbf{S4} = \mathbf{K4} \oplus \Box p \to p$$
$$\mathbf{Grz} = \mathbf{K} \oplus \Box(\Box(p \to \Box p) \to p) \to p, \qquad \mathbf{S5} = \mathbf{S4} \oplus \mathbf{D}p \to \Box\mathbf{D}p$$

As in Fig. 1, the Gödel-Mckinsey-Tarski translation G can embed **Int** into **S4** (Refs. 11,20), where a dotted line of Fig. 1 corresponds to an embedding G. **Grz** is the greatest extension of **S4** into which **Int** is embeddable (Ref. 21). Visser also proved that **FPL** is embeddable in modal logic **GL** via G

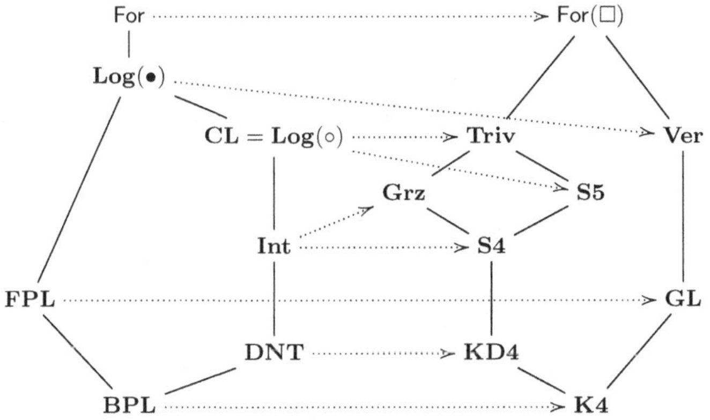

Fig. 1. Gödel-Mckinsey-Tarski translation G

in Ref. 1. Here we use the semantic method from Ref. 11 (pp. 96-98) to show some embedding results. Notice that some embedding results with a different translation can be found also in Ref. 10.

Definition 8.1. The Gödel-Mckinsey-Tarski translation $G : \mathsf{For} \to \mathsf{For}(\Box)$ is defined recursively as follows:

$$G(p) = \Box p, \qquad G(\bot) = \bot,$$
$$G(\alpha \wedge \beta) = G(\alpha) \wedge G(\beta), \qquad G(\alpha \vee \beta) = G(\alpha) \vee G(\beta),$$
$$G(\alpha \to \beta) = \Box(G(\alpha) \to G(\beta)).$$

Definition 8.2 (Skeleton frame). *Let $\mathfrak{F} = (W, R)$ be a transitive frame. Define a binary relation $\sim$ over W by:*

$$x \sim y \text{ iff } x = y \text{ or } (xRy \ \& \ yRx).$$

Then $\sim$ is an equivalence relation. Let $C(x) = \{y \in W \mid x \sim y\}$ and $W^\rho = \{C(x) \mid x \in W\}$. We say that $C(x)$ is a cluster *containing x. Define the relation R^ρ over W^ρ by:*

$$C(x) R^\rho C(y) \text{ iff } xRy.$$

The frame $\mathfrak{F}^\rho = (W^\rho, R^\rho)$ is called the skeleton *of $\mathfrak{F}$.*

Note that $\mathfrak{F}^\rho$ is transitive. Moreover, if every cluster in $\mathfrak{F}^\rho$ is *simple*, i.e., a singleton, then $\mathfrak{F}^\rho$ is isomorphic to $\mathfrak{F}$.

Definition 8.3 (Skeleton model). *Given any (possibly non-persistent) model* $\mathfrak{M} = (\mathfrak{F}, V)$ *where* $\mathfrak{F}$ *is a transitive frame, define a valuation* V^ρ *in* $\mathfrak{F}^\rho$ *by* $V^\rho(p) = \{C(x) \mid \mathfrak{M}, x \Vdash \Box p\}$ *for each* $p \in \mathsf{Prop}$. *Then the model* $\mathfrak{M}^\rho = (\mathfrak{F}^\rho, V^\rho)$ *is called the* skeleton *of* $\mathfrak{M}$.

It is clear that $\mathfrak{M}^\rho$ is persistent. Hence it is a model for **BPL**. Skeletons of transitive frames can be found in Refs. 11 (p.68) and 18. Here we use this construction to show some embedding results for superbasic logics.

Lemma 8.1. *For every model* $\mathfrak{M}$, *state* x *of* $\mathfrak{M}$, *and formula* $\alpha \in \mathsf{For}$, $\mathfrak{M}, x \Vdash G(\alpha)$ *iff* $\mathfrak{M}^\rho, C(x) \models \alpha$.

Proof. The proof by induction on α is omitted. (cf. Ref. 11 (p.96)). □

Corollary 8.1. *For any transitive* $\mathfrak{F}$ *and* $\alpha \in \mathsf{For}$, $\mathfrak{F} \Vdash G(\alpha)$ *iff* $\mathfrak{F}^\rho \models \alpha$.

Proof. Let $\mathfrak{F} = (W, R)$. Assume $\mathfrak{F} \nVdash G(\alpha)$. There exists a valuation V in $\mathfrak{F}$ and $w \in W$ such that $\mathfrak{F}, V, w \nVdash \alpha$. Let $\mathfrak{M} = (\mathfrak{F}, V)$. Define a valuation V^ρ in $\mathfrak{F}^\rho$ by $V^\rho(p) = \{C(x) \mid \mathfrak{M}, x \Vdash \Box p\}$ for each $p \in \mathsf{Prop}$. By Lemma 8.1, $\mathfrak{M}^\rho, C(x) \models \alpha$. Hence $\mathfrak{F}^\rho \not\models \alpha$. Conversely, assume $\mathfrak{F}^\rho \not\models \alpha$. Then there exists a persistent valuation U^ρ in $\mathfrak{F}^\rho$ such that $(\mathfrak{F}^\rho, U^\rho), w \not\models \alpha$. Define a valuation U in $\mathfrak{F}$ by $U(p) = \bigcup U^\rho(p)$ for each $p \in \mathsf{Prop}$. Let $\mathfrak{N} = (\mathfrak{F}, U)$. Then it is easy to verify that $\mathfrak{N}^\rho = (\mathfrak{F}^\rho, U^\rho)$. By Lemma 8.1, $(\mathfrak{F}, U), w \nVdash G(\alpha)$. Hence $\mathfrak{F} \nVdash G(\alpha)$. □

Theorem 8.1. *For every formula* $\alpha \in \mathsf{For}$, $\alpha \in \mathbf{BPL}$ *iff* $G(\alpha) \in \mathbf{K4}$.

Proof. Assume that $G(\alpha) \notin \mathbf{K4}$. Then by completeness of **K4**, there exists a transitive frame $\mathfrak{F} \nVdash G(\alpha)$. By Corollary 8.1, $\mathfrak{F}^\rho \not\models \alpha$. Hence $\alpha \notin \mathbf{BPL}$. Conversely, assume that $\alpha \notin \mathbf{BPL}$. By Theorem 4.2, there exists a finite antisymmetric transitive frame $\mathfrak{F} \not\models \alpha$. Note that every cluster in $\mathfrak{F}^\rho$ is simple since $\mathfrak{F}$ is finite and antisymmetric. Hence $\mathfrak{F}$ is isomorphic to $\mathfrak{F}^\rho$, and so $\mathfrak{F}^\rho \not\models \alpha$. Then $\mathfrak{F} \nVdash G(\alpha)$. Hence $G(\alpha) \notin \mathbf{K4}$. □

As we see above, the logic **BPL** is embedded into **K4**. To complete our observations on the lattice of superbasic logics, we apply the model-theoretic argument to prove that **FPL** is embedded into **GL**. Recall that for every frame $\mathfrak{F}$, it validates the modal Löb formula iff it is transitive and there is no infinite chain starting from one state (see Proposition 3.47 of Ref. 11). Moreover, $\mathbf{K4} \subseteq \mathbf{GL}$.

Lemma 8.2. *For every formula* $\alpha \in \mathsf{For}$, *if* $\alpha \in \mathbf{FPL}$, *then* $G(\alpha) \in \mathbf{GL}$.

Proof. Assume $\alpha \in \mathbf{FPL}$. By induction on the length of proof in **FPL**, it is easy to show $G(\alpha) \in \mathbf{GL}$. □

Theorem 8.2. *For every formula* $\alpha \in \mathsf{For}$, $\alpha \in \mathbf{FPL}$ *iff* $G(\alpha) \in \mathbf{GL}$.

Proof. The left-to-right direction is by Lemma 8.2. Assume that $\alpha \notin \mathbf{FPL}$. Then by completeness of **FPL**, there exists a finite irreflexive transitive frame $\mathfrak{F}$ such that $\mathfrak{F} \not\models \alpha$. Clearly, $\mathfrak{F}$ is also antisymmetric, and hence $\mathfrak{F}$ is isomorphic to $\mathfrak{F}^\rho$. Then $\mathfrak{F}^\rho \not\models \alpha$. Hence $\mathfrak{F} \nVdash G(\alpha)$ by Corollary 8.1. Since $\mathfrak{F}$ is frame for **GL**, we have $G(\alpha) \notin \mathbf{GL}$. □

It is already known that **GL** is embeddable into modal logic **K4** (Ref. 22). Hence **FPL** is embeddable into **K4**. Finally, we show that **Log(•)** is embeddable into **Ver** and the superbasic logic **DNT** is embeddable into **KD4** via the translation G.

Theorem 8.3. *For every formula* $\alpha \in \mathsf{For}$, $\alpha \in \mathbf{Log}(\bullet)$ *iff* $G(\alpha) \in \mathbf{Ver}$.

Proof. It is easy to prove that $G(\alpha) \in \mathbf{Ver}$ by induction on the length of proof of α in the natural deduction $\mathcal{N}_\bullet$ for **Log(•)**. Obviously, the rules $(\wedge I), (\wedge E_i), (\vee I_i), (\vee E)$ and $(\bot E)$ are preserved under the translation G. For $(\to)$, $G(\alpha \to \beta) = \Box(G(\alpha) \to G(\beta))$ which is a theorem of **Ver**. Therefore, the translation of each theorem in **Log(•)** is a theorem of **Ver**. Conversely, assume $\alpha \notin \mathbf{Log}(\bullet)$. Note that the skeleton of $\bullet$ is itself. Hence $\bullet^\rho \not\models \alpha$. Then $\bullet^\rho \nVdash G(\alpha)$. It follows that $G(\alpha) \notin \mathbf{Log}(\bullet)$. □

Theorem 8.4. *For every formula* $\alpha \in \mathsf{For}$, $\alpha \in \mathbf{DNT}$ *iff* $G(\alpha) \in \mathbf{KD4}$.

Proof. First, if $\alpha \in \mathbf{DNT}$, by induction on the length of proof in **DNT**, we can prove that $G(\alpha) \in \mathbf{KD4}$. Notice that the translation $G(\neg\neg\top)$ is equivalent to $\Box\mathbf{D}\top$ which is a theorem of **KD4**. Conversely, assume $\alpha \notin \mathbf{DNT}$. By Theorem 5.5, there exists a finite antisymmetric transitive serial frame $\mathfrak{F}$ such that $\mathfrak{F} \not\models \alpha$. Clearly, $\mathfrak{F}$ is isomorphic to $\mathfrak{F}^\rho$. Hence $\mathfrak{F}^\rho \not\models \alpha$. Then $\mathfrak{F}^\rho \nVdash G(\alpha)$. Moreover, $\mathfrak{F}^\rho$ is also transitive and serial. It follows that $G(\alpha) \notin \mathbf{KD4}$. □

9. Open Problems

Problem 9.1. *Does the operation* $\oplus$ *in* $\mathtt{Ext}(\mathbf{BPL})$ *preserve finite axiomatizability?*

Problem 9.2. *Provide a sufficient and necessary condition for the logic* $\mathbf{Log}(\mathbb{F})$ *of a frame class* $\mathbb{F}$ *to be a superbasic logic.*

Problem 9.3. *Can the syntax* $\mathcal{L}$ *such that* $\mathcal{L} \cap \mathbf{BPL} = \mathcal{L} \cap \mathbf{Int}$ *be characterized?*

Problem 9.4. *Which superbasic logics can be embedded into* **BPL***?*

Aghaei and Ardeshir gave an embedding from **Int** to **BPL** in Ref. 23.

Problem 9.5. *Are there logics* $\Lambda \neq \mathbf{Int}$ *such that the double negation translation can embed* **CL** *into* Λ*?*

Acknowledgement

We would like to give our thanks to the anonymous reviewer, who gave useful comments on this paper. We also thank the participants at the Asian Logic Conference 2013 who commented on our draft. The work of the first author was supported by China national funding of social sciences (grant no. 12CZX054) and the research funding of China ministry of education (grant no. 12YJC72040001) and the work of the second author was partially supported by JSPS KAKENHI, Grant-in-Aid for Young Scientists (B) 24700146.

References

1. A. Visser, A propositional logic with explicit fixed points, *Studia Logica* **40**, 155–175 (1981).
2. K. Ishii, R. Kashima and K. Kikuch, Sequent calculi for Visser's propositional logics, *Notre Dame Journal of Formal Logic* **42**, 1–22 (2001).
3. K. Ishii, New sequent calculi for Visser's formal propositional logic, *Mathematical Logic Quarterly* **49**, 525–535 (2003).
4. K. Kikuchi and K. Sasaki, A cut-free Gentzen formulation of basic propositional calculus, *Journal of Logic, Language and Information* **12**, 213–254 (2003).
5. M. Aghaei and M. Ardeshir, A Gentzen-style axiomatization for basic predicate calculus, *Archive for Mathematical Logic* **42**, 245–259 (2003).
6. Y. Suzuki and H. Ono, *Hilbert style proof systems for BPL*, Tech. Rep. RR-97-0040F, Japan Advanced Institute of Science and Technology (1997).
7. R. Ishigaki and K. Kikuchi, Tree-sequent methods for subintuitionistic predicate logics, *LNCS* **4548**, 149–164 (2007), TABLEAUX 2007, Proceedings of the 16th international conference on Automated Reasoning with Analytic Tableaux and Related Methods.

8. M. Alizadeh and M. Ardeshir, On Löb algebras, *Mathematical Logic Quarterly* **52**, 95–105 (2006).
9. M. Alizadeh and M. Ardeshir, Completions of basic algebra, in *WoLLIC 2009*, eds. M. K. H. Ono and R. de Queiroz (Springer pp. 72–83.
10. G. Corsi, Weak logics with strict implication, *Mathematical Logic Quarterly* **33**, 389–406 (1987).
11. A. Chagrov and M. Zakharyaschev, *Modal Logic* (Clarendon Press, 1997).
12. M. Ardeshir, Aspects of basic logic, PhD thesis, Department of Mathematics, Statistics and Computer Science, Marquette University, Milwaukee, 1995.
13. B. Davey and H. Priestley, *Introduction to Lattices and Order*, 2 edn. (Cambridge University Press, 2002).
14. M. Ardeshir and W. Ruitenburg, Basic propositional calculus II, *Archive for Mathematical Logic* **40**, 349–364 (2001).
15. K. Sasaki, Formalizations of the consequence relation of Vseer's propositional logic, *Reports on Mathematical Logic* **33**, 65–78 (1999).
16. W. Ruitenburg, Basic logic, **K4** and persistence, *Studia Logica* **63**, 343–352 (1999).
17. F. Bou, Strict weak languages. An analysis of strict implication, PhD thesis, University of Barcelona, 2004.
18. Y. Suzuki, F. Wolter and M. Zakharyaschev, Speaking about transitive frames in propositional languages, *Journal of Logic, Language and Information* **3**, 317–339 (1998).
19. D. Makinson, Some embedding theorems for modal logic, *Notre Dame Journal of Formal Logic* **12**, 252–254 (1971).
20. J. McKinsey and A. Tarski, The algebra of topology, *Annals of Mathematics* **45**, 141–191 (1944).
21. M. Zakharyaschev, The greatest extension of s4 into which intuitionistic logic is embeddable, *Studia Logica* **59**, 345–358 (1997).
22. P. Balbiani and A. Herzig, A translation from the modal logic of provability into K4, *Journal of Applied Non-classical Logics* **4**, 73–77 (1994).
23. M. Aghaei and M. Ardeshir, A bounded translation of intuitionistic propositional logic into basic propositional logic., *Mathematical Logic Quarterly* **46**, 199–206 (2000).

Large Cardinals and Higher Degree Theory

Xianghui Shi

School of Mathematical Sciences, Beijing Normal University
Key Laboratory of Mathematics and Complex Systems, Ministry of Education
Beijing 100875, P.R. China
Email: shi.bnu@gmail.com

This is a survey paper on some recent developments in the study of higher degree theory, the theory of degree structure of generalized degrees at uncountable cardinals, in particular at those with countable cofinality.

The study of generalized degree notions, known as α-recursion theory, was initiated by Sacks in the early 70s. But the early works were mainly from recursion theoretical perspective, and the results are mostly limited to Gödel's constructive universe. Some recent works (see Refs. 1,2) in set theory indicate that there is a profound connection between the complexity of degree structures at singular cardinals of countable cofinality and the large cardinal properties at the vicinity. This paper shall exam degree structures cross through L-like models for various mild large cardinals, or under large cardinal assumptions, by which hopefully a new line of research will be depicted.

Keywords: Turing degree, hyperarithmetic degree, inner model operator, jump, incomparable degrees, minimal degree, Posner-Robinson Theorem, Degree Determinacy, higher degree theory, Covering Lemma, large cardinal, core model, Axiom I_0

1. Organization of the paper

In this article, we survey some recent developments in degree theory, in particular, the study of degree structures of generalized degrees at singular cardinals. The paper consists of three parts. In **S2**, we give a brief account of various generalizations of classical recursion theory, in particular the two directions – one up in degree notion hierarchy and the other lifting to larger ordinals. In this part, we investigate these structures focusing on a particular set of structural questions. In **S3**, we present the latest discovery of the connection between the complexity of degree structures at countable cofinality singular cardinals and the large cardinal strength of relevant cardinals. The structure of Zermelo degrees at countable cofinality singular cardinals in various core models of large cardinals, or under strong large cardinal principles, are compared, according to that particular list of structural questions. In the last part (**S**4) we put down some remarks, proposing directions for further investigation.

Notations in this paper are very set theoretic, if not given explicitly, mostly follow Jech[3] and Kanamori[4].

2. Generalizing the classical recursion theory

Early in 1960s, efforts had already been made to generalize classical recursion theory to higher level or in broader context. In this section, we briefly recall various generalizations of classical recursion theory, in particular the two directions – one up in degree notion hierarchy and the other lifting to larger ordinals.

2.1. *Inner model operators*

In classical recursion theory, two subsets $A, B \subseteq \omega$, A is *Turing reducible to* B, denoted as $A \leq_T B$, if both A and the complement of A can be recursively computed given B as an oracle, or equivalently, A is Δ^0_1 definable in B (both A and $\omega - A$ are Σ^0_1 definable in B, using B as a predicate symbol). This partial ordering induces an equivalence relation – the notion of Turing degrees – on subsets of ω. Turing degrees may as well be called Δ^0_1-degrees. For $A \subseteq \omega$, let $[A]_T$[a] denote the degree represented by A, and $[A]'_T$ denote the Turing jump of $[A]_T$, the degree of the Σ^0_1-theory of $(\mathbb{N}, +, \times, 0, 1, A)$.

By replacing Δ^0_1 with larger collection of sets, one can define degree notions for higher levels of definability. For instance, a natural landmark in this hierarchy of definability degree notions is the hyperarithmetic degree. The hyperarithmetic degrees are the Δ^1_1-degrees, obtained by considering the class of Δ^1_1 subsets of ω; the hyperarithmetic jump of $[\emptyset]$ is the degree of Kleene's $\mathcal{O}$, a complete Π^1_1-set.

In Ref. 5, Hodes applied Jensen's fine structure theory to iterate the jump through transfinite up to $\aleph^L_1$.

Theorem 2.1 (Hodes[5]). *Use Jensen's J_α hierarchy for L.*

Whenever $(\Delta_{n+1}(J_\alpha) - \Delta_n(J_\alpha)) \cap \mathscr{P}(\omega) \neq \varnothing$, within this set there is a largest Turing degree, which contains $\Delta_{n+1}(J_\alpha)$-master code.

---

[a]In recursion theory, people normally use $\boldsymbol{a}$ or $\boldsymbol{a}_T$. Due to our excessive use of subscripts, we use the bracket form $[a]_T$, treating it as equivalence class.

The master code degrees are wellordered by the order of construction of their members in the J-hierarchy, and this order coincides with Turing reducibility on these degrees: for $\alpha < \aleph_1^L$*, the Turing jump of the* α*-th master code is an* $(\alpha+1)$*-th master code.*

Master code is a terminology from fine structure theory. We omit its definition. The degree of α-th master codes is now widely accepted as the α-th iterate of the Turing jump, denoted as $\emptyset^{(\alpha)}$. The hyperarithmetic degree of $\emptyset$ consists of exactly those $x \subseteq \omega$ such that $x \leq_T \emptyset^{(\alpha)}$ for some $\alpha < \omega_1^{\mathrm{CK}}$, and Kleene's $\mathcal{O} \equiv_T \omega_1^{\mathrm{CK}}$-th master code. But master code is only helpful for degrees below the constructible degrees (see the proposition on p.204).

In Ref. 6, Steel made the notion of degree operator precise.

Definition 2.1 (Steel[6]). *Let* $M : 2^\omega \to \mathscr{P}(2^\omega)$ *and*

$\forall x, y\,(x \equiv_T y \Rightarrow M(x) = M(y))$,
$\forall x\,(M(x)$ *is closed under join, Turing jump and Turing reducibility*$)$,
$\forall x, y\,(y \in M(x) \Rightarrow M(y) \subseteq M(x))$,
there is a relation $W(x,y,z)$ *so that* $\forall x\,(W_x = \{(y,z) \mid W(x,y,z)\}$ *is a wellorder of* $M(x))$,
for $\alpha < \mathrm{otp}(W_x)$ *and* $e \in \omega$*, let* $z_e = z$ *if there is a* y *such that* $y \equiv_T x$ *via* e *and* z *is the* α*-th element of* W_y*; and* $z_e = \emptyset$ *otherwise. Then there is a real in* $M(x)$ *coding the sequence* $\langle z_e : e < \omega\rangle$*.*[b]

Then we say that M *is an* inner model operator *(IMO).*

The intuition behind IMO is to consider $M(x)$ as $M_x \cap \mathscr{P}(\omega)$, where M_x is a transitive (set) model of ZFC, or natural fragment of ZFC. For the hyperarithmetic degree, the associated inner model operator is the map $x \mapsto L_{\omega_1^{\mathrm{CK}}}[x]$, the smallest Kripke-Platek model containing x. The constructible degree is given by $x \mapsto L[x]$. The readers can find more examples of "natural" inner model operators in Ref. 7.

For the next result, we assume AD. As we only work with (Turing) degree invariant sets and functions, we shall not distinguish $d \subseteq \omega$ and its degree $[d]$. Let μ be the cone measure on Turing degrees. Given two IMOs M, N, we say $M \leq N$ if $M(d) \subseteq N(d)$ for μ-a.e. d. A *jump operator* is a function $f : \mathscr{P}(\omega) \to \mathscr{P}(\omega)$ such that $f(d) \geq_T d$ for μ-a.e. d and f is uniformly (Turing) degree invariant, i.e. there is a $\pi : \omega \to \omega$ such that for all $e < \omega$, $x \equiv_T y$ via $e \Rightarrow f(x) \equiv_T f(y)$ via $\pi(e)$.

[b]This item is a uniformity requirement. $\mathrm{otp}(W_x)$ stands for the ordertype of W_x.

Theorem 2.2 (Steel[6]). [c]

$\leq$ *prewellorders inner model operators.*
If $f(d) \in L[d]$ for μ-a.e. d, then f is a jump operator iff $f(d)$ is a d-master code for μ-a.e. d.

In some sense, this says that degree notions defined via inner models of the form $L_{\gamma_d}[d] \cap \mathscr{P}(\omega)$ (for some $\gamma_d < \aleph_1^L$), where d is a master code, forms the initial segment of the hierarchies of degree notions. For IMOs of this form, the jump of the associated degree notions is the least d-master code not in $M(d)$ (for $d \subset \omega$), and this master code codes the relevant theory of $M(d)$:

Proposition 2.1 (Hodes[5]). *If $L_{\gamma_d}[d] \not\models \Delta_{n+1}$-CA, then $(\omega\alpha+n)$-th master code for d codes the n-quantifier theory of $(L_{\gamma_d}[d], \in, d)$.*

The theories (in the language of second-order arithmetic) related to these IMOs include Δ^1_n-CA ($n < \omega$), full CA, Σ^0_α-det ($\alpha < \omega_1^{\mathrm{CK}}$), Δ^1_1-det. Π^1_1-det is the least theory whose associated IMO is not of aforementioned form, as this theory is equivalent to the existence of sharps. For IMOs whose associated theories include full comprehension, the associated jump of $d \subset \omega$ naturally has to be the sharp of M_d, namely, the set coding the full (first-order) theory of $(M_d, \in, d)$.

Finding the "right" analogue of degree notions to higher levels, in particular generalizing Δ^1_1-degree to the second-order pointclasses in the context of projective determinacy, once had been in the focus of the interest of descriptive set theorists for some time around 1980s. Though many of the results are folklores among Cabal people, the reader can still find a good account on the development at the time in Ref. 7. Although this is a fascinating topic, in this paper we would like to focus on the structure of these degree structures, which did not get much attention of descriptive set theorists, at least judging by the literature.

2.2. *Degree structures I: The partial order*

In this part, we compile some of known facts about various degree structures. Most results in the literature are statements in $L(\mathbb{R})$, however, $L(\mathbb{R})^V$ may vary if V is different, thus the degree structures differ drastically in

[c]Steel's results on jump operators is part of his investigation on Martin's conjecture, and further developments on this topic can be found in Refs. 8 and 7.

these models. Later in this part, we will also briefly discuss certain degree structures in some inner models of set theory which do not have all the reals.

Given a definable reducibility notion $\leq$, let $\equiv$ denote the induced equivalence relation: $x \equiv y \Leftrightarrow x \leq y \wedge y \leq x$, and $[a]$ denote the equivalence class of a, i.e. the degree of a. For different degree notions, we use subscripts to distinguish them.

First, it's easy to see that the partial ordering $(\mathscr{D}, \leq)$, together with the join operator, is an upper semi-lattice. And secondly, as the set $\{y \subset \omega \mid y \leq x\}$ is countable for any $x \subset \omega$, $(\mathscr{D}, \leq)$ must have height ω_1. Then the next question is naturally about its width.

Post Problem is the question whether there are incomparable degrees, i.e. $[a], [b]$ such that $\neg([a] \leq [b] \vee [b] \leq [a])$. If yes, then the further question is how large the maximal antichains could be.

Another question regarding the order type of $(\mathscr{D}, \leq)$ is the *density question*: Is it always true that given two degrees $[a] < [b]$, there is always a $[c]$ such that $[a] < [c] < [b]$? The negation of this question is whether there is a gap, namely, a pair of degrees $[a] < [b]$ such that there is no c such that $[a] < [c] < [b]$. Such $[b]$ is called a *minimal cover* of $[a]$. If $[a] = [\emptyset]$, $[b]$ is called a *minimal degree.* One can also further ask the question whether the relation $\leq$ is well-founded, namely, if there is an infinite strictly $\leq$-decreasing sequence.

Let $(\mathscr{D}_T, \leq_T)$ denote the structure of Turing degrees and $(\mathscr{R}, \leq_T)$ the structure of recursively enumerable (r.e.) Turing degrees. Here are some of their properties:

Friedberg-Muchnik[9,10] showed that there are incomparable degrees in $(\mathscr{R}, \leq_T)$. The same is true in $(\mathscr{D}_T, \leq_T)$ (Kleene-Post[11]).
$(\mathscr{R}, \leq_T)$ is dense (Sacks[12]), while $(\mathscr{D}_T, \leq_T)$ has minimal degrees (Spector[13], Sacks[14]). In fact, there are 2^ω many minimal degrees, which are also pairwise incomparable degrees.
$(\mathscr{R}, \leq_T)$ is ill-founded as it is dense. Furthermore, Harrison[15] (see also Ref. 16, III.3.6) showed that in $(\mathscr{D}_T, \leq_T)$ there is an infinite sequence of degrees $\langle [a_i]_T : i < \omega \rangle$ such that $[a_{i+1}]'_T \leq [a_i]_T$, for all $i < \omega$.

There are abundant results on the structural properties of $(\mathscr{R}, \leq_T)$ and $(\mathscr{D}_T, \leq_T)$, for instance, minimal upper bound, exact pairs, poset-embedding problem, high/low hierarchies, decidability problems, etc. It's beyond our ability to discuss all of them here, we only mention some basic ones to make our points.

By a result of Spector[17] (see also Ref. 16, II.7.2), restricted to Π^1_1 reals, there are only two $\equiv_h$-degrees, thus the analogue of $(\mathscr{R}, \leq_T)$ for hyperarithmetic degrees is a trivial poset, only has the degrees of $\emptyset$ and Kleene's $\mathcal{O}$. Let $(\mathscr{D}_h, \leq_h)$ denote the structure of hyperarithmetic degrees. Adding two mutually generic reals Cohen generic over $L_{\omega_1^{CK}}$ produces two incomparable $\equiv_h$-degrees. And by adding ω Cohen reals that mutually generic over $L_{\omega_1^{CK}}$, one can arrange a set of $<_h$-decreasing sequence of reals. Thus $(\mathscr{D}_h, \leq_h)$ is not wellfounded. With perfect set forcing, Sacks[18] showed that Sacks reals have minimal degrees in $(\mathscr{D}_h, \leq_h)$. Again, there are also 2^ω many minimal degrees, and therefore pairwise incomparable degrees in $(\mathscr{D}_h, \leq_h)$.

These forcing arguments all work for constructibility degrees $(\mathscr{D}_c, \leq_c)$ and lead to the same conclusion. Below are some other relevant results.

(Friedman[19]) Assume $0^\sharp$ exists. There is a Π^1_2 singleton $a \subseteq \omega$ such that $0 <_c [a]_c <_c 0^\sharp$.[d] So the constructibility degrees restricted to Π^1_2 singletons is non-trivial.
(Friedman[19]) Let X be the set of Π^1_2 singletons that is $\leq_c$-comparable with every Π^1_2-singleton. Then $\leq_c$ restricted to X is pre-wellordered, and for every $x \in X$, the immediate $<_c$-successor of $[x]_c$ is $[x^\sharp]_c$.
(Harrington-Kechris[20]) If $d \subseteq \omega$ is a Π^1_2-singleton, then either $0^\sharp \leq_c a$ or $0^\sharp \equiv_c d^\sharp$. As a result of relativizing their argument, one can get that $[0^\sharp]_c, [0^{\sharp\sharp}]_c, [0^{\sharp\sharp\sharp}]_c, \ldots$ are the first ω degrees of sharps – $\leq_c$-jumps – of Π^1_2 singletons.

Next let us look at the structure of Δ^1_n-degrees. For that Projective Determinacy (PD) is always assumed for the sake of convenience, we leave it to the reader to figure out the necessary amount of "local" determinacy needed for each statement.

Theorem 2.3 (Kechris[21]). [e] *Suppose $n > 0$ is odd.*

There exists a largest thin Π^1_n set, denoted as C_n;
C_n is closed under Δ^1_n-jump, i.e. $[d]_n \subset C_n \Rightarrow [d]'_n \subset C_n$;
The Δ^1_n-degrees of members of C_n are wellordered by $[a]_n \leq_n [b]_n \Leftrightarrow a$ is Δ^1_n in b, in particular, the immediate $<_n$-successor of every Δ^1_n-degree is its Δ^1_n-jump.

---

[d] $0^\sharp$ is a Π^1_2 singleton.
[e] According to Ref. 21, the three results below in the case $n = 1$ were also proved independently by D. Guaspari and G. Sacks[18].

Let $(\mathscr{R}_n, \leq_n)$, $n > 0$ odd, be the analogue of $(\mathscr{R}, \leq_T)$ for Δ^1_n-degrees, namely the Δ^1_n-degrees for Π^1_n subset of ω. Then we have

Corollary 2.1. *$(\mathscr{R}_n, \leq_n)$, $n > 0$ odd, is a trivial poset, consisting of only the smallest and the largest elements – the degree of complete Π^1_n set of integers.*[f]

It is a well known fact in set theory that $\mathbb{R} \cap L$ is the largest countable Σ^1_2 set if $\aleph_1^L < \aleph_1$ (by Solovay[22]), and there exists a largest countable Σ^1_n set of reals for every even n (by Kechris-Moschovakis[23]). These largest countable sets are denoted as C_n for $n > 0$ even. Each C_{2n} $(n > 0)$ is the set of reals in an inner model of ZFC (for instance $L(C_{2n})$). However, C_{2n+1} is not the set of reals of any transitive model of ZFC. In Ref. 24, the authors invented the Q-set in order to develop the "right" theory generalizing that of hyperarithmetic degrees to odd levels of second order arithmetics. We omit the definition of Q_{2n+1}, $n > 0$, but only the following relevant facts about Q_{2n+1}.

Theorem 2.4 (Kechris-Moschovakis-Solovay[24]). *Assume* PD.

Q_{2n+1} is the maximal countable Π^1_{2n+1} set downward closed under Turing as well as Δ^1_{2n+1}-degrees;
Q_{2n+1} is closed under the Δ^1_{2n+1}-jump;
The Δ^1_{2n+1}-degrees restricted to Q_{2n+1} forms a proper initial segment of that restricted to C_{2n+1};
$Q_{2n+1} = \mathbb{R} \cap L(Q_{2n+1})$.

Let $(\mathscr{D}_n, \leq_n)$ denote the poset of Δ^1_n-degrees. Then we have a rather simple structure of degree in an inner model of ZFC.

Corollary 2.2 (PD). *In $L(Q_{2n+1})$, $n > 0$, $(\mathscr{D}_{2n+1}, \leq_{2n+1})$ is a wellordering of ordertype $\omega_1^{L(Q_{2n+1})}$.*

Of course, one can still do the forcing argument as before to produce complex degree structures, however, what interests us is the simplicity of these degree structures in inner models – notice that in $L(\mathbb{R})$ the poset $(\mathscr{D}, \leq_T)$ has all the properties discussed earlier using forcing arguments (see p.206). So the question is what causes, or when does, the change happen.

Very little is known about the degrees at the even levels at this point.

---

[f]For the case $n = 1$, Spector[17] (see also Ref. 16, II.7.2) showed that every Π^1_1 set of integers is either $\leq_1 \emptyset$ or $\geq_1 \mathcal{O}$. The case $n > 1$ is Theorem (3B-1) of Ref. 21.

2.3. *Degree structures II: Posner-Robinson problem and degree determinacy*

Before moving on to the next topic, we would like to discuss two more properties of degree structures. The first one, we called it *Posner-Robinson problem.*

In classical recursion theory, a fundamental task is to understand the jump operator. In the literature, there are quite a number of jump inversion theorems for that purpose. The Posner-Robinson theorems to be discussed here belongs to jump inversion problems, the basic theme is that every nontrivial real can be viewed as a jump of some other real (modulo that real itself).

For $X, Y, Z \subset \omega$, $X \leq_T (Y, Z)$ if X is recursive in the pair (Y, Z), i.e. there is a recursive bijection $\pi : \omega \to \omega \times \omega$ such that $X \leq_T \pi^{-1}[Y \times Z]$. (Y, Z) here is essentially the join of Y and Z.

The classical Posner-Robinson theorem (see Refs. 25,26) asserts that for any $A \not\leq_T \emptyset$, there is a real G such that A appears to be the Turing jump of G modulo G, more precisely, $G' \equiv_T (A, G)$. Shore-Slaman[27] generalize this to any α-REA operators, $\alpha < \omega_1^{\mathrm{CK}}$. More precisely, for any $A \notin I_{<\alpha}$, there is a real G such that $G^{(\alpha)} \equiv_T (A, G)$, where $I_{<\alpha} = \{X \subset \omega \mid \exists \beta < \alpha\, (X \leq_T \emptyset^{(\beta)})\}$. Woodin later proved (unpublished) the Posner-Robinson Theorem for hyperarithmetic jump as well as for the sharp.

> For any real $A \notin L_{\omega_1^{\mathrm{CK}}}$, there is a $G \subset \omega$ such that $\mathcal{O}^G \equiv_T (A, G)$, where $\mathcal{O}^G$ is the complete Π^1_1-in-G set.
> Assume $\forall x \subset \omega\, (x^\sharp$ exists). Then for any real $A \notin L$, there is a $G \subset \omega$ such that $G^\sharp \equiv_T (A, G)$, where $G^\sharp$ is the real coding the theory of $L[G]$.

We are more interested in the following less specific statement:

(PR) There are co-countable many reals A such that the Posner Robinson equation $x^\sharp \equiv_T (A, x)$ has a solution.

So $(\mathscr{D}_T, \leq_T)$, $(\mathscr{D}_h, \leq_h)$ and $(\mathscr{D}_c, \leq_c)$ all satisfies (PR). Note that in inner model $L(Q_{2n+1})$, the structure $(\mathscr{D}_{2n+1}, \leq_{2n+1})$, $n > 1$, is a wellordering, and the immediate $<_{2n+1}$-successor of every Δ^1_{2n+1}-degree is its Δ^1_{2n+1}-jump, so the Posner-Robinson equations fail to have solutions at limit degrees (i.e. the limit iterates of Δ^1_{2n+1}-jump of $\emptyset$), thus we have

Corollary 2.3. *Assume* $V = L(Q_{2n+1})$. PR *is false over* $(\mathscr{D}_{2n+1}, \leq_{2n+1})$.

The second one is the *Degree Determinacy Problem.* Given a degree structure $(\mathscr{D}, \leq)$. A set $A \subseteq \mathbb{R}$ is *degree invariant* if $x \equiv y \Rightarrow (x \in A \leftrightarrow y \in A)$. A function $f : \mathbb{R} \to \mathbb{R}$ is *degree invariant* if $x \equiv y \Rightarrow f(x) \equiv f(y)$. A *cone* of reals is a set of the form $C_x =_{\text{def}} \{u \in \mathbb{R} \mid x \leq u\}$. Let $C_{[x]} =_{\text{def}} \{[u] \mid u \in C_x\}$. *Degree Determinacy* is the assertion that $C_{[x]}$, $x \in \mathbb{R}$, generate an ultrafilter on the poset of degrees, in other word, every degree invariant set of reals either contains or is disjoint from a cone. *Turing Determinacy* (TD) is the Turing degree version of Degree Determinacy. And the Turing cone measure is often called *Martin measure*.

TD is a consequence of AD, many consequences of AD can also be derived from TD (see Ref. 28 for some examples). In fact, all versions of Degree Determinacy for reasonable definability degree notions all follow from AD. Unlike structure properties discussed before, the statement of Degree Determinacy connects second order objects (subsets of reals) of second order arithmetic to first order objects (bases of cones), it does not speaks about the partial ordering directly, however, it has fundamental impact on the global structure of degree functions. In response to Sacks' question regarding the existence of degree-invariant solution to Posts problem[29], Martin made a global conjecture that the only nontrivial definable Turing invariant functions are the Turing jump and its iterates through the transfinite. More precisely,

Conjecture 2.1 (Martin Conjecture, see Ref. 30, p.281). *Assume* ZF + DC + AD.

If f is a Turing degree invariant function and $x \not\leq_T f(x)$ for a cone of x, then $f(x) \equiv_T x$ for a cone of x.
Degree invariant functions on $\mathbb{R}$ are pre-wellordered by the relation $\leq_M$, where $f \leq_M g$ iff $f(x) \leq_T g(x)$ on a cone of x. Let f' be such that $f'(x) \equiv_T f(x)'$. Then $\text{rank}_{\leq_M}(f) = \alpha \Rightarrow \text{rank}_{\leq_M}(f') = \alpha + 1$.

Although this conjecture remains open, it has already been proven to be true when restricted to the class of uniformly Turing invariant functions. The conjecture were stated with AD, we believe that if it is true, TD should suffice. We refer interested readers to Refs. 6,8 and more recent Ref. 31. Aforementioned Posner-Robinson theorems for iterated jumps of Turing degrees played an important role in Slaman-Steel's proof of (1) for uniformly Turing invariant functions.

Back to the posets discussed earlier, Degree Determinacy holds in $(\mathscr{D}_T, \leq_T)$, $(\mathscr{D}_{2n+1}, \leq_{2n+1})$ $(n < \omega)$ and $(\mathscr{D}_c, \leq_c)$, while it is false in $(\mathscr{D}_{2n+1}, \leq_{2n+1})^{L(Q_{2n+1})}$, $n > 0$, as it is wellordered and any two disjoint unbound subsets of this ordering witness the failure of Degree Determinacy.

2.4. α-recursion theory

Another direction for generalization is to lift the notion of degrees to subset of α where α is an arbitrary limit ordinal $> \omega$. This is so called α-recursion theory. In order to preserve a good collection of results in classical recursing theory, it's necessary to consider ordinals with sufficient closure properties, in particular those are Σ_1-admissible. Many of the classical results lift to such α by means of recursive approximations and fine structure techniques.

A set is *admissible* if it is transitive and models KP set theory. This is the same as Σ_1-admissible, which is about to be defined later. An ordinal α is admissible if L_α is admissible. ω and ω_1^{CK} are the first two admissible ordinals. Let α be an admissible ordinal. Call a set *α-finite* if it belongs to L_α, *α-r.e.* (or *α-recursive*) if it is Σ_1-definable (Δ_1-definable, respectively) over $(L_\alpha, \in)$ (allowing parameters). The *α-jump* of $\emptyset$ is accordingly given by the complete $\Sigma_1(L_\alpha)$-set.[g] For readers not familiar to theory of admissible ordinals, Ref. 16 (Part C) and Ref. 32 are good places to look, Refs. 33,34 are good sources for advanced techniques in this area. The admissible initial segments of L provide natural settings for generalizing classical recursion theory, as such L_α admits an L_α-recursive bijection between L_α and α. L provides an ideal structure for developing higher recursion theory. The results in α-recursion theory cited in this paper all assume $V = L$.

Note that, although we follow the tradition and use the terminology α-degree, it should be called something like α-Δ_1-degree, at least in this paper, as it is the analogue of Turing degrees for subsets of α.

Let $\alpha \in \mathrm{Ord}$ be admissible. A set $X \subseteq \alpha$ is *regular*[h] if $A \cap \beta \in L_\alpha$ for all $\beta < \alpha$. For $A, B \subseteq \alpha$, write $A \leq_\alpha B$ if A is Δ_1-definable over $(L_{\alpha_1^B}[B], \in, B)$ (allowing parameters), where α_1^B is the least ordinal $\geq \alpha$ such that $L_\alpha[B]$ is admissible. A is regular iff $L_\alpha[A] = L_\alpha$. In recursion theory, amenability is equivalent to an important dynamic property in priority argument. In a 1966 paper, Sacks established the following basics of α-recursion theory.

Theorem 2.5 (Sacks[35]). *Let $\alpha \in \mathrm{Ord}$ be admissible.*

---

[g]It should be pointed out that although $[\emptyset]'$ is defined to the degree of some complete $\Sigma_1(L_\alpha)$-set, but in general $[\emptyset]^{(2)}$, the double jump $([\emptyset]')'$, does not have the same α-degree as some complete $\Sigma_2(L_\alpha)$-set.

[h]This is Sacks' terminology. Jensen call it *amenable*.

Every α-r.e. degree can be represented by an regular subset of α.
The poset of α-r.e. degrees, $(\mathscr{R}_\alpha, \leq_\alpha)$, is nontrivial, i.e. there exists a non-α-recursive, regular, α-r.e. set.

By exploiting the combinatoric power of admissibility and the techniques from fine structure theory, Sacks and his students manage to use Σ_1-admissibility to do the work of Σ_2. They lifted the classical finite injury argument to α-recursion theory and provided positive solutions to Post's problem in α-recursion theory

Theorem 2.6.

(Sacks-Simpson[36]) There exist two $\leq_\alpha$-incomparable α-r.e. subsets of α. Furthermore,
(Shore[37]) There is a uniform solution to Post's problem: There exist $m, n < \omega$ such that for all admissible α, the m-th and n-th lightface Σ_1 subsets of α are $\leq_\alpha$-incomparable.

With his Σ_2-blocking technique, Shore proved a splitting theorem at α, from which a positive solution to Post's problem also follows.

Theorem 2.7 (Shore[38]). *Let A be α-r.e. and regular. Then there exists α-r.e. B_0 and B_1 such that $A = B_0 \cup B_1$, $B_0 \cap B_1 = \varnothing$ and $A \nleq_\alpha B_i$ $(i < 2)$.*

Soon after the splitting theorem, Shore proved the density theorem at α, which is also the first instance of α-infinite injury.

Theorem 2.8 (Shore[39]). *Let A and C be α-r.e. sets such that $A <_\alpha C$. Then there exists an α-r.e. B such that $A <_\alpha B <_\alpha C$.*

This pretty much gives us the picture of $(\mathscr{R}_\alpha, \leq_\alpha)$. The picture in $(\mathscr{D}_\alpha, \leq_\alpha)$, the global poset of α-degrees, is complicated – in general it is not dense about $[\emptyset]'_\alpha$.

For an ordinal α, its Σ_n-*cofinality*, $n < \omega$, is the least $\rho \leq \alpha$ such that there is a $\Sigma_n(L_\alpha)$ function f mapping ρ into cofinally into α. An ordinal is Σ_n-*admissible* if its Σ_2-cofinality equals to itself. Regular cardinals are Σ_n-admissible for all $n < \omega$.

Theorem 2.9.

(MacIntyre[40]) There exists a minimal α-degree for every countable admissible ordinal.
(Shore[41]) Minimal α-degrees exist for Σ_2-admissible cardinal α.

These leave out the case for Σ_1-admissible but not Σ_2-admissible cardinals,[i] in particular $\alpha = \aleph_\omega$, is still open. Although the cardinal $\aleph_{\omega_1}$ is also a such cardinal, inspired by Silver's work[43] on Singular Cardinal Problem, using the combinatorics of stationary set, Sy Friedman[44] showed that $\aleph_{\omega_1}$-degree is wellordered above $[0]'_{\aleph_{\omega_1}}$. Given a singular cardinal λ and a degree notion at λ, we call a degree represented by a cofinal subset of λ of ordertype $\mathrm{cf}(\lambda)$ *a singularizing degree* (at λ).

Theorem 2.10 (Sy Friedman[44]). *For any singular cardinal λ with $\mathrm{cf}(\lambda) > \omega$, the λ-degrees are wellordered above every singularizing degree, and the immediate successor is given by the jump. In particular, $\aleph_{\omega_1}$-degree is wellordered above $[0]'_\alpha$.*

Therefore, minimal $\aleph_{\omega_1}$-cover exists for every degree above $[0]'_{\aleph_{\omega_1}}$. But the rest of the picture is still not quite clear. Chong[45] recently gives a partial answer: Minimal $\aleph_{\omega_1}$-degree if exists must be $< [0]'_{\aleph_{\omega_1}}$.

Let Δ_1^α*-degree Determinacy* be the statement that any Δ_1^α-degree invariant subset of $\mathscr{P}(\alpha)$ either contains or is disjoint from of a cone of subset of α. From Sy Friedman's theorem, immediately we have

Corollary 2.4. *Δ_1^λ-degree Determinacy fails at singular cardinal λ with uncountable cofinality.*

Proof. By Sy Friedman's theorem, the λ-degrees are wellordered above $[0]'_\lambda$. Let A be the set of all the odd iterates of λ-jumps of $[\emptyset]$ and B for all the even iterates. Then A, B are disjoint and $\leq_\lambda$-unbounded subsets of $\mathscr{D}_\lambda$, and none of them contains a cone of λ-degrees. □

Remark 2.1. Sy Friedman's argument works for any degree notions with large equivalence classes. If the degree notion under consideration is Δ_n-degree at $\aleph_{\omega_1}$, $n > 1$, for instance, then the Δ_n-degree of $\emptyset$ already contains a cofinal subset of $\aleph_{\omega_1}$, thus Δ_n-degree at $\aleph_{\omega_1}$, $n > 1$, are completely wellordered. But the α-degrees (take $\alpha = \aleph_{\omega_1}$) below $[\emptyset]'_\alpha$ is illfounded by Shore's Density Theorem for α-r.e. sets (see p.211).

The situation at $\aleph_\omega$ is different. According to Ref. 46, Harrington and Solovay independently proved that there are incomparable $\aleph_\omega$-degrees above $[0]'_\alpha$.[j] Later, we will discuss recently developments in degree theories

---

[i]Maass[42] improved Shore's result to a slight weaker assumption.

[j]Sy Friedman promised to give the proof in Ref. 47, but Ref. 47 seems never appeared.

at countable cofinality singular cardinals, but focusing on what we called Zermelo degrees, rather than Δ_1-degrees, at $\aleph_\omega$.

There aren't any research on the determinacy of α-degrees to date. Although lack of anything like determinacy axioms for large ordinals in general, our study of consequences of strong axioms like I_0, whose impact at the associated countable cofinality cardinal resembles a great deal to that of AD at ω, in degree theory supports the view that there is a deep connection with the complexity of degree structures and the strength of large cardinals of the universe carries.

Next consider the Posner-Robinson Problem. There are no Posner-Robinson results for α-degrees in the literature. Here is something close to it. The Simpson jump theorem below is a lifting of the Friedberg jump theorem of classical recursion theory.

Theorem 2.11 (Simpson[48]). *The following are equivalent.*

$\emptyset' \leq_\alpha D$ *and* D *has the same* α*-degree as some regular set.*
$C' \equiv_\alpha D$ *for some regular, hyperregular* C*.*

Informally speaking, this says that every degree $[d]_\alpha \geq [0]'_\alpha$ is an α-jump of some generalized low degree. For our PR statement on p.208, it only make sense to generalize it to cardinals. Let λ be a cardinal. PR_λ is the following assertion:

(PR_α) There are co-λ many reals A such that the Posner Robinson equation $x' \equiv_\lambda (A, x)$ has a solution.

In other word, the Posner-Robinson equations fail at no more than λ many places. As a consequence of Sy Friedman and Simpson's theorems, we have

Corollary 2.5. PR_λ *fails for* Δ_1*-degrees at singular cardinals* λ *with uncountable cofinality.*

Proof. In L, singular cardinals are strong limit, so every subset of λ that Δ_1 computes $\emptyset'$ is regular. By Simpson's theorem, every degree $[d]_\lambda \geq [\emptyset]'_\lambda$ is a λ-jump of some $[c]_\lambda$. Now suppose $[d]_\lambda$ is a limit iterate of λ-jump of $[\emptyset]_\lambda$, i.e. $d \equiv_\lambda c' \equiv_\lambda \emptyset^{(\alpha)}$, for some limit ordinal α.

We claim that the Posner-Robinson equation for $[c]_\lambda$ does not have a solution. Suppose NOT, $(c, g) \equiv_\lambda g'$ for some $g \subset \lambda$. Since $g' \geq_\lambda \emptyset'$, it must be that $g' \equiv_\lambda \emptyset^{(\beta)}$ for some β. Then

$$\emptyset^{(\beta+1)} \equiv_\lambda g'' \equiv_\lambda (c, g)' \equiv_\lambda (c', g') \equiv_\lambda (\emptyset^{(\alpha)}, \emptyset^{(\beta)}).$$

Then we would have $\alpha = \beta + 1$. Contradiction!

But there are at least λ^+ many such $[c]_\lambda$'s. This means that Posner-Robinson equations fails at more than λ many places, so PR_λ is false for Δ_1-degrees. □

It is also worth mentioning that, using Simpson's jump theorem together with another result of Shore[49], Sy Friedman were able to give a quick proof that the α-jump operator is definable in $(\mathscr{D}_\alpha, \leq_\alpha)$ (see Ref. 46, Theorem 6), in contrast to the sophisticated machinery used to prove the definability of Turing jump in $(\mathscr{D}_T, \leq_T)$ (see Ref. 27).

There are also works on α-degrees at inadmissible ordinals, which we avoid in this paper, as it is hard to organize and fit them into the theme we are laying out here.

Sacks and Slaman also did some ground work (see Ref. 50) for generalized hyperarithmetic theory, namely generalizing the Δ_1-degree at α to the analogue of hyperarithmetic degree at α. This combines the aforementioned two directions for generalizing classical recursion theory. At this point, as far as the questions we are interested here, not much can we say about these generalized hyperarithmetic degree structures.

There are also so called E-recursion theory, which from a different perspective extends the notion of computation from hereditarily finite sets to sets of arbitrary rank. We will not discuss these generalization due to its little relevance to the later part of this paper (maybe we just don't see yet).

The interests of recursion theorists and (descriptive) set theorists seem shift away from generalizing recursion theory after mid 1990s. Until very recent, some applications of large cardinals to degree structures emerged. In the rest of this paper, we report the recent developments in, what we prefer to call, *higher degree theory*.

3. Higher degree theory

3.1. *Preparation*

This section surveys the results in Ref. 2. First are some definitions.

Definition 3.1 (ZFC). *Suppose Γ is a fragment of* ZFC *such that* ZFC *proves the consistency of Γ. Suppose λ is an uncountable cardinal satisfying $2^{<\lambda} = \lambda$. Let H_λ be the collection of sets whose transitive closure has cardinality $< \lambda$. Fix a $\Delta \subset \lambda$ which codes a well ordering of H_λ of ordertype λ. For each $a \subseteq \lambda$, let α_a be the least ordinal $\alpha > \lambda$ such that*

$M_a =_{\mathrm{def}} L_\alpha[\Delta, a]$ models Γ. If there is a definable wellordering of H_λ (of ordertype λ) in V, then there is no need to mention Δ explicitly.

For any two subsets $a, b \subseteq \lambda$, set $a \leq_\Gamma b$ if and only if $M_a \subseteq M_b$. This gives rise to a degree notion, which we call Γ-degree. *To each $a \subseteq \lambda$, the* Γ-jump of a, *a'_Γ, is the theory of M_a, which is identified as a subset of γ.*

For instance, hyperarithmetic degree is KP-degree, where KP is Kirpke-platek set theory. For this section, we fix $\Gamma = \mathsf{Z}$, Zermelo set theory, i.e. ZF − Replacement. The point is that Z is sufficient for proving Covering lemmas for fine structure models. In this case, the ordinal α_a is called the *Zermelo ordinal for a*. The above definitions are given under ZFC. In general, suppose T_0 is our working theory, and T_1 is a consistently weaker fragment of T_0 (i.e. the existence of minimal models of T_1 can be derived from T_0), then one can define a degree notion for T_1 under T_0 as above. To illustrate our point, we shall only cross examine Z-degrees at countable cofinality singular cardinals in various large cardinal inner models.

Recall that in **S**2.2 and **S**2.3, we propose to consider the following structural properties of degree posets.

(*Post Problem*). Are there two incomparable degrees, i.e. two sets $a, b \subseteq \lambda$ such that $a \not\leq b$ and $b \not\leq a$? One can also ask a relativized question, i.e. incomparable degrees above a given degree. A set of pairwise incomparable degrees form an antichain. A related question is the size(s) of maximal antichains, if exist.
(*Minimal Cover*). Given a degree $[a]$, is there a minimal cover for $[a]$, i.e. a $[c] > [a]$ such that there is no $b \subseteq \lambda$ such that $[a] < [b] < [c]$?
(*Posner-Robinson*). Is it true that for almost all (co-λ many) $x \subseteq \lambda$, the Posner-Robinson equation for x has a solution, i.e. $\exists g \subseteq \lambda\, [(x, g) \equiv g']$?
(*Degree Determinacy*). Is it true that every degree invariant subset of $\mathscr{P}(\lambda)$ either contains or is disjoint from a cone?

Among the four questions, the first one is about antichains, therefore is related to the width of the degree poset, the second is about the organization of degrees, like whether the degrees are dense or discrete, the third is about the internal understanding of the degrees, such as what information do the degrees carry, and the last is more or less a question about the connection between the members of $\mathscr{P}(\lambda)$ and the subsets of $\mathscr{P}(\lambda)$, a bridge between these two types.

The first three problems are first order questions regarding $\mathscr{P}(\lambda)$ (more frequently $(V_{\lambda+1}, \in)$ in practice). However, for the degree determinacy problem, it makes more sense to state it for degree invariant subsets of $\mathscr{P}(\lambda)$ in $L(\mathscr{P}(\lambda))$, just like the situation of Turing Determinacy for sets of reals in $L(\mathbb{R})$. So it is more appropriate to think the degree determinacy problem as a question quantifying over second order sets in $L(\mathscr{P}(\lambda))$ (or $L(V_{\lambda+1})$ in practice). Since $V_{\lambda+1}$ varies in different universes, answers to degree theoretical questions, such as these four questions, often vary in different V's. We will give an example on this matter in **S**4.1.

These are certainly important degree theoretic questions. This list, however, is by no means meant to be comprehensive, it is merely a list of questions that at this point we are confident to answer.

At a strongly inaccessible cardinal λ (even regular cardinal satisfying $2^{<\lambda} = \lambda$), the degree notions in general are similar to their counterparts at ω, since most usual constructions for degrees at ω, priority argument, local forcing argument, et al, can be carried out at λ with very few changes. So the degree structures for an analogue degree notion at λ is very much like its counterpart at ω, not much new insight is obtained there. At these cardinals, the answers to the first three questions are all "Yes", as in the case of ω. However, for the Degree Determinacy question, the answer on the contrary is very likely to be "No". This will be discussed in **S**3.7. Also, as discussed in **S**2.4, Sy Friedman settled the situations at singular cardinals of uncountable cofinality, our focus will be mainly on degrees at singular cardinals of countable cofinality. We shall analyze Zermelo degree structures in some fine structure extender models. The reason for working with fine structure models will be discussed in **S**4.1.

Our main tool is Covering lemmas in various cases, Mitchell's handbook article (Ref. 51) is the main reference for that. Schimmering's introductory article (Ref. 52) is good enough for most fine structure contents in this paper, if the reader are not familiar with fine structures. We also assume familiarity of Prikry-type forcings, for which Gitik's handbook article (Ref. 53) is recommended.

3.2. *Zermelo degrees in L*

Let us start with Woodin's observation in L about Zermelo degrees at $\aleph_\omega$. The argument is a simple application of the Covering Lemma for L. First recall Jensen's Covering Lemma for L.

Lemma 3.1 (Covering Lemma for L, see Refs. 51,54). *Assume $0^\sharp$ does not exist. Then for every set x of ordinals, there is an $y \in L$ such that $x \subseteq y$ and $|x| = |y| + \omega_1$.*

Theorem 3.1 (Shi[2]). *Assume $V = L$. Let λ be a singular cardinal with $\mathrm{cf}(\lambda) > \omega$. Let $x \subset \lambda$ be cofinal in λ and* $\mathrm{otp}(x) = \mathrm{cf}(\lambda)$. *Then the Zermelo degrees at λ above $[x]_Z$ are well ordered. In particular, Zermelo degrees at $\aleph_\omega$ are well ordered.*

Proof. The argument is a simple application of the Covering Lemma for L. Suppose $a \subseteq \lambda$ and $a \geq_Z x$. Consider $M(a)$, the minimal Zermelo model containing a. Notice that $M(a)$ has the same reals as V and sharps are absolute between transitive models containing ω_1, therefore M_a contains no sharps. In M_a, as $x \in M_a$, every $z \subset \lambda$ can be identified with a countable subset of λ. By Covering in M_a, a is covered by a $b \in L^{M_a} = L_{\alpha_a}$ such that $|b| = |a| + \omega_1$. As M_a and L_{α_a} agree on $\mathscr{P}(\omega_1)$, $|b|^{L_{\alpha_a}} = \omega_1$. Let $\pi : b \to |b|^{L_{\alpha_a}}$ be a bijection in L_{α_a}. Then $\pi[a] \in \mathscr{P}(\omega_1) \subseteq L_{\alpha_a}$. It follows that a can be computed from b in L_{α_a}. (This is essentially the proof of Theorem V.5.4 in Devlin[55].) Thus $M_a = L_{\alpha_a}$. This means that the mapping $[a] \mapsto \alpha_a$ is injective. Therefore Zermelo degrees at λ are well ordered above $[x]$. Since the sequence $\{\aleph_n \mid n < \omega\}$ is Σ_1-definable over $L_{\aleph_\omega}$, Zermelo degrees at $\aleph_\omega$ are fully well ordered. □

Immediately, we have the following answers to the four questions. Enumerate $\{\alpha_x \mid x \subset \lambda\}$ in increasing order, let α_η be its η-th member. Say $[x]$ is a *successor* (resp. *limit*) *degree* if $\alpha_x = \alpha_\eta$ for some successor (resp. limit) ordinal η.

Corollary 3.1. *Assume $V = L$. Let λ be a singular cardinal and* $\mathrm{cf}(\lambda) = \omega$. *Then*

There are no incomparable Zermelo degrees above any singularizing degree. In particular, every two Zermelo degrees at $\aleph_\omega$ are comparable.
Every Zermelo degree above any singularizing degree has a unique minimal cover. Again this holds for every Zermelo degree at $\aleph_\omega$.
PR_λ *is false for Zermelo degrees at λ.*
Degree Determinacy fails for Zermelo degrees at λ.

Proof. (1) and (2) are immediate from this wellordered structure, so the answers are "No" for the first question and "Yes" for the second. However, for the multi-minimal-cover question, the answer is "No".

In this wellordered structure, Posner-Robinson equation is equivalent to the jump inversion equation, namely, $(\exists G)(x \equiv_Z J_Z(G))$. Notice that whenever there is a new subset of λ is constructed, say in $L_{\alpha+1} \setminus L_\alpha$, we have $L_{\alpha+1} \models$ "$|L_\alpha| = \lambda$". Therefore, a successor degree knows that the minimal

Zermelo model associated to its (immediate) predecessor degree has size λ, therefore can compute the jump of its predecessor. So the jump operator in the degree structure coincides with the successor operator in the well-order. Thus limit degrees can not be the jump of any degree. There are λ^+ many limit degrees, therefore the answer for (3), the Posner-Robinson question, is "No".

(4) follows from the fact that there are two disjoint sets of degrees that are unbounded in this wellorder, which witness the failure of degree determinacy. □

Here L is viewed as the core model for the negation of the large cardinal axiom that $0^\sharp$ exists. The core of the argument is the Covering lemma for L. The same form of Covering Lemma holds for inner models between L and $L[\mu]$ (the inner model for one measurable cardinal), which include models like $L(0^\sharp)$, $L(0^{\sharp\sharp})$ etc., and Dodd-Jensen's core model K^{DJ}, the core model below a measurable cardinal. For instance,

Lemma 3.2 (Covering Lemma for K^{DJ}, see Refs. 51,56). *Assume that there is no inner model for one measurable cardinal and the Dodd-Jensen core model K^{DJ} exists. Then for every set x of ordinals, there is a $y \in K^{\mathrm{DJ}}$ such that $x \subseteq y$ and $|x| = |y| + \omega_1$.*

These models can be obtained by (proper) partial measures using Steel's construction. For these models, the same covering argument works. The point is that in these inner models, the minimal model of the form M_a, $a \subseteq \lambda$, λ a countable cofinality singular cardinal, is always an initial segment of the core model. Thus Zermelo degrees for countable cofinality singular cardinal are always well ordered above the singularizing degree, the same as in L.

3.3. *Zermelo degrees in $L[\mu]$*

Next, consider $L[\mu]$, the canonical model for one measurable cardinal. The Covering Lemma for $L[\mu]$ starts to be different.

Lemma 3.3 (Covering Lemma for $L[\mu]$, see Refs. 51,57). *Assume that $0^\dagger$ does not exist but there is an inner model with a measurable cardinal, and that the model $L[\mu]$ is chosen so that $\kappa = \mathrm{crit}(j_\mu)$, the least ordinal moved by the elementary embedding j_μ given by μ, is as small as possible. Here j_μ denote the canonical embedding associated to μ. Then one of the following two statements holds:*

For every set x of ordinals there is a set $y \in L[\mu]$ with $x \subseteq y$ and $|y| = |x| + \omega_1$.
There is a sequence $C \subseteq \kappa$, which is Prikry generic over $L[\mu]$, such that for all sets x of ordinals there is a set $y \in L[\mu, C]$ such that $x \subseteq y$ and $|y| = |x| + \omega_1$. Furthermore, the sequence C is unique up to finite initial segments.

But the difference does not affect the structure of Zermelo degrees at cardinals other than κ.

Theorem 3.2 (Shi^2). *Assume $V = L[\mu]$. Zermelo degrees at countable cofinality singular cardinals are well ordered above any singularizing degree. Moreover, the successor of a degree above any singularizing degree is its jump.*

Proof. Reorganize $L[\mu]$ as $L[E]$ using Steel's construction, where E is a sequence of (possibly partial) measures. Here we omit the predicate for the extender: When we say $L_\alpha[E]$ ($\alpha \in \mathrm{Ord}$ or $\alpha = \mathrm{Ord}$), we often refer to the structure $\langle L_\alpha[E\restriction\alpha], \in, E\restriction\alpha\rangle$ or $\langle L_\alpha[E\restriction\alpha], \in, E\restriction\alpha, E(\alpha)\rangle$. A crucial point of using Steel construction is the acceptability condition, which says for any $\gamma < \alpha$,

$$(L_{\alpha+1}[E] \setminus L_\alpha[E]) \cap \mathscr{P}(\gamma) \neq \varnothing \quad \Longrightarrow \quad L_{\alpha+1}[E] \models |\alpha| = \gamma.$$

Here are some benefits of having the acceptability condition:

$\mathscr{P}(\gamma) \cap L[E] = \mathscr{P}(\gamma) \cap L_{\gamma^+}[E]$ for any cardinal γ.
Suppose λ is a cardinal, a, b are unbounded subsets of λ. If $b \leq_Z a$ and $\alpha_b < \alpha_a$, then $[b]'_Z \in M_a$, hence $[b]'_Z \leq_Z a$.

Fix an $a \subset \lambda$ in $L[E]$, consider M_a in $L[E]$. First suppose $\lambda > \kappa$. As M_a and $L[E]$ agree up to λ, arguing as in L, we get that M_a, $a \subseteq \lambda$, are initial segments of $L[E]$. Now suppose $\lambda < \kappa$. As M_a has the same reals as V, M_a does not have $0^\dagger$, so M_a could have at most one full measure.

Case 1. M_a has no full measure. In this case, as the Covering Lemmas for inner models below one measurable have the same form as the Covering Lemma for L (for instance, see the Covering Lemma for K^{DJ} on page 218), applying the corresponding covering lemma, we get that the minimal models M_a, $a \subset \lambda$, are the same as its own core model, namely $K^{M_a} = M_a$. Run the comparison process for M_a against $K^V = L[E]$. M_a is iterable and since M_a agrees with $L[E]$ up to λ, iteration maps used during the comparison do not move M_a, thus M_a is an initial segment of $K^V = L[E]$.

CASE 2. M_a has one full measure, say

$$M_a \models \mu' \text{ is a measure at some } \gamma > \lambda.$$

M_a satisfies the hypothesis of Covering Lemma for one measurable. Apply the Covering in M_a, then there are two cases, either

a is covered by a set $y \in (L[\mu'])^{M_a}$ with $|y| = |a| + \aleph_1$, or
a is covered by a set $y \in (L[\mu', C])^{M_a}$ with $|y| = |a| + \aleph_1$, where C is a Prikry generic over $(L[\mu'])^{M_a}$. Such C is unique up to finite differences.

Notice that $\lambda < \gamma$ and Prikry generics do not add new bounded subsets. a is a bounded subset of γ, so it must be case 1 – the covering set y for a is in $(L[\mu'])^{M_a}$. Thus $M_a = L[\mu']^{M_a} = L_{\alpha_a}[\mu']$. Run the comparison for M_a against K^V. Arguing as before, M_a is an initial segment of $K^V = L[E]$.

So either case, we have that M_a, $a \subset \lambda$, are initial segments of $L[E]$, exactly the same picture as in L – Zermelo degrees at λ in $L[E]$ are well ordered above every singularizing degrees via their Zermelo ordinals. The "moreover" clause follows from the acceptability condition. This completes the proof. □

It is not difficult to see that this argument can be adapted to show the result for core models of finitely many measurable cardinals. So Corollary 3.1 should also include large cardinal core models beyond L up to core models for finitely many measurable cardinals.

3.4. *Zermelo degrees in Mitchell models for an ω-sequence of measures.*

New picture starts to emerge in the canonical model for ω many measurable cardinals, $L[\bar{\mu}]$, where $\bar{\mu} = \langle \mu_n : n < \omega \rangle$ and each μ_n is a measure on κ_n and $\kappa_n < \kappa_{n+1}$, $n < \omega$.

Consider $V = L[\bar{\mu}]$. Again we view $L[\bar{\mu}]$ as built with (partial) measures using Steel's construction. Let $\kappa_\omega = \sup_n \kappa_n$. Let λ be a countable singular cardinal. It is not difficult to see that, when $\lambda > \kappa_\omega$ or $\lambda < \kappa_\omega$, arguing as in $L[\mu]$, Zermelo degrees at λ is well ordered above the singularizing degree. The new picture appears at $\lambda = \kappa_\omega$. The Covering Lemma for $L[\bar{\mu}]$ is similar to that of $L[\mu]$, except that C in the second case now is a system of indiscernibles $C = \langle C_n : n < \omega \rangle$ with the following property:

Each $C_n \subset \kappa_n$ is either finite or a Prikry sequence;

C as a whole is a uniform system of indiscernibles, i.e.

$$(\forall \bar{x} \in L[\bar{\mu}])\,(\forall n < \omega)(x_n \in \mu_n) \quad \Longrightarrow \quad |\bigcup\{C_n - x_n \mid n < \omega\}| < \omega.$$

In fact, for any function $f : \omega \to \omega \cup \{\omega\}$ with infinite support, i.e. the set $\mathrm{supp}(f) =_{\mathrm{def}} \{i \in \omega \mid f(i) > 0\}$ is infinite, one can use the following variation of diagonal Prikry forcing $\mathbb{P}^f_{\bar{\mu}}$ to produce an indiscernible system such that $|C_n| = f(n)$:

- The conditions of $\mathbb{P}^f_{\bar{\mu}}$ are pairs $(\bar{a}, \bar{A})$ such that each $a_i \subset \kappa_i$, and $|a_i| \leq f(i)$, each $A_i \in \mu_i$, for $i < \omega$, moreover, $\bigcup_i a_i$ is finite.
- The order is defined by $(\bar{a}, \bar{A}) \leq (\bar{a}', \bar{A}')$ iff $a(i) \supseteq a'(i)$, $A'(i) \subset A(i)$, and $a_i - a'_i \in A_i$ for $i < \omega$.

These discussion about system of indiscernibles can be found in **S**4 of Mitchell's handbook article (see Ref. 51). The proof of classical Mathias condition for characterizing diagonal Prikry sequence for $\mathbb{P}_{\bar{\mu}}$ can be easily adapted to show the $\mathbb{P}^f_{\bar{\mu}}$-version of Mathias condition.

Proposition 3.1 (Mathias Condition for $\mathbb{P}^f_{\bar{\mu}}$). *Suppose M is an inner model of ZC, Zermelo set theory plus choice, $\bar{\mu} \in M$, f and $\mathbb{P}^f_{\bar{\mu}}$ are defined in M as above. Suppose $G \in \prod_{n \in \omega}(\kappa_n)^{f(n)}$. Then G is a generic sequence for $\mathbb{P}^f_{\bar{\mu}}$ over M if and only if for any sequence $\bar{A} \in M$ such that $A_n \in \mu_n$ for $n < \omega$, there is an $m < \omega$ such that $G(n) \subset A_n$, for $n \geq m$.*

To simplify the presentation of our next theorem, we use the standard diagonal Prikry poset, namely $\mathbb{P}_{\bar{\mu}} = \mathbb{P}^f_{\bar{\mu}}$ with the constant function $f(n) = 1$, for $n < \omega$. With this diagonal Prikry forcing, one can use a single diagonal Prikry sequence C in case (2) of the Covering Lemma for $L[\bar{\mu}]$.

Lemma 3.4 (Covering Lemma for $L[\bar{\mu}]$, Ref. 51). *Assume the sharp of $L[\bar{\mu}]$ does not exist and there is an inner model containing ω measurable cardinals. Let $L[\bar{\mu}]$ be such that every $\kappa_\omega = \sup_{n<\omega} \kappa_n$, where each $\kappa_n = \mathrm{crit}(j_{\mu_n})$, is as small as possible. Then one of the following two statements holds:*

For every set x of ordinals there is a set $y \in L[\bar{\mu}]$ with $x \subseteq y$ and $|y| = |x| + \omega_1$.
There is a sequence $C \subseteq \kappa$, which is $\mathbb{P}_{\bar{\mu}}$-generic over $L[\bar{\mu}]$, such that for all sets x of ordinals there is a set $y \in L[\bar{\mu}, C]$ such that $x \subseteq y$ and $|y| = |x| + \omega_1$. Furthermore, the sequence C is unique up to finite differences.

Using generics for the standard diagonal Prikry forcing as the system of indiscernibles, we describe the structures of Zermelo degrees at countable cofinality singular cardinals in $L[\bar{\mu}]$ as follows. For this subsection, we say an ordinal $\alpha > \lambda$ is a *Zermelo ordinal for* $a \subseteq \lambda$ if $L_\alpha[\bar{\mu}, a] \models$ Zermelo set theory. For each ordinal η, let β_η denote the η-th Zermelo ordinal (for $\varnothing$).

Theorem 3.3 (Shi[2]). *Assume $V = L[\bar{\mu}]$, where $\bar{\mu} = \langle \mu_n : n < \omega \rangle$, each μ_n is a measure on κ_n. Let $\kappa_\omega = \sup_n \kappa_n$. Suppose λ is a singular cardinal of countable cofinality.*

If $\lambda \neq \kappa_\omega$, then the Zermelo degrees at λ are wellordered above any singularizing degree;
If $\lambda = \kappa_\omega$, consider only Zermelo degrees at λ above the degree of $\bar{\mu}$, identifying $\bar{\mu}$ as a subset of λ. Then

$\{\alpha_a \mid a \subset \lambda\} = \{\beta_\eta \mid \eta < \lambda^+ \wedge \beta_\eta > \lim_{\xi<\eta} \beta_\xi\}$.
Therefore Zermelo degrees at λ above the degree of $\bar{\mu}$ are prewellordered via their Zermelo ordinals, i.e. for $a, b \subset \lambda$, the ordering $\preccurlyeq$ given by $[a] \preccurlyeq [b] \Leftrightarrow \alpha_a \leq \alpha_b$ prewellorders the Zermelo degrees at λ above the degree of $\bar{\mu}$.
For each $\eta < \lambda^+$, let α_η be the η-th member of $\{\beta_\eta \mid \beta_\eta > \lim_{\xi<\eta} \beta_\xi\}$, let A_η be a subset of λ that codes the sequence $\langle \alpha_\xi : \xi < \eta \rangle$, and $\mathcal{C}_\eta$ be the set of diagonal Prikry generic sequences for $\mathbb{P}_{\bar{\mu}}$ that are $L_{\alpha_\eta}[\bar{\mu}]$-generic. Then Zermelo degrees at λ (above the degree of $\bar{\mu}$) whose Zermelo ordinals equal to α_η are exactly the degrees given by

$$A_\eta \oplus \mathcal{C}_\eta = \{(A_\eta, C) \mid C \in \mathcal{C}_\eta \cup \{\varnothing\}\}.$$

Although we use the standard diagonal Prikry sequence to state this theorem, the argument works for every $\mathbb{P}^f_{\bar{\mu}}$. So for each f as on p.221, every Zermelo degree can be represented by a diagonal Prikry sequence for $\mathbb{P}^f_{\bar{\mu}}$.

Compared with previous pictures, though not eventually well ordered, this is still a rather simple structure. We have definite answers to the four questions.

Corollary 3.2. *Assume $V = L[\bar{\mu}]$, and $\bar{\mu}, \bar{\kappa}, \lambda$ be as in Theorem 3.3. Let $\lambda = \sup_n \kappa_n$. Consider the Zermelo degrees at λ above the degree of $\bar{\mu}$.*

There are incomparable Zermelo degrees.
No Zermelo degree has a minimal cover.
Posner-Robinson Theorem for Zermelo degrees at λ is false.

Degree determinacy for Zermelo degrees at λ is false.

The proof is rather sophisticated, we refer the reader to Ref. 2 Corollary 3.1 for the details.

For (1), a further question is whether there is a size λ^ω antichain of Zermelo degrees. As there are no minimal degrees, the usual way of getting 2^ω many incomparable degree at ω by constructing 2^ω many minimal degrees no longer works here.

Note that if $C_i \in \mathcal{C}_\eta$, $i = 0, 1$, are such that $C_0(n) \subset C_1(n)$ for all $n < \omega$, and $C_1(n) \setminus C_0(n) \neq \varnothing$ infinitely often, then $C_0 <_Z C_1$. As all the models of the form $M_{A_\eta, C}$, $C \in \mathcal{C}_\eta$ have the same reals, it follows that the poset $(\mathscr{P}(\omega)/\mathrm{Fin}, \subseteq^*)$, where $[a] \subseteq^* [b]$ iff $a \setminus n \subseteq b \setminus n$ for some $n < \omega$, can be embedded into Zermelo degrees. In particular, there are infinite descending sequences of Zermelo degrees. However, as Zermelo-jump increases associated Zermelo ordinals, there is no Harrison-type (see Ref. 15, or Ref. 16 III.3.6) descending sequence, i.e.

Corollary 3.3 (Shi[2]). *Assume $V = L[\bar{\mu}]$, and $\bar{\mu}, \bar{\kappa}, \lambda$ be as in Theorem 3.3. At λ, there are infinite descending sequences of Zermelo degrees, but there is no infinite sequence $\langle [a_i]_Z : i < \omega \rangle$ such that $[a_{i+1}]'_Z \leq_Z [a_i]_Z$.*

The theorem below says that over the structure $(\mathscr{D}_Z, \leq_Z)$, the set of degrees represented by the sets coding the Zermelo ordinals, and the relation that two sets share the same Zermelo ordinal, are definable.

Theorem 3.4 (Shi[2]). *Assume $V = L[\bar{\mu}]$, and $\bar{\mu}, \bar{\kappa}, \lambda$ be as in Theorem 3.3. The following are definable over the structure $(\mathscr{D}_Z, <_Z)$:*

$\mathcal{I} = \{[A_\eta]_Z \mid A_\eta \subset \lambda \text{ codes } \langle \alpha_i : i < \eta \rangle, \eta < \lambda^+\}$.
$\mathcal{R} = \{([a]_Z, [b]_Z) \mid a, b \subset \lambda, \alpha_a = \alpha_b\}$.

3.5. *Zermelo degrees in models beyond ω many measurable cardinals.*

Let us look at Mitchell models with more measurable cardinals. In Ref. 58 (Theorem 4.1) Mitchell showed that if there is no inner model with an inaccessible limit of measurable cardinals then, as in the Dodd-Jensen covering lemma, for each minimal Zermelo model M_a, there is a single *maximal* system of indiscernibles C which can be used to cover any set $x \subset \lambda$ in M_a. A fair amount of analyses above can be carried out at ω-limits of measurable cardinals below the least inaccessible limit of measurable cardinals, if there exists one. Therefore, the pictures at those places are rather similar to the one at κ_ω in $L[\bar{\mu}]$.

Once we past models with inaccessible limit of measurable cardinals, the systems of indiscernibles are no longer unique – may depend on the set of ordinals to be covered (see Ref. 51, p.1555) – and are extremely difficult to analyze. However, Yang proved the existence of minimal covers at ω-limit of certain measurable cardinals.

Theorem 3.5 (Yang[1]). *Suppose $\langle\kappa_n : n < \omega\rangle$ is an increasing sequence of measurable cardinals such that each κ_{n+1} carries κ_n different normal measures, $n \in \omega$, and $\lambda = \sup_n \kappa_n$. Let $\mathcal{U}$ denote this matrix of normal measures, and let W be any subset of λ that codes $\langle V_\lambda, \in, \Delta, \{\kappa_i \mid i < \omega\}, \mathcal{U}\rangle$, where $\Delta \subset \lambda$ and codes a well ordering of V_λ of ordertype λ. Then there is a minimal cover for the Zermelo degree of W.*

Yang's result holds for Δ_1-degrees and any larger degree notions at λ, here we only state it for Zermelo degrees. This result can be relativized to any degrees above that of W. This implies that for instance, in the Mitchell model for $o(\kappa) = \kappa$,[k] a new picture appears at the λ in the hypothesis – there are minimal covers (over almost every degree). Yang's forcing in fact produces a large perfect set (has size λ^ω) of subsets of λ that are minimal above W. As every degree contains only at most λ many of them, thus the size of antichains of Zermelo degrees at this λ can be as large as possible. So we have "Yes" to the first two questions. We don't know the answers to Posner-Robinson and Degree Determinacy at this λ, but speculate "No" for both of them.

3.6. *The picture from I_0*

The analyses above relies heavily on the fine structure theory, especially covering and comparison. Once past Mitchell models, we are out of comfort zone. Though there are still some variations of covering lemmas for inner models past Mitchell models, very little have we derived from them for the structures of Zermelo degrees in those models. But the emerging new pictures suggest that larger cardinals give us more power to create rich degree structures.

In the later part of this paper, we consider the degree structures at countable cofinality singular cardinals from the other extreme – looking

---

[k]This is not the minimal inner model for Yang's hypothesis, however it has the "shortest" o-expression.

at the strongest large cardinal, Axiom I_0. I_0 asserts the existence of an elementary embedding $j : L(V_{\lambda+1}) \to L(V_{\lambda+1})$ with critical point below λ. This λ is an ω-limit of very large cardinals, it satisfies Yang's hypothesis. Therefore at this λ, there are a large perfect set of minimal covers for every degree (above the degree of Yang's W set). Let $\mathcal{E}(L(V_{\lambda+1}))$ denote the set of all elementary embeddings that witness I_0 at λ.

Corollary 3.4. *Assume* ZFC *and* $\mathcal{E}(L(V_{\lambda+1})) \neq \varnothing$. *Let* W *be as in Theorem 3.5. Consider the Zermelo degrees at* λ *above the degree of* W.

There are incomparable Zermelo degrees. In fact, there are antichains (of Zermelo degrees) of size λ^ω.
No Zermelo degree has a minimal cover. In fact, every Zermelo degree has λ^ω *many minimal covers.*

Moreover, as applications of Generic Absoluteness Theorem in I_0 theory (see Refs. 59,60) we have the following following results regarding Posner Robinson problem and Degree determinacy at λ for Zermelo degrees.

Theorem 3.6 (Shi[2]). *Assume* ZFC *and* $\mathcal{E}(L(V_{\lambda+1})) \neq \varnothing$. *Then for every* $A \in V_{\lambda+1}$, *and for almost all (i.e. except at most* λ *many)* $B \geq_Z A$, *the Posner-Robinson equation for* B *has a solution, i.e. there exists a* $G \in V_{\lambda+1}$ *such that* $(B, G) \equiv_Z G'$, *where* G' *denote the Zermelo jump of* G.

The proof in fact shows something stronger, $\leq_Z$ here can be replaced by Δ_1-reducibility for subsets of λ. The argument works if G' is replaced by any reasonable jump operator at λ. This theorem says that PR_λ holds for Zermelo degrees of subsets of λ above any $A \subset \lambda$.

For the degree determinacy problem, we have an almost negative answer.

Theorem 3.7 (Shi[2]). *Assume* ZFC *and* $j \in \mathcal{E}(L(V_{\lambda+1})) \neq \varnothing$. *Let* $\kappa = \mathrm{crit}(j)$ *and suppose* $V_\lambda \models$ *"the supercompactness of* κ *is indestructible by* κ*-directed closed posets". Then* $L(V_{\lambda+1}) \models$ *Degree Determinacy fails for Zermelo degrees at* λ.

Although the theorem uses an additional indestructibility requirement (see Ref. 61), the hypothesis of this theorem is equiconsistent with ZFC + I_0 (see Ref. 2).

3.7. *A conjecture*

We continue the discussion on degree determinacy problem in this subsection. We have seen the structures of Zermelo degrees at countable cofinality strong limit singular cardinals in the early subsections. Now consider the situations for singular (strong limit) cardinals of uncountable cofinalities as well as for regular cardinals. The case that λ is a strong limit singular cardinal of uncountable cofinality follows from the following result of Shelah (see Ref. 62).

Theorem 3.8 (Shelah[62]). *Assume* ZFC. *Then for every strong limit singular cardinal λ of uncountable cofinality, $L(\mathscr{P}(\lambda)) \models$ Axiom of Choice.*

Using the choice, one can select in $L(\mathscr{P}(\lambda))$ two disjoint unbounded set of degrees, which witness the failure of degree determinacy for Zermelo degrees at λ.

On page 216, we mentioned that at regular cardinals, answers to the degree determinacy questions are alway "No". Here we discuss this matter. For regular cardinals, we first look at the case that λ is regular and satisfies the weak power condition, i.e. $2^{<\lambda} = \lambda$. Jensen's lemma (see Ref. 63) can be generalized to such λ – Jensen's proof in the context of ω can be literally adapted for such λ. More precisely, we have the following lemma.

Lemma 3.5 (Jensen[63]). *Assume* $\mathsf{ZF} + \lambda^+\text{-}\mathsf{DC}$. *Suppose $\lambda > \omega$ is a regular cardinal and $2^{<\lambda} = \lambda$. Suppose $a \subset \lambda$ and $A \subset (\lambda, \lambda^+)$ is a scattered set (i.e. $\alpha > \sup(A \cap \alpha)$ for every $\alpha \in A$) such that every $\alpha \in A$ is a Zermelo ordinal for a. Suppose $\mathrm{otp}(A) \leq \lambda$. Suppose $B \subseteq A$ and $\mathrm{otp}(B) = \mathrm{otp}(A)$. Then there is a $b \subset \lambda$ such that $b \geq_{\mathsf{Z}} a$ and B is the set of the α-th Zermelo ordinal for b, $\alpha < \mathrm{otp}(A)$.*

Let $\mathrm{Det}_\lambda(\mathscr{D}_{\mathsf{Z}})$ denote the statement of degree determinacy for Zermelo degrees. If $\mathrm{Det}_\lambda(\mathscr{D}_{\mathsf{Z}})$ were true in $L(\mathscr{P}(\lambda))$, then applying this lemma to $A \subset \lambda$ that is scattered and $\mathrm{otp}(A) = \omega_1$, one would get a countably additive coherent measure on $[\lambda^+]^{\omega_1}$. This implies the determinacy for sets of reals that are ω_1-Suslin, hence there is no sequence of distinct reals of length ω_1, contradicting to the assumption that V_λ is wellordered. So in ZFC models, $\mathrm{Det}_\lambda(\mathscr{D}_{\mathsf{Z}})$ fails in $L(\mathscr{P}(\lambda))$ for regular cardinal λ such that $2^{<\lambda} = \lambda$. Let $\mathsf{ZFC}^{-\epsilon}$ be a fragment of ZFC sufficient for proving this.

Now consider the case that λ is regular but $2^{<\lambda} > \lambda$. If degree determinacy for Zermelo degrees at λ were true in $L(\mathscr{P}(\lambda))$, then there would be a (in fact a cone of) $u \subset \lambda$ and an $\eta_u < \lambda^+$ such that

$$L_{\eta_u}[u] \models \mathsf{ZFC}^{-\epsilon} + \text{“}L(\mathscr{P}(\lambda)) \models \mathrm{Det}_\lambda(\mathscr{D}_{\mathsf{Z}})\text{”}.$$

However $L_{\eta_u}[u]$ "thinks" $2^{<\lambda} = \lambda$. This is because if $x \subset \delta < \lambda$, then there are $\alpha, \beta < \lambda$ such that $x \in L_\alpha[u \cap \beta]$; then it follows that $|\mathscr{P}(\delta)| \leq$

λ. Therefore $L_{\eta_u}[u] \models$ "$2^{<\lambda} = \lambda$". But according to the discussion in the last paragraph, it must be that $L_{\eta_u}[u] \models$ "$L(\mathscr{P}(\lambda)) \models \neg\,\mathrm{Det}_\lambda(\mathscr{D}_\mathsf{Z})$". Contradiction!

This concludes the case that λ is regular. So at least, we know that

Theorem 3.9. *If $\lambda > \omega$ is a regular cardinal or a strong limit singular cardinal of uncountable cofinality, then $L(\mathscr{P}(\lambda)) \models \neg\,\mathrm{Det}_\lambda(\mathscr{D}_\mathsf{Z})$.*

In light of Shelah's result that $L(\mathscr{P}(\lambda))$ is a model of choice if λ is a strong limit singular cardinal and $\mathrm{cf}(\lambda) > \omega$, together with evidences for degree structures at other cardinals, the author (see Ref. 2) makes the following conjecture

Conjecture 3.1 (ZFC). *Let λ be any uncountable cardinal. Then Degree Determinacy for Zermelo degrees at λ is false in $L(\mathscr{P}(\lambda))$.*

At this point, very little is known about singular cardinals that are not strong limit.

4. Remarks

Now is the time for final comments.

4.1. *Why inner models?*

The first remark is regarding the question why we focus on degree structures in inner models.

Yang's theorem and those I_0 results are stated under large cardinal assumption, it seems to be natural to study the consistency strength of those degree theoretical properties. For instance,

- What is the consistency strength of having minimal covers as in Yang's Theorem (see page 224)?
- What is the consistency strength of having Posner-Robinson result as in Theorem 3.6?

These are interesting questions in its own, especially for set theorists. Properties regarding generalized recursive degree are subjects of α-recursion theory, which concerns only degrees in L (see Ref. 16). While we investigate structural properties of higher level degree notions, it also makes more sense to consider them in canonical settings such as fine structure extender models. This is because ZFC alone, even plus large cardinal assumption, though may decide certain individual properties, can hardly determine the structure of degree posets. For instance, consider the structure of Zermelo degrees at $\aleph_\omega$.

Example 4.1. Assume $\mathsf{ZFC} + \mathsf{GCH}$ and plus some large cardinal assumption, say a measurable cardinal κ of Mitchell order $o(\kappa) = \kappa^{++}$ plus a measurable cardinal $\kappa' > \kappa$. With a small forcing, one can arrange that in the generic extension $\kappa = \aleph_\omega$, GCH remains true below $\aleph_\omega$, $2^{\aleph_\omega} = \aleph_{\omega+2}$ while the measurability of κ' is preserved (This combines results of Woodin and Gitik, see Ref. 64). But then the Zermelo degree posets at $\aleph_\omega$ can not be well ordered (even prewellordered) in the generic extension, as every degree has only $\aleph_\omega$ many predecessors in the degree partial ordering. This is in contrast to the pictures in $L[\mu]$ (see Theorem 3.2, p.219).

It is the well organized structure of $L[\mu]$ (organized using Steel's construction) that forces the degrees to line up in a well ordered fashion. The existence of measurable cardinals alone (more precisely, without appealing to forcings) is not strong enough to create "untamed" degrees – incomparable degrees, unless we go up to the ω-limit of measurable cardinals (see Corollary **S**3.2) and beyond.

4.2. *Degree structures in canonical models*

In **S**3, we analyzed Zermelo degree structures in several canonical models or under some stronger large cardinals. An immediate conclusion is that larger cardinals create more complicated Zermelo degree structures at some critical cardinals (more precisely, ω-limit of certain large cardinals). In other words, in these models the complexity of the Zermelo degree structure at these critical cardinals reflects the strength of the relevant cardinals.

The next natural step is to look into larger cardinal axioms and hope to find more complicated degree structures. For instance, what degree structures can one see at an ω-limit of strong cardinals, or Woodin cardinals, or supercompact cardinals, etc.? During the process, it would be interesting in itself to extend the question list on page 215 to differentiate these degree structures, in a way that the natural order of large cardinals sorts these structural properties into layers.

At the mean time, the pictures of Zermelo degree structures in L and through up to the core models for finitely many measurable cardinals strongly suggest that in any reasonable inner model, at every singular cardinal λ with $\mathrm{cf}(\lambda) = \omega$ and below the least measurable, the Zermelo degrees are wellordered above some degree. In particular,

Conjecture 4.1. *In all fine structure extender models the Zermelo degree structures at (their) $\aleph_\omega$ are all (eventually) wellordered and the immediate successor is given by Zermelo jump.*

Combining these remarks, one can see that the complexity of a particular degree structure does not necessarily gives the large cardinal strength of the core model, but it does indicate the levels of the associated cardinals that the structure resides. In other words, from the variety of the types of degree structures that appear in a core model one can tell the lower bound of the large cardinals the given core model carries. This is a complete new perspective for looking into large cardinal axioms.

4.3. *New techniques are needed*

Our proofs for L and up to $L[\bar{\mu}]$ use heavily one particular form of covering lemmas, we expect that that analysis will work as far as that form of Covering Lemma holds, namely at least up to Mitchell models for sequences of measures.

Next key step is to check whether in M_1, the minimal iterable class model for one Woodin cardinal, the scenarios described above continue.

Conjecture 4.2. *In M_1, the Zermelo degrees at $\aleph_\omega$ are well ordered.*

Moreover, we expect that this to be true for singular countable cofinality λ's that are above or in-between critical large cardinals, as this fits well with the intuition that universes of small large cardinals are initial segment of universes of larger cardinals.

Climbing up the cardinal ladder, although new pictures may appear at certain cardinals, as well as degree structures in between these special cardinals, as what we have just discussed about $\aleph_\omega$, it seems reasonable to conjecture that the structure at a particular cardinal once appear in a core model for certain large cardinal axiom, will stay unchanged as we move up to core models for larger cardinals, assuming they exist.

However, as the classical form of Covering is not available for M_1 and larger core models, deeper understanding of their structures[1] and new techniques are necessary for the investigation of degree structures in these models.

[1]So far the best result on constructing core models is due to Neeman (see 65), who produces a core model for a Woodin limit of Woodin cardinals.

Besides the classical fine structure models, recent developments in descriptive inner model theory (see for example Sargsyan's survey paper[66]) suggest a much advanced and daring path of investigation – looking into higher degrees in the HODs of determinacy models, as determinacy gives a whole family of canonical models – the ones given by Solovay hierarchy. Assume $\mathsf{AD}^+ + V = L(\mathscr{P}(\mathbb{R}))$, it's believed that HOD is a canonical model. Although it's still an open question whether $\mathsf{AD}^+ + V = L(\mathscr{P}(\mathbb{R}))$ proves that HOD is a fine structure model, HOD of $\mathsf{AD}^+ + V = L(\mathscr{P}(\mathbb{R}))$ models are believed to be fine structure models at least all the way up to $\Theta =_{\text{def}} \sup\{\alpha \mid \exists f : \mathbb{R} \xrightarrow{\text{onto}} \alpha\}$ (see Steel[67]). Based on this understanding, the first test question would be

Question 4.1. Assume $\mathsf{AD}^+ + V = L(\mathscr{P}(\mathbb{R}))$. Look at Zermelo degrees within HOD at $(\aleph_\omega)^{\mathsf{HOD}}$, are they (eventually) well ordered?

This is a *great* question! One can not expect to solve this problem with only Covering, one would need mouse analysis for arbitrary AD^+ models. But the mouse analysis technique is still in its development, there is very little on this matter that is valuable to say at this point.

4.4. *Evidences of the impact of large cardinals on structures of degrees*

Next we leave the canonnical models, look at the impact of large cardinal alone on the structures of degrees. The theme is that stronger large cardinal yields more complicated degree structures at certain strong limit, countabe cofinality, singular cardinals. We have seen two evidences, one is the prewellordered degree structure in Theorem 3.3, where you can find incomparable degrees (see Corollary 3.2), the other is Yang Sen's minimal cover result quoted on page 224.

In Ref. 2, it is shown that I_0, one of the strongest large cardinal hypotheses, entails a richer degree structures at a certain strong limit λ with countable cofinality – it gives positive answers to Post problem, minimal covver problem and Posner-Robinson problem. Furthermore, it proves that I_0 together with a mild indestructibility assumption imply the failure of degree determinacy in $L(\mathscr{P}(\lambda))$ for Zermelo degrees at a particular strong limit, countable cofinality, singular cardinal (see Theorem 3.7) by exploiting the richness of the degree structure provided by the large cardinal axiom. As part of the global conjecture (see page 227), it is also conjectured that the failure of degree determinacy in $L(\mathscr{P}(\lambda))$ for Zermelo degrees at λ, for countable cofinality λ, is a theorem of ZFC. But as our analysis indicates, if one wants to prove this conjecture, the proof has to be very subtle: In early stages of canonical inner models, the degree structures are very simple, the degree determinacy fails due to that simplicity and our approach for proving the failure of degree determinacy by exploiting the richness of degree structures does not work there.

4.5. *Structures of other degrees*

In **S3**, we have been focusing on Zermelo degrees, only compare structures of Zermelo degrees crossing over inner models. But there is a whole spectrum of degree notions one can explore, as we have seen in **S2**. And certainly there are many more questions one can pursue if structures of different degree notions are compared.

In fact, the newly discovered connection between the complexity of degree structures and the large cardinal strength of relevant cardinals lead us to review some old results with new perspective.

Recall that there are incomparable $\aleph_\omega$-degrees (see p.212). Comparing it with the fact that the Zermelo degree at $\aleph_\omega$ is wellordered, one may draw a conclusion that smaller degree operator exhibit rich degree structures at early stage of inner models. This is not something exciting, as the larger degree operator often absorbs part of structure induced by the smaller degree operator. And this is well supported by examples discussed in **S2** – structures of larger degree notions are always simpler than those given by smaller degree operators. For instance, while the poset $(\mathscr{R}, \leq_T)$ is dense and has incomparable degrees, it analogue for hyperarithmetic degrees – the poset of hyperdegrees restricted to Π^1_1 subsets of ω – is trivial, consisting of only $[\emptyset]_h$ and $[\mathcal{O}]_h = [\emptyset]'_h$. Here is a question on the spot:

Question 4.2. Assume $V = L$. Are there incomparable generalized hyperdegrees at $\aleph_\omega$?

Let us take a closer look. In $V = L$, although it is still open whether there are minimal $\aleph_\omega$-degrees, if move to a Σ_2-admissible ordinal α, one starts to see minimal α-degrees. Yang's argument gives minimal α-degrees at α which is the ω-limit of certain large cardinals there are, a little adaption gives minimal Zermelo degrees as well. So these large cardinals create not only minimal Δ_1-degrees at α, but any reasonable degree operator at α. This is beyond Σ_2-admissibility. Just as admissible ordinals are "recursively" regular, Δ_1-ly speaking, Σ_2-admissible ordinals behave like a very large "recursive" large cardinals. To extend this similarity, a sample question would be to find the ordinal for minimal hyperdegrees.

Question 4.3. Assume $V = L$. At what ordinal α can there be minimal generalized hyperdegrees at α?

Acknowledgement

The author is indebted to W. Hugh Woodin for his insightful advices on this project. The author would also like to thank Professor Chong Chi-Tat and Yang Yue of National University of Singapore for their friendship and help throughout this project. This project is partially supported by NSFC (No. 11171031) and the Fundamental Research Funds for the Central Universities.

References

1. S. Yang, Prikry-type forcings defined from many measures and an application, Preprint (April, 2013).
2. X. Shi, Axiom I_0 and higher degree theory, preprint, 2014, to appear in JSL.
3. T. Jech, *Set theory*Springer Monographs in Mathematics, Springer Monographs in Mathematics (Springer-Verlag, Berlin, 2003), The third millennium edition, revised and expanded.
4. A. Kanamori, *The higher infinite*Springer Monographs in Mathematics, Springer Monographs in Mathematics, second edn. (Springer-Verlag, Berlin, 2003), Large cardinals in set theory from their beginnings.
5. P. Erdős, Méthodes probabilistes en théorie des nombres, in *Séminaire Delange-Pisot-Poitou (15e année: 1973/74), Théorie des nombres, Fasc. 1, Exp. No. 1*, (Secrétariat Mathématique, Paris, 1975) p. 4.
6. J. R. Steel, A classification of jump operators, *J. Symbolic Logic* **47**, 347 (1982).
7. H. Friedman and R. Jensen, Note on admissible ordinals. Syntax Semantics infinit. Languages, Symposium Ucla 1967, Lect. Notes Math. 72, 77-79 (1968).
8. T. A. Slaman and J. R. Steel, Definable functions on degrees, in *Cabal Seminar 81–85*, Lecture Notes in Math. Vol. 1333 (Springer, Berlin, 1988) pp. 37–55.
9. R. M. Friedberg, Two recursively enumerable sets of incomparable degrees of unsolvability (solution of Post's problem, 1944), *Proc. Nat. Acad. Sci. U.S.A.* **43**, 236 (1957).

10. A. A. Muchnik, Negative answer to the problem of reducibility of the theory of algorithms (russian), *Dokl. Akad. Nnauk SSSR* **108**, 194 (1956).
11. S. C. Kleene and E. L. Post, The upper semi-lattice of degrees of recursive unsolvability, *Ann. of Math. (2)* **59**, 379 (1954).
12. G. E. Sacks, The recursively enumerable degrees are dense, *Ann. of Math.* **80**, 300 (1964).
13. C. Spector, On degrees of recursive unsolvability, *Ann. of Math. (2)* **64**, 581 (1956).
14. G. E. Sacks, A minimal degree less than $0'$, *Bull. Amer. Math. Soc.* **67**, 416 (1961).
15. J. Harrison, Recursive pseudo-well-orderings, *Trans. Amer. Math. Soc.* **131**, 526 (1968).
16. G. E. Sacks, *Higher recursion theory*Perspectives in Mathematical Logic, Perspectives in Mathematical Logic (Springer-Verlag, Berlin, 1990).
17. C. Spector, Recursive well-orderings, *J. Symb. Logic* **20**, 151 (1955).
18. G. E. Sacks, Countable admissible ordinals and hyperdegrees, *Advances in Math.* **20**, 213 (1976).
19. S. D. Friedman, The Π^1_2-singleton conjecture, *J. Amer. Math. Soc.* **3**, 771 (1990).
20. L. Harrington and A. S. Kechris, Π^1_2 singletons and $0^{\#}$, *Fund. Math.* **95**, 167 (1977).
21. A. S. Kechris, The theory of countable analytical sets, *Trans. Amer. Math. Soc.* **202**, 259 (1975).
22. R. M. Solovay, On the cardinality of Σ^1_2 sets of reals, in *Foundations of Mathematics (Symposium Commemorating Kurt Gödel, Columbus, Ohio, 1966)*, (Springer, New York, 1969) pp. 58–73.
23. A. S. Kechris and Y. N. Moschovakis, Two theorems about projective sets, *Israel J. Math.* **12**, 391 (1972).
24. A. S. Kechris, D. A. Martin and R. M. Solovay, Introduction to Q-theory, in *Cabal seminar 79–81*, , Lecture Notes in Math. Vol. 1019 (Springer, Berlin, 1983) pp. 199–282.
25. D. B. Posner and R. W. Robinson, Degrees joining to $\mathbf{0}'$, *J. Symbolic Logic* **46**, 714 (1981).
26. T. A. Slaman and J. R. Steel, Complementation in the Turing degrees, *J. Symbolic Logic* **54**, 160 (1989).
27. R. A. Shore and T. A. Slaman, Defining the Turing jump, *Mathematical Research Letter* **6**, 711 (1999).

28. R. L. Sami, Turing determinacy and the continuum hypothesis, *Arch. Math. Logic* **28**, 149 (1989).
29. G. E. Sacks, *Degrees of unsolvability* (Princeton University Press, Princeton, N.J., 1963).
30. A. S. Kechris and Y. N. Moschovakis (eds.), *Cabal Seminar 76–77*, Lecture Notes in Mathematics Vol. 689, (Springer, Berlin, 1978).
31. A. Marks, T. Slaman and J. Steel, Martin's conjecture, arithmetic equivalence, and countable borel equivalence relations (September 2011).
32. J. Barwise, *Admissible sets and structures* (Springer-Verlag, Berlin, 1975), An approach to definability theory, Perspectives in Mathematical Logic.
33. S. G. Simpson, Short course on admissible recursion theory, in *Generalized recursion theory, II (Proc. Second Sympos., Univ. Oslo, Oslo, 1977)*, , Stud. Logic Foundations Math. Vol. 94 (North-Holland, Amsterdam, 1978) pp. 355–390.
34. C. T. Chong, *Techniques of admissible recursion theory*, Lecture Notes in Mathematics, Vol. 1106 (Springer-Verlag, Berlin, 1984).
35. G. E. Sacks, Post's problem, admissible ordinals, and regularity, *Trans. Amer. Math. Soc.* **124**, 1 (1966).
36. G. E. Sacks and S. G. Simpson, The α-finite injury method, *Ann. Math. Logic* **4**, 343 (1972).
37. R. A. Shore, Σ_n sets which are Δ_n-incomparable (uniformly), *J. Symbolic Logic* **39**, 295 (1974).
38. R. A. Shore, Splitting an α-recursively enumerable set, *Trans. Amer. Math. Soc.* **204**, 65 (1975).
39. R. A. Shore, The recursively enumerable α-degrees are dense, *Ann. Math. Logic* **9**, 123 (1976).
40. J. M. Macintyre, Minimal α-recursion theoretic degrees, *J. Symbolic Logic* **38**, 18 (1973).
41. R. A. Shore, Minimal α-degrees, *Ann. Math. Logic* **4**, 393 (1972).
42. W. Maass, On minimal pairs and minimal degrees in higher recursion theory, *Arch. Math. Logik Grundlagenforsch.* **18**, 169 (1976).
43. J. Silver, On the singular cardinals problem, in *Proceedings of the International Congress of Mathematicians (Vancouver, B. C., 1974), Vol. 1*, (Canad. Math. Congress, Montreal, Que., 1975).
44. S. D. Friedman, Negative solutions to Post's problem. I, in *Generalized recursion theory, II (Proc. Second Sympos., Univ. Oslo, Oslo, 1977)*, Stud. Logic Foundations Math. Vol. 94 (North-Holland, Amsterdam,

1978) pp. 127–133.
45. C. T. Chong, The minimal α-degree revisited, Unpublished (May, 2013).
46. S. D. Friedman, Negative solutions to Post's problem. II, *Ann. of Math.* **113**, 25 (1981).
47. S. D. Friedman, Negative solutions to Post's problem. III, (1981).
48. S. G. Simpson, Degree theory on admissible ordinals, in *Generalized recursion theory (Proc. Sympos., Univ. Oslo, Oslo, 1972)*, (North-Holland, Amsterdam, 1974) pp. 165–193. Studies in Logic and Foundations of Math., Vol. 79.
49. R. A. Shore, α-recursion theory, in *Handbook of mathematical logic, Part C*, (North-Holland, Amsterdam, 1977) pp. 653–681. Studies in Logic and the Foundations of Math., Vol. 90.
50. G. E. Sacks and T. A. Slaman, Generalized hyperarithmetic theory, *Proc. London Math. Soc.* **60**, 417 (1990).
51. W. J. Mitchell, The covering lemma, in *Handbook of set theory. Vols. 1, 2, 3*, (Springer, Dordrecht, 2010) pp. 1497–1594.
52. E. Schimmerling, The ABC's of mice, *Bull. Symbolic Logic* **7**, 485 (2001).
53. M. Gitik, Prikry-type forcings, in *Handbook of set theory. Vols. 1, 2, 3*, (Springer, Dordrecht, 2010) pp. 1351–1447.
54. R. B. Jensen, The fine structure of the constructible hierarchy, *Ann. Math. Logic* **4**, 229 (1972), With a section by Jack Silver.
55. K. J. Devlin, *Constructibility*Perspectives in Mathematical Logic, Perspectives in Mathematical Logic (Springer-Verlag, Berlin, 1984).
56. T. Dodd and R. Jensen, The covering lemma for K, *Ann. Math. Logic* **22**, 1 (1982).
57. A. J. Dodd and R. B. Jensen, The covering lemma for $L[U]$, *Ann. Math. Logic* **22**, 127 (1982).
58. W. J. Mitchell, On the singular cardinal hypothesis, *Trans. Amer. Math. Soc.* **329**, 507 (1992).
59. W. H. Woodin, Suitable extender models I, *J. Math. Log.* **10**, 101 (2010).
60. W. H. Woodin, Suitable extender models II: beyond ω-huge, *J. Math. Log.* **11**, 115 (2011).
61. R. Laver, Making the supercompactness of κ indestructible under κ-directed closed forcing, *Israel J. Math.* **29**, 385 (1978).
62. S. Shelah, Pcf without choice, *ArXiv e-prints* (2010), [Sh:835]. arXiv:math/0510229.

63. R. B. Jensen, Forcing with classes of conditions Chapter 6 of Admissible Sets, URL: `http://www.mathematik.hu-berlin.de/~raesch/org/jensen/pdf/AS_6.pdf`.
64. M. Gitik, The negation of the singular cardinal hypothesis from $o(\kappa) = \kappa^{++}$, *Ann. Pure Appl. Logic* **43**, 209 (1989).
65. I. Neeman, Inner models in the region of a Woodin limit of Woodin cardinals, *Ann. Pure Appl. Logic* **116**, 67 (2002).
66. G. Sargsyan, Descriptive inner model theory, *Bull. Symbolic Logic* **19**, 1 (2013).
67. J. R. Steel, $\mathrm{HOD}^{L(\mathbf{R})}$ is a core model below Θ, *Bull. Symbolic Logic* **1**, 75 (1995).

Degree Spectra of Equivalence Relations

Liang Yu

Institute of Mathematical Science
Nanjing University, Jiangsu Province 210093
P. R. of China
E-mail: yuliang.nju@gmail.com

In Refs. 2 and 3, Greenberg, Montalbán and Slaman investigated both hyperarithmetic and constructibility degree spectra of countable structures. Inspired by their results, we push them to a more general setting by investigating degree spectra of equivalence relations.

Keywords: Hyperarithmeticity, constructibility, equivalence relation, spectrum.

1. Introduction

Definition 1.1. For any equivalence relation E, reduction $\leq_r$ over 2^ω and real $x \in 2^\omega$, let

$$\mathrm{Spec}_{E,r}(x) = \{y \mid \exists z \leq_r y(E(z,x))\}$$

be the (E,r)-*spectrum* of x.

In Refs. 2 and 3, Greenberg, Montalbán and Slaman investigated $\mathrm{Spec}_{\cong,h}(x)$ and $\mathrm{Spec}_{\cong,L}(x)$, where $\cong$ denotes the isomorphism relation, and $\leq_h$ and $\leq_L$ denote hyperarithmetic and constructibility reducibility respectively. They prove the following result.

Theorem 1.1 (Greenberg, Montalbán and Slaman,[2] and[3]).

(1) *For the countable structures of partial ordering language, there is a linear ordering structure* $\mathcal{M}$ *so that* $\mathrm{Spec}_{\cong,h}(\mathcal{M}) = \{y \mid y$ *is not hyperarithmetic*$\}$.
(2) *Assume that* ω_1 *is inaccessible. For any recursive language and any countable structure* $\mathcal{M}$ *of the language, if* $\mathrm{Spec}_{\cong,L}(\mathcal{M})$ *contains all the nonconstructible reals, then it contains all the reals.*

The motivation to classify degree spectra of an equivalence relation is, as in the introduction in Ref. 3, to find which recursion theoretical aspects of a set x of natural numbers are reflected in the equivalence class of x. Moreover, pushing the results to the general setting may give some clearer explanation why the argument used in the classical setting works. For example, (1) in Theorem 1.1 can be viewed as a result for Σ_1^1-equivalence relations. One may wonder whether there is a Π_1^1-equivalence relation so that the conclusion of (1) remains true. We refute this by showing Proposition 2.1. And the genius method used in the proof of (2) of Theorem 1.1 by Greenberg, Montalbán and Slaman is much more powerful than it looked. Actually the method in the proof can be viewed as a generalization of the proof of the classical result that every nontrivial upper cone of Turing degrees is null. We show that, in Theorem 3.1, the conclusion remains true for any Σ_2^1-equivalence relation under a fairly weak set theoretical assumption. Moreover, we prove that the relativization of the conclusion does not require any large cardinal assumption by showing Corollary 3.1 (The existence of an inaccessible cardinal seems necessary to relativize their original proof to arbitrary countable language and structures of (2) of Theorem 1.1). Both the proofs use some ideas from Ref. 2.

Mostly we follow the notations from Ref. 7. Readers should be familiar with higher recursion and set theory.

We enumerate some classical results which are needed later.

We say that a real x codes a well ordering if the relation $R(n,m) \leftrightarrow x(2^n \cdot 3^m) = 1$ is a well ordering of ω.

For $n \in \mathcal{O}$, H_n^x is a $\Pi_2^0(x)$-singleton. Actually it is the $|n|$-th Turing jump relative to x. If $\omega_1^x = \mathrm{CK}$, then each real hyperarithmetic in x is recursive in H_n^x for some $n \in \mathcal{O}$. If a real x codes a well ordering of order type α, then we use $|x|$ to denote α.

For any $\sigma \in 2^{<\omega}$, $[\sigma] = \{x \in 2^\omega \mid x \succ \sigma\}$.

Theorem 1.2 (Sacks[6]). *Let μ be the Lesbegue measure, then $\mu(\{x \mid \omega_1^x = \mathrm{CK}\}) = 1$.*

Theorem 1.3 (Sacks[6]). *For any Π_1^1 set $B \subseteq 2^\omega \times 2^\omega \times \omega$, the set $\{(y,p) \mid p$ is a rational $\wedge\, \mu(\{x \mid (x,y) \in B\}) > p\}$ is Π_1^1.*

Theorem 1.4 (Sacks[6] and Tanaka[8]). *Every Π_1^1 positive measure set $A \subseteq 2^\omega$ contains a hyperarithmetic real.*

For the set theory notions, we follow from the book Ref. 4. We use $\dot{x}, \dot{y}, \cdots$ to denote names over a forcing language.

Others can be found in Refs. 4,5,7, and the forthcoming book. [1]

2. On Π^1_1-equivalence relations

Proposition 2.1. *For any Π^1_1-equivalence relation E and real x, if $\mathrm{Spec}_{E,h}(x) \supseteq \{z \mid z \notin \Delta^1_1\}$, then $\mathrm{Spec}_{E,h}(x) = 2^\omega$.*

Proof. Suppose that E is a Π^1_1-equivalence relation. Fix a real x so that $\mathrm{Spec}_{E,h}(x) \supseteq \{z \mid z \notin \Delta^1_1\}$. So $\mu(\mathrm{Spec}_{E,h}(x)) = 1$. For any $n \in \mathcal{O}$ and Turing oracle functional Φ^*_e with $e \in \omega$, the set

$$A_{n,e} = \{z \mid E(\Phi_e^{H_n^z}, x)\}$$

is a $\Pi^1_1(x)$ subset of $\mathrm{Spec}_{E,h}(x)$ and so measurable. By Theorem 1.2,

$$\mu(\bigcup_{n\in\mathcal{O}, e\in\omega} A_{n,e}) = 1.$$

So there must be some $n \in \mathcal{O}$ and e so that the set $A_{n,e}$ has positive measure. By the Lesbegue density theorem, there must be some $\sigma \in 2^{<\omega}$ so that

$$\mu(A_{n,e} \cap [\sigma]) > \frac{3}{4} \cdot 2^{-|\sigma|}.$$

Let

$$B_{n,e} = \{y \succ \sigma \mid \mu(\{z \succ \sigma \mid E(\Phi_e^{H_n^z}, \Phi_e^{H_n^y})\}) > \frac{3}{4} \cdot 2^{-|\sigma|}\}.$$

Then by Theorem 1.3, $B_{n,e}$ is a Π^1_1 set. Moreover $B_{n,e} = A_{n,e} \cap [\sigma]$. So $B_{n,e}$ has positive measure. Then by Theorem 1.4, $B_{n,e}$ contains a hyperarithmetic real. Thus $\mathrm{Spec}_{E,h}(x) = 2^\omega$. □

Note that Proposition 2.1 fails for Σ^1_1-equivalence relations due to (1) of Theorem 1.1. Here we give a much simpler example. Let $E(x,y)$ if and only if $x = y$ or $x \notin \Delta^1_1$ and $y \notin \Delta^1_1$. Then E is a Σ^1_1-equivalence relation and for any nonhyperarithmetic real x, $\mathrm{Spec}_{E,h}(x) = \{z \mid z \notin \Delta^1_1\}$.

3. On Σ^1_2-relations

For any real x, let $\mathbb{P}_x = (\mathbf{P}_x, \leq)$ be the random forcing over $L[x]$, the constructible universe relative to x.

Theorem 3.1. *Assume that $\mu(\{x \mid x \text{ is } L\text{-random }\}) = 1$. Then for any Σ^1_2-relation E and real x, if $\mathrm{Spec}_{E,L}(x) \supseteq \{z \in 2^\omega \mid z \notin L\}$, then $\mathrm{Spec}_{E,L}(x) = 2^\omega$.*

Proof. Note that by the assumption, for almost every real x, $\mu(\{y \mid y \text{ is } L[x]\text{-random }\}) = 1$.

Let E be a Σ^1_2-relation and x be a real so that $\mathrm{Spec}_{E,L}(x) \supseteq \{z \in 2^\omega \mid z \notin L\}$. Since E is Σ^1_2, there must be some Π^1_1-relation $R_0 \subseteq (2^\omega)^3$ so that

$$\forall y \forall z (E(y,z) \leftrightarrow \exists s R_0(y,z,s)).$$

By the Shoenfield absoluteness theorem,

$$\forall y \forall z (E(y,z) \leftrightarrow \exists s \in L_{\omega_1^{L[y\oplus z]}}[y \oplus z] R_0(y,z,s)).$$

In particular,

$$\forall y (E(y,x) \leftrightarrow \exists s \in L_{\omega_1^{L[y\oplus x]}}[y \oplus x] R_0(y,x,s)).$$

Whence

$$z \in \mathrm{Spec}_{E,L}(x) \leftrightarrow$$
$$\exists t \exists y \exists s (t \text{ codes a well ordering } \wedge y \in L_{|t|}[z] \wedge s \in L_{|t|}[y \oplus x] \wedge R_0(y,x,s)).$$

Note that, by the assumption, the set $\mathrm{Spec}_{E,L}(x)$ is $\Sigma^1_2(x)$ and conull.

By the assumption, there are conull many L-random reals. Since random forcing does not collapse cardinals, the set $\{y \mid \omega_1^L = \omega_1^{L[y]}\}$ is conull. If $\mathrm{Spec}_{E,L}(x)$ contains all non-constructible reals, then there exists x_0 so that $E(x_0,x)$ and $\omega_1^{L[x_0]} = \omega_1^L$; so by passing to x_0 if necessary, we may assume that $\omega_1^{L[x]} = \omega_1^L$ and the set of $L[x]$-random reals is of measure 1. So the set $\{y \mid \omega_1^{L[x\oplus y]} = \omega_1^L\}$ is also conull.

For any real t coding a well ordering, let

$$z \in R_{1,t} \leftrightarrow \exists y \in L_{|t|}[z] \exists s \in L_{|t|}[y \oplus x](R_0(y,x,s)).$$

Then $R_{1,t} \subseteq \mathrm{Spec}_{E,L}(x)$ is a $\Pi^1_1(t \oplus x)$-set and so measurable. Moreover, if z is $L[x]$-random, then $z \in \mathrm{Spec}_{E,L}(x)$ if and only if $z \in R_{1,t}$ for some real $t \in L$ coding a well ordering. Since $\mu(\mathrm{Spec}_{E,L}(x)) = 1$ and the set of $L[x]$-random reals is of measure 1, there must be some $L[x]$-random real z and a real $t \in L$ coding a well ordering so that $z \in R_{1,t}$. Then there must be some condition $p \in \mathbf{P}_x$ so that $z \in p$ and

$$p \Vdash_{\mathbb{P}_x} \exists \dot{y} \in L_{|t|}[\dot{z}] \exists \dot{s} \in L_{|t|}[\dot{y} \oplus x](R_0(\dot{y},x,\dot{s})).$$

Since $\mu(p) > 0$ and almost every real in p is $L[x]$-random, we have that $\mu(R_{1,t}) > 0$. Fix such a real $t \in L$ to be t_0. By the countable additivity of Lesbegue measure, there must be some formula φ in the set theory language so that the set

$$R_{1,t_0,\varphi} = \{z \mid \exists y \exists s \in L_{|t_0|}[y \oplus x](\forall n(n \in y \leftrightarrow L_{|t_0|}[z] \models \varphi(n)) \wedge R_0(y,x,s))\}$$

has positive measure. Then there must be some $\sigma \in 2^{<\omega}$ so that

$$\mu(R_{1,t_0,\varphi} \cap [\sigma]) > \frac{3}{4} \cdot 2^{-|\sigma|}.$$

Now we try to get rid of the parameter x.

Let

$$S = \{r \mid \mu(\{z \succ \sigma \mid \exists y \exists s \in L_{|t_0|}[y \oplus r] \\ (\forall n(n \in y \leftrightarrow L_{|t_0|}[z] \models \varphi(n)) \wedge R_0(y,r,s))\}) > \frac{3}{4} \cdot 2^{-|\sigma|}\}.$$

Then S is a $\Pi^1_1(t_0)$-set and every real in S is E-equivalent to x. Since $x \in S$, we have that S is not empty. Thus there must be some t_0-constructible, and so constructible, real in S.

This completes the proof. □

Corollary 3.1.

(1) *Assume $\omega_1^L < \omega_1$. Then for any Σ^1_2-relation E and real x, if* $\mathrm{Spec}_{E,L}(x) \supseteq \{z \in 2^\omega \mid z \notin L\}$, *then* $\mathrm{Spec}_{E,L}(x) = 2^\omega$.

(2) *Assume that $MA + 2^{\aleph_0} > \aleph_1$, where MA is Martin's axiom. Then for any reals x_0, x and $\Sigma^1_2(x_0)$-relation E, if* $\mathrm{Spec}_{E,L}(x) \supseteq \{z \in 2^\omega \mid z \notin L[x_0]\}$, *then* $\mathrm{Spec}_{E,L}(x) = 2^\omega$.

Proof. (1). If $\omega_1^L < \omega_1$, then the set of L-random random reals is conull. Thus the assumption of Theorem 3.1 is satisfied.

(2). We prove the lightface version. The boldface version follows immediately by a relativization. By $MA + 2^{\aleph_0} > \aleph_1$, a union $\aleph_1$-many null sets is null. So for any real x, the set of $L[x]$-random reals is of measure 1. Thus the assumption of Theorem 3.1 is satisfied. □

Note the conclusion of Theorem 3.1 cannot be proved under ZFC. If $V = L[g]$ for a Sacks generic real g, then there are only two constructible degrees in V. Let $E(x,y)$ if and only if $x = y$. Then for any nonconstructible real x, $\mathrm{Spec}_{E,L}(x) = \{z \mid z \notin L\}$.

References

1. Chong CT and Yu L. *Recursion Theory: The Syntax and Structure of Definability.* http://math.nju.edu.cn/ yuliang/course.pdf.
2. Noam Greenberg, Antonio Montalbán, and Theodore A. Slaman. The Slaman-Wehner theorem in higher recursion theory. *Proc. Amer. Math. Soc.*, 139(5):1865–1869 (2011).
3. Noam Greenberg, Antonio Montalbán, and Theodore A. Slaman. Relative to any non-hyperarithmetic set,. *Journal of Mathematical Logic.*, to appear.
4. Thomas Jech. *Set theory.* Springer Monographs in Mathematics. Springer-Verlag, Berlin (2003).
5. Yiannis N. Moschovakis. *Descriptive set theory*, volume 155 of *Mathematical Surveys and Monographs.* American Mathematical Society, Providence, RI, second edition (2009).
6. Gerald E. Sacks. Measure-theoretic uniformity in recursion theory and set theory. *Trans. Amer. Math. Soc.*, 142:381–420 (1969).
7. Gerald E. Sacks. *Higher recursion theory.* Perspectives in Mathematical Logic. Springer-Verlag, Berlin (1990).
8. Hisao Tanaka. A basis result for $\Pi_1{}^1$-sets of postive measure. *Comment. Math. Univ. St. Paul.*, 16:115–127 (1967/1968).

Printed in the United States
By Bookmasters